# Lecture Notes in Bioinformatics

T0238310

Subseries of Lecture Notes in Computer Science

Sarah Cohen-Boulakia  Val Tannen (Eds.)

# Data Integration in the Life Sciences

4th International Workshop, DILS 2007
Philadelphia, PA, USA, June 27-29, 2007
Proceedings

 Springer

Series Editors

Sorin Istrail, Brown University, Providence, RI, USA
Pavel Pevzner, University of California, San Diego, CA, USA
Michael Waterman, University of Southern California, Los Angeles, CA, USA

Volume Editors

Sarah Cohen-Boulakia
University of Pennsylvania, Department of Computer and Information Science
303 Levine Hall, 3330 Walnut St., Philadelphia, PA 19104, USA
E-mail: sarahcb@seas.upenn.edu

Val Tannen
University of Pennsylvania, Department of Computer and Information Science
570 Levine Hall, 3330 Walnut St., Philadelphia, PA 19104, USA
E-mail: val@cis.upenn.edu

Library of Congress Control Number: 2007928915

CR Subject Classification (1998): H.2, H.3, H.4, J.3

LNCS Sublibrary: SL 8 – Bioinformatics

ISSN      0302-9743
ISBN-10   3-540-73254-3 Springer Berlin Heidelberg New York
ISBN-13   978-3-540-73254-9 Springer Berlin Heidelberg New York

Springer is a part of Springer Science+Business Media

springer.com

© Springer-Verlag Berlin Heidelberg 2007
Printed in Germany

Typesetting: Camera-ready by author, data conversion by Scientific Publishing Services, Chennai, India
Printed on acid-free paper      SPIN: 12081769      06/3180      5 4 3 2 1 0

# Preface

Understanding the mechanisms involved in life (e.g., discovering the biological function of a set of proteins, inferring the evolution of a set of species) is becoming increasingly dependent on progress made in mathematics, computer science, and molecular engineering. For the past 30 years, new high-throughput technologies have been developed generating large amounts of data, distributed across many data sources on the Web, with a high degree of semantic heterogeneity and different levels of quality. However, one such dataset is not, by itself, sufficient for scientific discovery. Instead, it must be combined with other data and processed by bioinformatics tools for patterns, similarities, and unusual occurrences to be observed. Both data integration and data mining are thus of paramount importance in life science.

DILS 2007 was the fourth in a workshop series that aims at fostering discussion, exchange, and innovation in research and development in the areas of data integration and data management for the life sciences. Each previous DILS workshop attracted around 100 researchers from all over the world. This year, the number of submitted papers again increased. The Program Committee selected 19 papers out of 52 full submissions. The DILS 2007 papers cover a wide spectrum of theoretical and practical issues including scientific workflows, annotation in data integration, mapping and matching techniques, and modeling of life science data. Among the papers, we distinguished 13 papers presenting research on new models, methods, or algorithms and 6 papers presenting implementation of systems or experience with systems in practice. In addition to the presented papers, DILS 2007 featured two keynote talks by Kenneth H. Buetow, National Cancer Institute, and Junhyong Kim, University of Pennsylvania.

The workshop was held at the University of Pennsylvania, in Philadelphia, USA. It was kindly sponsored by the School of Engineering and Applied Science of the University of Pennsylvania, the Penn Genomics Institute, and Microsoft Research, who also made available their conference management system. As editors of this volume, we thank all the authors who submitted papers, the Program Committee members, and the external reviewers for their excellent work. Special thanks go to Susan Davidson, General Chair, Chris Stoeckert, PC Co-chair, as well as Olivier Biton, Tara Betterbid, and Howard Bilowsky. Finally, we are grateful for the cooperation and help of Springer in putting this volume together.

June 2007

Sarah Cohen-Boulakia
Val Tannen

# Organization

## Executive Committee

General Chair
> Susan Davisdon, University of Pennsylvania, USA

Program Chairs
> Chris Stoeckert, University of Pennsylvania, USA
> Val Tannen, University of Pennsylvania, USA

## Program Committee

| | |
|---|---|
| Judith Blake | Jackson Laboratory, USA |
| Sarah Cohen-Boulakia | University of Pennsylvania, USA |
| Marie-Dominique Devignes | LORIA, Nancy, France |
| Barbara Eckman | IBM |
| Christine Froidevaux | LRI, University of Paris-Sud XI, France |
| Cesare Furlanello | ITC-irst, Trento, Italy |
| Jim French | University of Virginia, USA |
| Floris Geerts | University of Edinburgh, UK |
| Amarnath Gupta | University of California San Diego, USA |
| Ela Hunt | ETH Zurich, Switzerland |
| Jacob Koehler | Rothamsted Research, UK |
| Anthony Kosky | Axiope Inc. |
| Hilmar Lapp | NESCENT |
| Ulf Leser | Humboldt-Universität zu Berlin, Germany |
| Bertram Ludäscher | University of California Davis, USA |
| Victor Markowitz | Lawrence Berkeley Labs |
| Peter Mork | MITRE |
| Tom Oinn | European Bioinformatics Institute, UK |
| Meral Ozsoyoglu | Case Western Reserve University, USA |
| John Quackenbush | Harvard, USA |
| Louiqa Raschid | University of Maryland, USA |
| Fritz Roth | Harvard Medical School, USA |
| Susanna-Assunta Sansone | European Bioinformatics Institute, UK |
| Kai-Uwe Sattler | Technical University of Ilmena, Germany |
| Chris Stoeckert | University of Pennsylvania, USA |
| Val Tannen | University of Pennsylvania, USA |
| Olga Troyanskaya | Princeton University, USA |

## External Reviewers

Jérôme Azé
Jana Bauckmann
Julie Bernauer
Shawn Bowers
William Bug
Amy Chen

Adnan Derti
Francis Gibbons
Philip Groth
Timothy McPhillips
Joe Mellor
Krishna Palaniappan

Norbert Podhorszki
Murat Tasan
Weidong Tian
Silke Trißl
Daniel Zinn

## Sponsorship Chair

Howard Bilofsky, University of Pennsylvania

## Sponsoring Institutions

School of Engineering and Applied Science at the University of Pennsylvania
http://www.seas.upenn.edu/

Penn Genomics Institute
http://www.genomics.upenn.edu/

Microsoft Research
http://research.microsoft.com/

## Web Site and Publicity Chairs

Olivier Biton            University of Pennsylvania
Sarah Cohen-Boulakia     University of Pennsylvania

DILS 2007 Web site       http://dils07.cis.upenn.edu/

# Table of Contents

## Mapping and Matching Techniques

## Modeling of Life Science Data

## Annotation in Data Integration

# Enabling the Molecular Medicine Revolution Through Network-Centric Biomedicine
## (Keynote Presentation)

Kenneth H. Buetow

National Cancer Institute
2115 East Jefferson Street
Suite 6000, MSC 8505
Bethesda, MD 20892, USA
buetowk@nih.gov

To deliver on the promise of next generation treatment and prevention strategies in cancer, we must address its multiple dimensions. The full complement of the diverse fields of modern biomedicine are engaged in the assault on this complexity. These disciplines are armed with the latest tools of technology, generating mountains of data. Each surpasses the next in their unprecedented and novel view of the fundamental nature of cancer. Each contributes a vital thread of insight. Information technology provides a promising loom on which the threads of insight can be woven.

Bioinformatics facilitates the electronic representation, redistribution, and integration of biomedical data. It makes information accessible both within and between the allied fields of cancer research. It weaves the disparate threads of research information into a rich tapestry of biomedical knowledge. Bioinformatics is increasingly inseparable from the conduct of research within each discipline. The linear nature of science is being transformed into a spiral with bioinformatics joining the loose ends and facilitating progressive cycles of hypothesis generation and knowledge creation.

To facilitate the rapid deployment of bioinformatics infrastructure into the cancer research community the National Cancer Institute (NCI) is undertaking the cancer Biomedical Informatics Grid, or caBIG$^{TM}$. caBIG$^{TM}$, is a voluntary virtual informatics infrastructure that connects data, research tools, scientists, and organizations to leverage their combined strengths and expertise in an open environment with common standards and shared tools. Effectively forming a World Wide Web of cancer research, caBIG$^{TM}$ promises to speed progress in all aspects of cancer research and care including etiologic research, prevention, early detection, and treatment by breaking down technical and collaborative barriers.

Researchers in all disciplines have struggled with the integration of biomedical informatics tools and data; the caBIG$^{TM}$ program demonstrates this important capability in the well-defined and critical area of cancer research, by planning for, developing, and deploying technologies which have wide applicability outside the cancer community. Built on the principles of open source, open access, open development, and federation, caBIG$^{TM}$ infrastructure and tools are open

S. Cohen-Boulakia and V. Tannen (Eds.): DILS 2007, LNBI 4544, pp. 1–2, 2007.

and readily available to all who could benefit from the information accessible through its shared environment. caBIG$^{TM}$ partners are developing or providing standards-based biomedical research applications, infrastructure, and data sets. The implementation of common standards and a unifying architecture ensures interoperability of tools, facilitating collaboration, data sharing, and streamlining research activities across organizations and disciplines.

The caBIG$^{TM}$ effort has recognized that in addition to new infrastructure new information models are required to capture the complexity of cancer. While the biomedical community continues to harvest the benefits of genome views of biologic information, it has been clear from the founding of genetics that biology acts through complex networks of interacting genes. Information models and a new generation of analytic tools that utilizing these networks are key to translating discover to practical intervention.

# Phyl-O'Data (POD) from Tree of Life: Integration Challenges from Yellow Slimy Things to Black Crunchy Stuff

## (Keynote Presentation)

Junhyong Kim

Department of Biology
Penn Center for Bioinformatics
Penn Genomics Institute
415 S. University Ave.
Philadelphia, PA 19104, USA
junhyong@sas.upenn.edu

## 1 Background

The AToL (Assembling the Tree of Life) is a large-scale collaborative research effort sponsored by the National Science Foundation to reconstruct the evolutionary origins of all living things. Currently 31 projects involving 150+ PIs are underway generating novel data including studies of bacteria, microbial eukaryotes, vertebrates, flowering plants and many more. Modern large-scale data collection efforts require fundamental infrastructure support for archiving data, organizing data into structured information (e.g., data models and ontologies), and disseminating data to the broader community. Furthermore, distributed data collection efforts require coordination and integration of the heterogeneous data resources. In this talk, I first introduce the general background of the phylogenetic estimation problem followed by an introduction to the associated data modeling, data integration, and workflow challenges.

## 2 Phylogeny Estimation and Its Utility

While ideas about genealogical reconstruction have been around since Darwin, quantitative algorithmic approaches to the problem have been developed only in the last 50 years. The basic structure of the problem involves considering all possible tree graph structures compatible with an organismal genealogy and measuring their fits to observed data by various objective functions. There are now many algorithms based on various inferential principles including maximum information, maximum likelihood, Bayesian posterior, etc. Many flavors of the phylogeny reconstruction problem have been shown to be NP-hard and there is a considerable body of literature on associated computational and mathematical problems. Phylogenetic methods provide the temporal history of biological diversity and have been used in many applications. For example: to track the

S. Cohen-Boulakia and V. Tannen (Eds.): DILS 2007, LNBI 4544, pp. 3–5, 2007.
© Springer-Verlag Berlin Heidelberg 2007

history of infectious diseases; to reconstruct ancestral molecules; to reveal functional patterns in comparative genomics; and even in criminal cases, to infer the relatedness of biological criminal evidence. But, a grand challenge for phylogenetic research is to reconstruct the history of all extent organismal life-the so-called Tree of Life.

## 3   Problems and Challenges

In modern times, much of phylogenetic estimation is carried out using molecular sequences, some explicitly gathered for phylogenetic research; others, systematically collected and deposited in public databases. There are many data management problems associated with using molecular sequence data even from public databases-which are not discussed here. But, for the frontline researcher the problem starts at the stage of actually collecting the biological material for experimentation. That is, the animal must be captured, preserved, measured, and recorded along with associated metadata (e.g., capture location). Such specimens must be physically archived (called voucher specimen) and identified if possible and given a name. All of these activities are sometimes called *alpha-taxonomy*.

Once a specimen has been obtained and named then it (or presumed identical specimens) must be measured for relevant traits including extracting molecular sequences. This action of obtaining "relevant measurements" involve gathering characteristics that will be broadly comparable amongst different varieties of organisms-thus require a prior data model of what is or is not relevant. Once the relevant measurements are recorded, the next important step is deriving an equivalence relationship between measurements on different organisms such that the measurements are considered to be evolutionarily comparable to each other. This activity is called "establishing homology relationships" and is a critical prelude to further analysis. An example of such homologizing activity is the alignment of molecular sequences whereby equivalent relations of individual sequence letters are established. This activity of defining relevant characters and homologizing their assembly is called *Systematics* and the final product of this activity is the "data matrix" that encapsulates the data model of relevant measurements and the relational maps of sets of such measurements. Biologists widely disagree on details of such matrices-for example, whether a particular measurement should be described as present or absent; thus, these matrices are best seen as a "data view" of the primary objects. It is common in the literature and in public databases to have available only the fixed data matrix. Given the complicated and uncertain ontology of such matrices, **a critical challenge is to endow the phylogenetic matrices with their provenance information as well as to provide a facility to change the "views" as biologists' assumptions change.**

Notwithstanding the fundamental problems described above, there are continuing activities to collect empirical measurements and generate data matrices and place them into a structured information source. For example, there are current database efforts within the AToL projects where all projects have sub-aims targeting data storage, access, and sharing; and, a small number of projects have been funded to develop domain specific data models and analysis tools such as databases for 3D morphological data, web-based information storage and dissemination, and mining molecular databases for phylogenetic information. As a simple fact the magnitude of the AToL efforts is insufficient to meet the real-world needs of the AToL projects that will become critical as each empirical project matures. More importantly, there is little or no coordination between these efforts and there is a critical need to enable the integration of distributed, heterogeneous, and changing data sources; provide for reliable data archiving and maintenance of data provenance; and, help manage the complex data collection and analysis processes. Many of the projects are already very mature and domain-specific problems, cultural problems, and legacy problems make it difficult to develop a single solution to the problems. Therefore, another **critical challenge is to provide ways to post-hoc integrate the extremely dispersed and heterogeneous phylogenetic data sources in a scalable manner.**

The ultimate end product of phylogenetic reconstruction is the tree graph depicting the genealogical history and associated data. The associated data is usually mapped to substructures within the tree graph-say the nodes of the graph, usually from a secondary analysis (so-called post-analysis). For example, once the phylogeny is known, there are algorithms available for reconstructing the measurements of putative ancestors; thus, we may assign data matrices to interior nodes of the tree. In actual practice, the researcher may try many different algorithms to reconstruct the tree, each algorithm may generate multiple trees (e.g., because of equivalent optimality score), and given some preliminary tree estimate one may want to modify the data matrix (try a different view) and re-estimate the tree. Furthermore, there is often a large battery of post-analysis routines that involve other calculations including calculations on substructures of estimated tree. As is typical in complicated data analysis, the total analysis may involve a large number of steps, some steps recursive, cyclic, or branching. Thus, **a final challenge is to develop a workflow for phylogenetic analysis that automatically tracks analysis flow and helps manage the complexity in such a way that is useful to the primary researcher and helps other researchers recapitulate analyses carried out by third parties.**

# Automatically Constructing a Directory of Molecular Biology Databases

Luciano Barbosa, Sumit Tandon, and Juliana Freire

School of Computing, University of Utah

**Abstract.** There has been an explosion in the volume of biology-related information that is available in online databases. But finding the *right* information can be challenging. Not only is this information spread over multiple sources, but often, it is hidden behind form interfaces of on-line databases. There are several ongoing efforts that aim to simplify the process of finding, integrating and exploring these data. However, existing approaches are not scalable, and require substantial manual input. Notable examples include the NCBI databases and the NAR database compilation. As an important step towards a scalable solution to this problem, we describe a new infrastructure that automates, to a large extent, the process of locating and organizing online databases. We show how this infrastructure can be used to automate the construction and maintenance of a Molecular Biology database collection. We also provide an evaluation which shows that the infrastructure is scalable and effective—it is able to efficiently locate and accurately identify the relevant online databases.

## 1 Introduction

Due to the explosion in the number of online databases, there has been increased interest in leveraging the high-quality information present in these databases [1, 2, 7, 10, 19]. However, finding the right databases can be challenging. For example, if a biologist needs to locate databases related to molecular biology and searches on Google for the keywords "molecular biology database" over 27 million documents are returned. Among these, she will find pages that contain databases, but the results also include a very large number of pages from journals, scientific articles, etc.

Recognizing the need for better mechanisms to locate online databases, there have been a number of efforts to create online database collections such as the NAR database compilation [9], a manually created collection which lists databases of value to biologists. Given the dynamic nature of the Web, where new sources are constantly added, manual approaches to create and maintain database collections are not practical. But automating this process is non-trivial. Since online databases are sparsely distributed on the Web, an efficient strategy is needed to locate the forms that serve as entry points to these databases. In addition, online databases do not publish their schemas and since their contents are often hidden behind form interfaces, they are hard to retrieve. Thus, a

S. Cohen-Boulakia and V. Tannen (Eds.): DILS 2007, LNBI 4544, pp. 6–16, 2007.

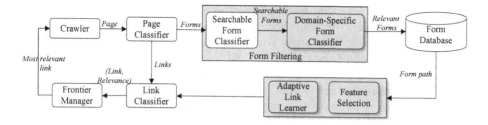

**Fig. 1.** Architecture of *ACHE*

scalable solution must determine the relevance of a form to a given database domain just by examining information that is available in and around forms.

In previous work [2,3,4], we proposed *ACHE* (**A**daptive **C**rawler for **H**idden-Web **E**ntry Points), a new scalable framework that addresses these problems. We showed, experimentally, that *ACHE* is effective for a representative set of commercial databases. In this paper, we describe a case study we carried out to investigate the effectiveness of this framework for different domains, and in particular, for non-commercial online databases. We chose to focus on databases related to molecular biology for two key reasons: these are often academic databases; and there is already a sizeable collection of these databases [9] which can serve as a basis for comparison.

The remainder of the paper is organized as follows. In Section 2, we give a brief overview of the *ACHE* framework. In Section 3, we describe in detail the process we followed to customize *ACHE* to the molecular biology domain. We discuss the issues we faced in the process, and show that, because the different components of *ACHE* use learning-based techniques, they can be easily adapted to a different domain. We present our experimental evaluation in Section 4. The results indicate that *ACHE* is effective: it is able to efficiently locate and accurately identify online databases related to molecular biology. We conclude in Section 6, where we outline directions for future work.

## 2   Searching and Identifying Online Databases

*ACHE* provides an end-to-end solution to the problem of locating and organizing online databases. The high-level architecture of the system is shown in Figure 1. *ACHE* uses a focused crawler to locate online databases. Similar to topic-specific crawlers, *ACHE* also uses Web page contents to focus its search on a given topic. But to deal with the sparseness of online databases on the Web, it prioritizes links that are more likely to lead to forms in the database domain sought. *ACHE* also uses a form-filtering process to select the relevant forms among the set of forms retrieved by the crawler. This form-filtering process is required because even a focused crawler invariably retrieves a diverse set of forms, including searchable forms (i.e., forms used to search over a database) from multiple database domains, and non-searchable forms that do not represent database queries such as, for example, forms for login, mailing list subscriptions, Web-based email forms. Consider for example, the Form-Focused Crawler (FFC) [2] which is optimized for locating searchable Web forms. For a set of representative database domains, on average, only 16% of the forms retrieved by the FFC are actually relevant— for some domains this percentage can be as low as 6.5%. These numbers are even lower for less focused crawlers [6, 8].

In what follows, to make this paper self-contained, we briefly describe the components of *ACHE*. For a detailed description, the reader is referred to [3, 4].

## 2.1 Searching for Online Databases

Each page retrieved by the crawler is sent to the *Page Classifier*, which is trained to identify pages that belong to a particular topic based on their contents. It uses the same strategy as the best-first crawler of [6]. The page classifier analyzes a page $P$ and assigns to it a score which reflects the probability that $P$ belongs to the focus topic. A page is considered relevant if this probability is greater than a certain threshold (0.5 in our case).

If a page is determined to be relevant, its links are extracted and used as inputs to the *Link Classifier*.

The Link Classifier learns to estimate the distance between a link and a target page based on link patterns: given a link, the link classifier assigns a score to the link which corresponds to the estimated distance between the link and a page that contains a relevant form. The Frontier Manager uses this estimate to prioritize *promising* links, including links that have *delayed benefit*—links which belong to paths that will *eventually* lead to pages that contain searchable forms. As we discuss in [3], considering links with delayed benefit is essential to obtain high harvest rates while searching for sparse concepts such as online databases on the Web. Since searchable forms are sparsely distributed on the Web, by prioritizing only the links that bring immediate return, i.e., links whose patterns are similar to those of links pointing to pages containing searchable forms, the crawler may miss target pages that can only be reached with additional steps.

The Link Classifier is constructed as follows. Given a set of URLs of pages that contain forms in a given database domain, paths to these pages are obtained by crawling backwards from these pages. *ACHE* uses two different approximations of the Web graph to perform a backward crawl: it uses the `link:` facility provided by search engines [5] at the beginning of the crawling process; and it uses the Web subgraph collected during the crawler execution. The backward crawl proceeds in a breadth-first manner. Each level $l+1$ is constructed by retrieving all documents that point to the documents in level $l$. From the set of paths gathered, the *best* features of the links are automatically selected. These features consist of the highest-frequency terms extracted from text in the neighborhood of the link, as well as from the URL and anchor. Using these features, the classifier is trained to estimate the distance between a given link (from its associated features) and a target page that contains a searchable form. Intuitively, a link that matches the features of level *1* is likely to point to a page that contains a form; and a link that matches the features of level $l$ is likely $l$ steps away from a page that contains a form.

The *Frontier Manager* controls the order in which pages are visited. It creates one queue for each level of the backward crawl. Links are placed on these queues based on their similarity to the features selected for the corresponding level of the link classifier. Intuitively, the lower the level of the link classifier, the higher is the priority of the queue. When the crawler starts, all seeds are placed in queue 1. At each crawling step, the crawler selects the link with the highest relevance score from the first non-empty queue. If the page it downloads belongs to the target topic, its links are classified by link classifier and added to the most appropriate queue.

The focused crawler learns new link patterns during the crawl and automatically adapts its focus based on these new patterns. As the crawler navigates through Web pages, successful paths are gathered, i.e., paths followed by the crawler that lead to relevant forms. Then, the *Feature Selection* component automatically extracts the patterns of these paths. Using these features and the set of path instances, the *Adaptive Link Learner* generates a new Link Classifier that reflects these newly-learned patterns.[1] The Adaptive Link Learner is invoked periodically, after the crawler visits a pre-determined number of pages. Experiments over real Web pages in a representative set of commercial domains showed that online learning leads to significant gains in harvest rates—the adaptive crawler retrieve up to three times as many forms as a crawler that use a fixed focus strategy [3].

## 2.2   Identifying Relevant Databases

The *Form Filtering* component is responsible for identifying relevant forms gathered by *ACHE*, and it does so by examining the visible content in the forms. The overall performance of the crawler is highly-dependent on the accuracy of the form filtering process, which assists *ACHE* in obtaining high-quality results and also enables the crawler to adaptively update its focus strategy. If the Form Filtering process is inaccurate, crawler efficiency can be greatly reduced as it drifts way from its objective through unproductive paths.

Instead of using a single, complex classifier, our form filtering process uses a sequence of simpler classifiers that learn patterns of different subsets of the form feature space [4]. The first is the *Generic Form Classifier* (*GFC*), which uses structural patterns to determine whether a form is searchable. Empirically, we have observed that these structural characteristics of a form are a good indicator as to whether the form is searchable [2]. The second classifier in the sequence identifies searchable forms that belong to a given domain. For this purpose, we use a more specialized classifier, the *Domain-Specific Form Classifier* (*DSFC*). The DSFC uses the textual content of a form to determine its domain. Intuitively, the form content is often a good indicator of the database domain—it contains metadata and data that pertain to the database.

---

[1] The length of the paths considered depends on the number of levels used in the link classifier.

By partitioning the feature space of forms, not only can simpler classifiers be constructed that are more accurate and robust, but this also enables the use of learning techniques that are more effective for each feature subset. Whereas decision trees [14] gave the lowest error rates for determining whether a form is searchable based on structural patterns, SVM [14] proved to be the most effective technique to identify forms that belong to the given database domain based on their textual content.

## 3  Constructing the Molecular Biology Database Directory

In this section we describe the process we followed to build a collection of molecular biology online databases. This process consists of customizing three components of the *ACHE* framework: Page Classifier, Link Classifier and Form Filtering.

*Page Classifier.* The Page Classifier defines the broad search topic for the crawler: based on the page content (words in the page), the Page Classifier predicts whether a given page belongs to a topic or not. We used Rainbow [12], a freely-available Naïve Bayes classifier, to build the Page Classifier. To train it, we crawled the biology-related Web sites listed in dmoz.org and gathered 2800 pages to serve as positive examples. Because of the great variety of pages that can be visited during the crawl, constructing a set of representative negative examples is more challenging. To select negative examples, we ran the crawler with the selected positive examples and an initial set of negative examples taken from a corpus that comes with the Rainbow classifier. We then added the misclassified pages to the set of negative examples. Examples of such misclassified pages included non-English pages and pages from foreign porn sites. A total of 4671 negative examples were collected.

The Page Classifier was then constructed using the 50 terms that led to the highest information gain. For the Molecular Biology domain, these terms included: biology, molecular, protein, genome, ncbi, length, substring, structure, gene, genomics, nih, parent, sequence, pubmed, entrez, nlm, fellows, postdoctoral, research, dna.

*Link Classifier.* We created the Link Classifier from a backward crawl of depth 3. The set of seeds chosen to train the Link Classifier comprised 64 relevant Web forms manually selected from NAR collection. For each of the feature spaces of links (url, anchor and text in link neighborhood), the 5 most frequent words are selected. To build the classifier we used WEKA [18], an open source data mining tool. The classification algorithm (Naïve Bayes) is used to estimate the probabilities of a link being 1, 2, or 3 steps away from a form page.

*Form Filtering.* As discussed above, the Form Filtering uses two classifiers: the GFC (based on form structure) and DSFC (based on form content). In our initial experiment, we used the GFC we had constructed for identifying searchable

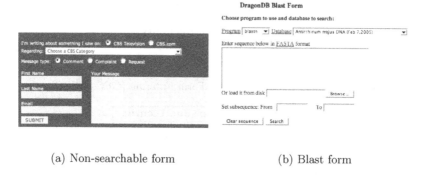

(a) Non-searchable form                    (b) Blast form

**Fig. 2.** Similarity of a non-searchable form in a commercial domain and a searchable form in the molecular biology domain

commercial databases [4]. An inspection of the misclassified forms showed that some searchable forms in the molecular biology domain are structurally similar to non-searchable forms of commercial sites. The presence of features such as text areas, buttons labeled with the string "submit", file inputs, are good indicators that a (commercial) form is non-searchable. However, these features are also present in many searchable forms in the molecular biology domain (e.g., Blast search forms). This is illustrated in Figure 2. To address this problem, we added a sample of the misclassified forms to the pool of positive examples, and generated a new instance of the classifier. The GFC was then able to correctly classify forms like the Blast forms as searchable—its accuracy improved to 96%.

To generate the DSFC, we manually gathered 150 positive examples of forms from the NAR collection [9]. The negative examples were obtained as follows: we ran the crawler and filtered the searchable forms using the GFC; then, from these searchable forms we manually selected 180 forms that did not belong to the molecular biology domain. These forms included, e.g., forms related to chemistry and agriculture, as well forms for searching for authors and journals related to molecular biology. Using these training examples, we generated the first version of DSFC. This version, however had a very low precision: only 16%. The problem was due to false positives. Unlike the commercial domains, the crawler retrieved a large number of non-English pages. As the DSFC was not trained to handle non-English terms it incorrectly classified these forms as relevant. After we added the misclassified (non-English) forms to the set of negative examples, the accuracy of the DSFC increased substantially, to 65%. To try and further improve the accuracy, we added additional false positives misclassified by the second version of the DSFC to the pool of negative examples and, once again, constructed a new instance of the classifier. The third version obtained 89% accuracy. The top 20 terms used to build the DSFC were: name, type, select, result, keyword, gene, sequenc, databa, enter, ani, option, page, help, titl, protein, number, advanc, onli, format, word.

**Table 1.** Quality measurement for GFC

|          | Recall | Specificity |
|----------|--------|-------------|
| Adaptive | 0.82   | 0.96        |

**Table 2.** Quality measurement for Form Filtering (GFC+DSFC)

|          | Recall | Precision | Accuracy |
|----------|--------|-----------|----------|
| Adaptive | 0.73   | 0.93      | 0.96     |

## 4   Experimental Evaluation

In this section we first assess the effectiveness of *ACHE* for constructing a high-quality set of molecular biology databases. We then compare our results with those of the manually constructed NAR collection.

To verify the effectiveness of the adaptive crawler in this domain, we executed the following crawler configurations:

- Baseline Crawler: A variation of the best-first crawler [6]. The page classifier guides the search and the crawler follows all links of a page whose contents are classified as being on-topic;
- Static Crawler: Crawler operates using a fixed policy which remains unchanged during the crawling process;
- Adaptive Crawler: ACHE starts with a pre-defined policy, and this policy is dynamically updated after crawling 10,000 pages.

All configurations were run using 35 seeds obtained from dmoz.org and crawled 100,000 pages. The Link Classifier was configured with three levels.

Since the goal of *ACHE* is to find *relevant* online databases in the molecular biology domain, we measured the effectiveness of the crawler configurations in terms of the total number of relevant forms gathered. We manually inspected the output of Form Filtering to calculate the values for: accuracy, recall; precision and specificity. *Accuracy* is a suitable measure when the input to the classifier contains similar proportions of positive and negatives examples; *recall* captures the number of relevant items retrieved as fraction of all relevant items; *precision* represents the number of relevant items as a fraction all the items predicted as positive by the classifier; and *specificity* is the proportion of actual irrelevant items predicted as irrelevant.

The results obtained by the GFC (see Table 1) confirm that it can identify most of the relevant forms (high recall) and to filter out most of the irrelevant forms (high specificity). As Table 2 shows, the combination of the GFC and DSFC leads to a very high recall, precision and accuracy. This indicates that the Form Filtering process is effective and that a high-quality (homogeneous) collection of databases can be generated by *ACHE*.

Figure 3 shows the number of relevant forms retrieved by the three crawler configurations over time. The Adaptive Crawler outperforms both the Static

**Table 3.** Features of the Link Classifiers used at the beginning and at the end of the crawl process

| Field | Initial features | Final features |
|-------|------------------|----------------|
| URL | link, search, full, genom, index | search, blast, genom, form, bioinfo |
| Anchor | data, genom, for, text, full | search, blast, gene, databas, sequenc |
| Around | bio, data, info, genom, gene | search, databas, gene, genom, sequenc |

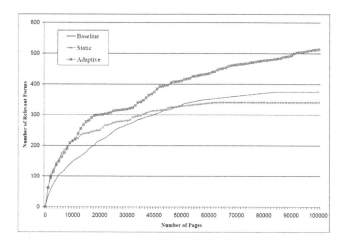

**Fig. 3.** Behavior of different crawler configurations over time

and the Baseline configurations, retrieving 513 relevant forms after crawling 100,000 pages versus 341 and 376 relevant forms retrieved by Static and Baseline, respectively. This shows that the feedback from Form Filtering is effective in boosting the crawler performance. Table 3 shows the features used by the initial Link Classifier and the features learned during the crawl that are used by the final classifier. A similar behavior has been observed for crawls over commercial domains [3]. This indicates that the adaptive strategy is effective regardless of the domain.

*NAR Collection.* The NAR collection lists 968 databases. But their concept of databases is more generic than ours: they consider as a database both pages that contains tables with information about genes and proteins, and pages that contain forms (or links to pages that contain forms). In our evaluation, we consider only the searchable forms accessible through the NAR collection. To extract the searchable forms directly or indirectly accessible through the NAR collection, we crawled the links provided (using `wget` with depth 1). Among the 20,000 pages retrieved by `wget`, 700 relevant forms were identified. Although *ACHE* obtained 513 forms, we should note that the NAR collection has been maintained for over 7 years—the earliest reference we found dates back to 1999—and it has become a very popular resource. Once *ACHE* was configured, the 513 forms were automatically gathered in 4 hours. This shows that such a collection can be efficiently maintained over time. In addition, these forms were obtained in a relatively small crawl (only 100,000 pages). The positive slope for the Adaptive Crawler graph in Figure 3 indicates that additional forms can be obtained in larger crawls. This is an issue we plan to investigate in future work.

## 5   Related Work

BioSpider [11] is a system that integrates biological and chemical online databases. Given a biological or chemical identifier, BioSpider produces a report containing physico-chemical, biochemical and genetic information about the identifier. Although the authors mention BioSpider performs a crawl to locate the underlying sources, no details are given about the crawling process. Also, the number of sources they integrate is very small—only about 20 databases are listed on their Web site.

Ngu et al. [15] proposed an approach to classify search interfaces by probing these interfaces and trying to match the control flow of the interface against a standard control flow. Thus, for a specific type of form (which they refer to as a service class), e.g., a Blast search, they create a corresponding flow graph pattern from a sample of known interfaces and try to match new interfaces against that pattern. An important limitation of this solution comes from its reliance on the ability to automatically fill out structured forms. The difficulties in automatically filling out structured Web forms are well-documented in the literature [7, 16].

InfoSpiders [17] is a multi-agent focused crawler specialized for biomedical information whose goal is to fetch information about diseases when given information about genes. A study by Menczer et al. [13] comparing several topic-driven crawlers (including InfoSpiders) found that the best-first approach (the Baseline configuration in Section 4) leads to the highest harvest rate among the crawlers in the study. As we discuss in Section 4, our adaptive crawler outperforms the best first crawler by a large margin.

## 6   Conclusion and Discussion

In this paper we described a case study we carried out to evaluate the extensibility and effectiveness of the *ACHE* framework for constructing a high-quality online database directories. We described the process of customizing the framework for molecular biology databases; and performed an evaluation which showed that *ACHE* is able to efficiently locate and accurately identify databases in this domain. The number of relevant forms automatically gathered (after a 4-hour crawl) is very close to the number of forms listed in a manually created collection that has been maintained for over 7 years. This indicates that *ACHE* provides a scalable solution to the problem of automatically constructing high-quality, topic-specific online database collections. These results also reinforce our choice of applying learning techniques. Because we use learning classifiers in the different components of *ACHE*, with some modest tuning, the system can be customized for different domains.

It is well-known, however, that the performance of machine learning techniques, such as the classifiers used in our framework, is highly-dependent on the choice of training examples used to construct them. And building a representative sample of forms is difficult due to the large variability in form content and structure, even within a well-defined domain. We are currently investigating strategies that simplify the process of gathering positive and negative examples.

To help users locate relevant databases, we are designing intuitive and expressive query interfaces that support both simple keyword-based queries and structured queries (e.g., find forms that contain an attribute with a given label). Although our focus has been on databases accessible through forms, in future work we plan to investigate extensions to our infrastructure for handling more general notions of online databases, such as for example, pages that contain tables with biology-related information.

**Acknowledgments.** This work was partially supported by the National Science Foundation and a University of Utah Seed Grant.

# References

1. Barbosa, L., Freire, J.: Siphoning Hidden-Web Data through Keyword-Based Interfaces. In: Proceedings of SBBD, pp. 309–321 (2004)
2. Barbosa, L., Freire, J.: Searching for Hidden-Web Databases. In: Proceedings of WebDB, pp. 1–6 (2005)
3. Barbosa, L., Freire, J.: An Adaptive Crawler for Locating Hidden-Web Entry Points. In: Proceedings of WWW (To appear 2007)
4. Barbosa, L., Freire, J.: Combining Classifiers to Organize Online Databases. In: Proceedings of WWW (To appear 2007)
5. Bharat, K., Broder, A., Henzinger, M., Kumar, P., Venkatasubramanian, S.: The connectivity server: Fast access to linkage information on the Web. Computer Networks 30(1-7), 469–477 (1998)
6. Chakrabarti, S., van den Berg, M., Dom, B.: Focused Crawling: A New Approach to Topic-Specific Web Resource Discovery. Computer Networks 31(11-16), 1623–1640 (1999)
7. Chang, K.C.-C., He, B., Zhang, Z.: Toward Large-Scale Integration: Building a MetaQuerier over Databases on the Web. In: Proceedings of CIDR, pp. 44–55 (2005)
8. Diligenti, M., Coetzee, F., Lawrence, S., Giles, C.L., Gori, M.: Focused Crawling Using Context Graphs. In: Proceedings of VLDB, pp. 527–534 (2000)
9. Galperin, M.: The molecular biology database collection: 2007 update. Nucleic Acids Res, vol. 35 (2007)
10. Hsieh, W., Madhavan, J., Pike, R.: Data management projects at Google. In: Proceedings of ACM SIGMOD, pp. 725–726 (2006)
11. Knox, C., Shrivastava, S., Stothard, P., Eisner, R., Wishart, D.S.: BioSpider: A Web Server for Automating Metabolome Annotations. In: Pacific Symposium on Biocomputing, pp. 145–156 (2007)
12. McCallum, A.: Rainbow http://www-2.cs.cmu.edu/mccallum/bow/rainbow/

13. Menczer, F., Pant, G., Ruiz, M., Srinivasan, P.: Evaluating topic-driven Web crawlers. In: Proceedings of SIGIR, pp. 241–249 (2001)
14. Mitchell, T.: Machine Learning. McGraw-Hill, New York (1997)
15. Ngu, A.H., Rocco, D., Critchlow, T., Buttler, D.: Automatic discovery and inferencing of complex bioinformatics web interfaces. World Wide. Web. 8(3), 463–493 (2005)
16. Raghavan, S., Garcia.-Molina, H.: Crawling the Hidden Web. In: Proceedings of VLDB, pp. 129–138 (2001)
17. Srinivasan, P., Mitchell, J., Bodenreider, O., Pant, G., Menczer, F.: Web Crawling agents for Retrieving Biomedical Information. In: Workshop on Agents in Bioinformatics (NETTAB-02) (2002)
18. Witten, I.H., Frank, E.: Data Mining: Practical Machine Learning Tools and Techniques. Morgan Kaufmann, San Francisco (2005)
19. Wu, W., Yu, C., Doan, A., Meng, W.: An Interactive Clustering-based Approach to Integrating Source Query interfaces on the Deep Web. In: Proceedings of ACM SIGMOD, pp. 95–106 (2004)

# The Allen Brain Atlas: Delivering Neuroscience to the Web on a Genome Wide Scale

Chinh Dang, Andrew Sodt, Chris Lau, Brian Youngstrom, Lydia Ng,
Leonard Kuan, Sayan Pathak, Allan Jones, and Mike Hawrylycz

The Allen Institute for Brain Science, 551 North 34th Street, Seattle, WA, USA
`chinhda@alleninstitute.org`, `mikeh@alleninstitute.org`

**Abstract.** The Allen Brain Atlas (ABA), publicly available at http://
www.brain-map.org, presents the expression patterns of more than
21,500 genes in the adult mouse brain. The project has produced more
than 600 Terabytes of cellular level in situ hybridization data whose images have been reconstructed and mapped into whole brain 3D volumes
for search and viewing. In this application paper we outline the bioinformatics, data integration, and presentation approach to the ABA and
the creation a fully automated high-throughput pipeline to deliver this
data set to the Web.

## 1 Introduction

There are several high-throughput projects underway to systematically analyze
gene expression patterns in the mammalian central nervous system [1,2,3,4].
These projects strive to gain insight into temporal and spatial expression of
specific genes throughout development and in the adult brain. Advances in
genomic sequencing methods, high-throughput technology, and bioinformatics
through robust image processing now enable the neuroscience community to
study nervous system function at the genomic scale [5,6,7]. Central to these
neuro-genomics efforts is understanding gene transcription in the context of the
spatial anatomy and connectivity of the central nervous system [8].

The recent completion of the Allen Brain Atlas (ABA, www.brain-map.org)
offers a technology platform for viewing, browsing, and searching cellular level
resolution in situ hybridization gene expression data in the brain for over 21,500
genes of the standard laboratory C56Bl/6J mouse genome. The value of this
dataset for basic neuroscience research, medicine, and pharmaceutical drug target development is enormous and becoming increasingly recognized [9,10]. In the
ABA expressing genes can be viewed with tools that can pan/zoom from a whole-
brain section down to a single cell and retrieve data in a multi-resolution format.
The full technology solution of the ABA consists of an automated informatics
mapping platform capable of starting with two dimensional histological and in
situ hybridization images and ending with the construction of three-dimensional
maps or atlases for each gene. Each of these genes is comparable to a standard
de facto reference brain for anatomic comparison and localization.

S. Cohen-Boulakia and V. Tannen (Eds.): DILS 2007, LNBI 4544, pp. 17–26, 2007.

The ultimate data set of the ABA is on the order of 600 terabytes and will require a substantial data mining effort to understand and integrate with existing biomedical knowledge. In addition to a variety of informatics tools, such as organizing the data for anatomic search and implementation of expression detection filters, a three-dimensional gene atlas viewer is available that maps the location and intensity of gene expression onto a reference atlas. While this is a substantial step in browsing gene expression pattern summaries, much work remains to be done in connecting the data to the community.

In this paper we describe the architecture and methods of the ABA pipeline and web application. The aim is to provide an overview of this important resource so that in addition to the tools and services already presented, the data integration and bioinformatics community may offer solutions toward best how to integrate this community resource for maximum benefit to neurobiology.

## 2    Technology Requirements/Challenges of the ABA

Leveraging the success of the Human Genome Project, the Allen Brain Atlas is one of the first projects to utilize a genomic scale approach to neuroscience. The ABA employs high throughput methods to automate the in situ hybridization process in the standard C56Bl/6J adult mouse brain [10]. These streamlined processes include animal and tissue processing, riboprobe generation, in situ hybridization (ISH), Nissl staining, and anatomic and gene expression quantification. Details of these processes can be found in [10]. At capacity, 600 genes each at 200 um spaced sagittal sections were processed each week. Higher density coronal replicates were generated for a subset of genes with restricted or interesting gene expression patterns. Imaging of each tissue section at 10X magnification with resolution at 1.07 $\mu m^2$/pixel generated 80-120 tiles having 3-8MB per image tile. Data generation was approximately 1 terabyte per day with data processing done on more than 300 genes per day.

The throughput of the ABA requires a data management and analysis system capable of the significant throughput of the laboratory processes. Additionally, the open source nature of the project required an efficient mechanism to deliver data to the public. Our approach to addressing these challenges can be divided into four primary areas: bioinformatics, laboratory information management, data processing, and data presentation. In each area, we will detail the strategy as well as describe the technologies used to address the above challenges.

### 2.1    Bioinformatics

Probe generation requires bioinformatics support in the area of primer and template design. The majority of the genes used in the ABA are catalogued in the NCBI's Reference Sequence (RefSeq) collection, providing a non-redundant set of transcript sequences. Other external sources for gene templates not represented in RefSeq included TIGR, Celera, and the Riken FANTOM3 clone collection. Using a standardized PostgreSQL RDBMS [11], these external sources are uploaded into a staging environment before an automated application is run to update public gene and transcript metadata as well as importing new genes into a set of tables within the Laboratory Information Management System (LIMS) described below. Templates used to design probes come from different clone collections including MGC and Riken. Metadata for these clones and their sequences are also imported into the database via the same strategy as the gene metadata information. An internal custom web-based application has been created to support end-user querying of gene/transcript information, integration of BLAST and Primer3 algorithms for alignment and primer design, and upload of design and probe information into the database. These are available in LIMS for use in the ISH process.

## 2.2   Laboratory Information Management (LIMS)

A custom Java web-based LIMS was modeled to the ISH process used in the ABA. At the most general level, it is designed around the set of discrete tasks that define particular workflows. All tasks are recorded and object identifiers are assigned to all inputs and outputs of tasks. Three additional design items in the LIMS include quality control, usage of a controlled set of vocabularies, and reporting. In addition to generating outputs from each task, success and failures are recorded as well as their reasons for failures. Reports, using Crystal Reports [12] web API, are integrated into the LIMS for ease of creating report templates and allows end-user access to a number of reports relating to task plans and status, object states, and quality statistics. A web application allows users to conveniently access computer terminals throughout the laboratory areas. Since all tasks and objects are in a database, reliance on spreadsheets is eliminated and further allows for subsequent automated processing.

## 2.3   Data Processing

**Image Capture.** Tissue processing from the laboratory, once completed, is imaged using an automated microscopy system called the Image Capture System (ICS). The system consists of 10 Leica DM6000M microscopes, each equipped with a DC500 camera, a custom stage, and a barcode scanner mounted. A custom Scope Controller Application was created to control the movement of the stage and image tile acquisition. Barcoded microscope slides are read by the Scope Controller as they are loaded onto the stage. As image tiles are acquired, they are deposited to the Storage Area Network along with the barcode number.

**Software Architecture.** The data processing component of the ABA involves careful development and integration of software and hardware components. The Informatics Data Processing pipeline (IDP) is a data processing engine with its architecture centering around three components: scheduler, computing modules, and business logic.

*Scheduler.* This layer of the IDP engine was developed to monitor, queue computing jobs, manage load balancing, and failure recovery. To develop this system we used a combination of custom software development leveraging existing open source applications, mainly Torque and Maui [13]. Within the IDP, the engine submits jobs through the use of PBS scripts to manage the different states of jobs as it progresses through the pipeline. The system is also configurable to manage the number of job batches being run and the number of different job queues. Automated re-submission of jobs due to failure is incorporated as well. Torque and Maui are used to schedule jobs to the computing servers and manage the load balancing of the system.

*Computing Modules.* The IDP was designed to accept a variety of different Informatics computation modules. The system is not as generically configurable as systems such as Taverna [14], or the LONI pipeline [15] but rather optimized for the computational environment particular to the ABA. Each different computing method is developed separately and packaged as a module in the IDP. Communication between the modules and the IDP follows a simple input and output XML scheme. The ABA informatics automated pipeline consists of modules supporting the following functions:

- Image preprocessing, including tile stitching and direct compression into JPEG2000 format [16]
- Image storage and indexing,
- Access to a novel online digital reference atlas for the adult C56Bl/6J mouse brain [17],
- 3-D image reconstruction and deformable registration to bring the ISH images into a common anatomic framework,
- Signal detection and estimation for segmentation of expressing cells and tissues,
- Compilation of gene expression results over 3-D regions and presentation in an online searchable database,
- Visualization tools for examining 3-D expression patterns of multiple genes in anatomic regions.

The details of these algorithms and methods are given in [10] and [18].

*Business Logic.* Workflows are defined in the IDP by specifying which computing modules to run for a particular project data type. This layer in the IDP also interacts with other systems such as the LIMS and the internal web application hosting the ABA via a series of web services calls. Upon identification of a valid barcode on the directory file system produced by the ICS, a web service call, implemented using the XFire Codehaus API [19], is initiated to the LIMS. Metadata passed from the LIMS instructs the IDP engine which workflow to be used. As the data progresses to the end of the pipeline, the IDP packages all required information to present for visualization and passes it off to the web application for end-user usage.

**Hardware Architecture.** In addition to the head node, the computing modules of the IDP are run using a Fedora Linux cluster consisting of 106 CPUs, 32 HP BL35 Blades with dual AMD 2.4 Ghz, 4 GB RAM and 21 IBM HS20 Blades with dual Intel 2.8 Ghz, 4GB RAM. Once raw tiles are stitched and compressed, they are deep archived using the SpectraLogic T120 robotic tape library system. Stitched images, compressed 16:1 using a wavelet based JPEG2000 format are stored on the Hitachi Thunder 9585V Storage Area Network along with other intermediate files produced by the Informatics modules in the IDP. Figure 1 shows the architecture of the IDP including the computing modules of the ABA workflow.

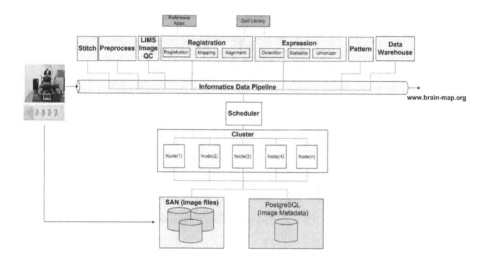

**Fig. 1.** The software and hardware architecture of the Informatics Data Pipeline, including the Informatics computation workflow of the ABA project

The Allen Brain Atlas hosting architecture is designed for performance, redundancy and ease of deployment. While servers can be added to any of the silos based on usage needs, the use of F5 Big IP allows for load balancing as well as fail over. This mutiple silos architecture has several advantages. Performance wise, it allows us to add additional servers to the silos and new silos can also be added as required. Since servers within each of the silos are independent of the other silos, this scheme provides additional redundancy within the system. During deployment upgrades and maintenance, single silos can be taken down without affecting the public availability of the website.

## 3   Data Presentation and Visualization

Successfully processed data are housed in a web application containing the primary image data as well as other summary data such as 2D expression quantification images, metadata, etc.. Two and three dimensional visualization of the data can be performed within the main application or Brain Explorer (described below), and both viewing mechanisms can be accessed from our public site at http://www.brain-map.org. In addition, computed data are made available through a number of different gene-based and structure-based searches within the application. Figure 2 shows the search interface with return results. In this instance a search was performed for genes expressing simultaneously high level and density in the cortex and hypothalamus. The ABA interface provides the return list with metadata as well as genes whose expression patterns are most correlated with the given gene image series.

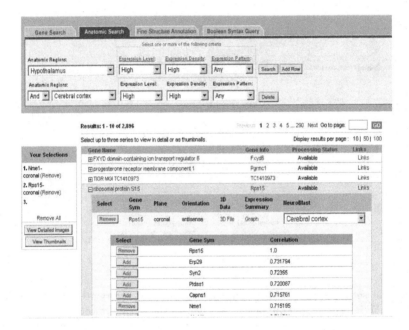

**Fig. 2.** ABA search interface with return list and metadata. Search can be based on expression level, density, or uniformity of signal. Genes can be individually loaded into detailed two or three dimensional viewing applications.

## 3.1 The Allen Brain Atlas Web Portal

In addition to serving images within the application, the database architecture of the Allen Brain Atlas is built using a warehousing approach to allow fast retrieval of textual image metadata information and their computed values. Metadata destined to be displayed externally are retrieved from the LIMS through a web service call from the IDP as successfully processed data are passed from the IDP to the web application along with informatics computed values. Informatics value contents and all associated metadata are flattened and warehoused in a simple schema centered on the concept of an image series.

It is now relatively common to present and manage high resolution imagery in bioinformatics applications. While our primary images are compressed in the

JPEG2000 format [18], delivering images through a browser in this format is infeasible due to the lack of native browser support without a plug-in. While some work has been done [20] in streaming images within the browser, ABA image sizes and the number of images to be displayed disfavor this approach. Using a JPEG multi-resolution tiled pyramid [21] allows browsing ease and is comparatively size effective. We created a Linux based custom modification of this approach to serve JPEG based pyramid images directly converted from the JPEG2000 originals. Additional development was done to integrate the Flash based user inferface within the web application as well as add additional end-user functionality.

The result of this architecture, shown in Figure 3, enables the development of a detailed image viewer that can support browsing of the reference atlas, individual ISH data sections, quantitation masks, as well as thumbnail selection and supporting metadata and a host of image manipulation options.

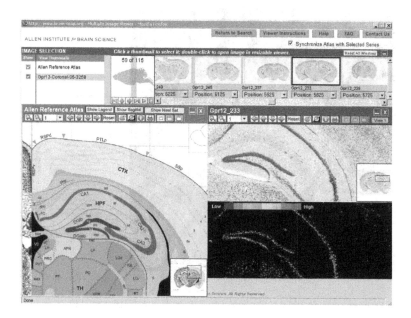

**Fig. 3.** The multiple image viewer of the ABA showing G-protein coupled receptor Gpcr-12 with closest reference plane section and expression pattern in the hippocampus. The right side images show the original ISH section and signal detection mask for expression.

## 3.2   3-D Viewing Tools

With gene expression data mapped into a common coordinate space, it is also possible to visualize and compare expression patterns in 3D. However, 3D expression visualization of brain data has not commonly been applied on a very large scale largely due to manual input requirements of validation and the task of accurate

3D reconstruction. Additionally limiting the usage of 3D visualization has been the necessary compromise in resolution relative to the resolution of 2D sections.

We developed a 3D viewing application, Brain Explorer, to visualize the Allen Reference Atlas and Allen Brain Atlas in three-dimensional space. The application has utility for both examining small sets of genes in detail as well as navigating the entire atlas of more than 21,500 genes. Our application also links the 3D representation of gene expression with the original full resolution 2D tissue sections. An area of interest in the 3D model can instantly link to the full resolution image (shown in Figure 3) for corroboration with the 3D model as well as detailed examination of subtle expression patterns. Figure 4 shows the gene protein kinase-C, prkcd implicated in the regulation of cell growth and death viewed in 3-D in Brain Explorer.

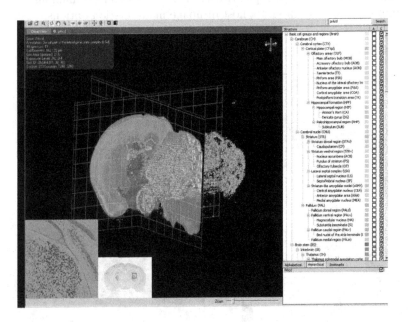

**Fig. 4.** The gene protein kinase-C,viewed in 3-D in Brain Explorer. The user has access to the full anatomy, original ISH section, maneuverable planes, and summary statistics.

## 4   Discussion

With the entire ABA data set now publicly available, current development focus is on additional features within the application for easier navigation, enabling users to download images, and more advanced mining tools. Because the ABA data set was calculated entirely by an automated pipeline, additional data integration and annotation can make the data set even more powerful. One way this can be achieved is by allowing the user community to bookmark genes and/or images of interest and assign a tag to describe the bookmark. Most popular and related tags, combined with the automated calculations can be retrieved to discover additional interesting annotations associated with the data. When implemented, other online content could integrate the information of the ABA with current data knowledge. Similarly, scientific information exposed using this framework could be retrieved by the ABA and returned to the users as added annotation bookmark tags and links to the existing ABA data. While this can be viewed as a tool to help link dynamic unstructured web data, there is great potential to create new knowledge from this integration strategy.

A detailed view of expression content in the brain can also be achieved by exposing the ABA's expression grid data, rather than summaries at the anatomic structural level. Achieving this type of API and providing structural ontology descriptions, as well as user community annotation bookmarks to the external world may best be done in the context of the Semantic Web, a novel paradigm for web information exchange and integration. Technology standards such as XML or JSON [22], Resource Description Framework (RDF) and Web Ontology Language (OWL) can be combined to provide a template for information exchange architecture.

Neurobiological content in the ABA data set can best be leveraged by effective integration with other online scientific data sources. It is widely recognized that lack of successful data integration is one of the most significant barriers to comprehensive understanding of the biological significance of available data. Ultimately, effective data integration, particularly for genomic and proteomic content, should contribute to more specific understanding of biological processes from fundamental science to the selection of pharmaceutical drug targets. One of the most challenging goals for the Allen Brain Atlas, or any large scale image database in the life sciences, is to expose its content to maximum advantage with respect to our present knowledge.

## Acknowledgements

This work was sponsored by the Allen Institute for Brain Science. The authors would like to thank the Institute Founders, Paul G. Allen and Jody Patton for their vision and support; The ABA Scientific Advisory Board for overall leadership of the ABA project; Carey Teemer, Kirk Larsen, Reena Kawal, Bryan Smith, Kevin Brockway, Rob Young, Tim Fliss, Mei Chi Chin, and Cliff Frensley for the development of www.brain-map.org; Jimmy Chong, Madhavi Sutram, and Tim Dolbeare for the development of the LIMS and internal tools; and Cliff Slaughterbeck for the development of the ICS system.

## References

1. Su, A.I., et al.: Large-scale analysis of the human and mouse transcriptomes. In: Proc. Natl. Acad. Sci. USA 99, 4465–4470 (2002)
2. Gong, S., et al.: A gene expression atlas of the central nervous system based on bacterial artificial chromosomes. Nature 425, 917–925 (2003)

3. Sandberg, R., et al.: Regional and strain-specific gene expression mapping in the adult mouse brain. In: Proc. Natl. Acad. Sci. USA 97, 11038–11043 (2000)

4. Carninci, P., et al.: The transcriptional landscape of the mammalian genome. Science 309, 1559–1563 (2005)

5. Siddiqui, A.S., et al.: A mouse atlas of gene expression: Large-scale digital gene-expression profiles from precisely defined developing C57BL/6J mouse tissues and cells. In: Proceedings of the National Academy of Sciences 102, 18485 (2005)

6. MacKenzie-Graham, A., et al.: The informatics of a C57BL/6J mouse brain atlas. Neuroinformatics 1, 397–410 (2003)

7. Schwartz, E.L., et al.: Applications Of Computer-Graphics And Image-Processing To 2D And 3D Modeling Of The Functional Architecture Of Visual-Cortex. IEEE Computer Graphics And Applications 8, 13–23 (1988)

8. Boguski, M.S., Jones, A.R.: Neurogenomics: at the intersection of neurobiology and genome sciences. Nature Neuroscience 7, 429–433 (2004)

9. Markram, H.: Industrializing Neuroscience, Nature 445 (January 11, 2007)

10. Lein, E., Hawrylycz, M., et al.: Genome-wide atlas of gene expression in the adult mouse brain. Nature 445(7124), 168–176 (2007)

11. http://www.postgresql.org

12. http://www.businessobjects.com

13. http://www.clusterresources.com

14. Hull, D., Wolstencroft, K., Stevens, R., Goble, C., Pocock, M.R., Li, P., Oinn, T.: Taverna: a tool for building and running workflows of services, Nucleic Acids Res 34 (July 1, 2006)

15. Rex, D.E., Ma, J.Q., Toga, A.W.: The LONI Pipeline Processing Environment

16. http://www.kakadusoftware.com

17. Dong, H.W.: Allen Reference Atlas Wiley (June 2007)

18. Ng, L., Pathak, S., Kuan, L., Lau, C., Dong, H.-W., Sodt, A., Dang, C., Avants, B., Yushkevich, P., Gee, J., Haynor, D., Lein, E., Jones, A., Hawrylycz, M.: Neuroinformatics for Genome-wide 3-D Gene Expression Mapping in the Mouse Brain, IEEE Transactions on Computational Biology and Bioinformatics (in press)

19. http://xfire.codehaus.org

20. Janosky, J., Witthus, R.: Using JPEG, for Enhanced Preservation and Web Access of Digital Archives-A Case Study (2000) http://charlesolson.uconn.edu

21. http://www.zoomify.com

22. http://www.json.org/

# Toward an Integrated RNA Motif Database

Jason T.L. Wang[1], Dongrong Wen[1], Bruce A. Shapiro[2],
Katherine G. Herbert[3], Jing Li[4], and Kaushik Ghosh[4]

[1] Bioinformatics Program and Department of Computer Science
New Jersey Institute of Technology, Newark, NJ 07102
[2] Center for Cancer Research Nanobiology Program
National Cancer Institute, Frederick, MD 21702
[3] Department of Computer Science
Montclair State University, Montclair, NJ 07043
[4] Applied Statistics Program and Department of Mathematical Sciences
New Jersey Institute of Technology, Newark, NJ 07102

**Abstract.** In this paper we present the design and implementation of an RNA structural motif database, called RmotifDB. The structural motifs stored in RmotifDB come from three sources: (1) collected manually from biomedical literature; (2) submitted by scientists around the world; and (3) discovered by a wide variety of motif mining methods. We present here a motif mining method in detail. We also describe the interface and search mechanisms provided by RmotifDB as well as techniques used to integrate RmotifDB with the Gene Ontology. The RmotifDB system is fully operational and accessible on the Internet at http://datalab.njit.edu/bioinfo/

## 1 Introduction

Post-transcriptional control is one of the mechanisms that regulate gene expression in eukaryotic cells. RNA elements residing in the UnTranslated Regions (UTRs) of mRNAs have been shown to play various roles in post-transcriptional control, including mRNA localization, translation, and mRNA stability [16]. RNA elements in UTRs can be roughly divided into three groups: elements whose functions are primarily attributable to their sequences, elements whose functions are attributable to their secondary or tertiary structures, and elements whose functions are attributable to both of their sequences and structures. Well-known sequence elements include AU-rich elements (AREs), some of which contain one or several tandem AUUUA sequences and are involved in regulating mRNA stability [2], and miRNA target sequences, which are partially complementary to cognate miRNA sequences and are involved in regulating translation or mRNA stability [14].

Well-known structure elements (or structural motifs) include the histone 3′-UTR stem-loop structure (HSL3) and the iron response element (IRE) [16]. Both sequence and structure are important for the functions of the structural motifs. HSL3 is a stem-loop structure of about 25 nucleotides (nt) that exists in

S. Cohen-Boulakia and V. Tannen (Eds.): DILS 2007, LNBI 4544, pp. 27–36, 2007.
© Springer-Verlag Berlin Heidelberg 2007

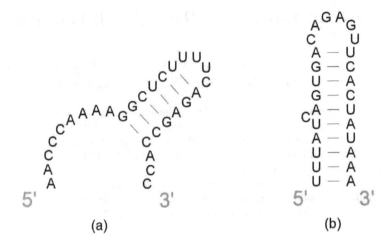

**Fig. 1.** (a) An example of the HSL3 motif. (b) An example of the IRE motif.

3′-UTRs of most histone genes. In Figure 1(a), an HSL3 motif is portrayed using the XRNA tool (http://rna.ucsc.edu/rnacenter/xrna/xrna.html). The HSL3 structure is critical for both termination of the transcription of mRNAs and stability of mRNAs. These functions are exerted by the stem-loop binding protein (SLBP) that interacts with HSL3.

IRE is a stem-loop structure of ∼30 nt with a bulge or a small internal loop in the stem (Figure 1(b)). IREs have been found in both 5′-UTRs and 3′-UTRs of mRNAs whose products are involved in iron homeostasis in higher eukaryotic species. IREs bind to the iron regulatory proteins (IRPs) of these species, which control translation and stability of IRE-containing mRNAs.

HSL3 and IRE are similar in several aspects: both are small simple RNA structures with less than 40 nt; both exist in UTRs of several genes with related functions; and both bind to cellular proteins and are involved in post-transcriptional gene regulation. The regulations via HSL3 and IRE constitute a distinct mode of gene regulation, whereby expression of several genes can be modulated via a common RNA structure in UTRs.

Functional sequence motifs in genomes have been heavily studied in recent years, particularly for the promoter region and sequences involved in splicing. In contrast, RNA structure elements have been investigated to a much lesser extent, largely due to the difficulties in predicting correct RNA structures and conducting RNA structure alignments, where huge computing costs are involved. While some success has been achieved using phylogenetic approaches [1] and sequence alignments [3,4] to gain accuracy in RNA structure prediction, large-scale mining for conserved structures in eukaryotic UTRs has been studied to a lesser extent. In addition, current methods for finding common stem-loop structures solely rely on the detection of structural similarities [9]. Gene Ontology information has not been used in the study of RNA structure, though integrating ontologies with other biological data has been studied extensively.

Here we present a database, called RmotifDB, that contains structural motifs found in 5′ and 3′ UTRs of eukaryotic mRNAs. The structural motifs are linked with Gene Ontology and PubMed entries concerning the motifs. A wide variety of motif mining methods are developed. In particular, we present in the paper a histogram-based method for discovering motifs in eukaryotic UTRs. In Section 2 we describe the histogram-based method in detail. Section 3 describes RmotifDB as well as its interface and search mechanisms. Section 4 presents techniques used to integrate RmotifDB with the Gene Ontology. Section 5 concludes the paper and points out some directions for future research.

## 2   A Motif Mining Method

We have developed several structural motif mining methods based on different RNA representation models. For example, in [19], we represented an RNA secondary structure using an ordered labeled tree and designed a tree matching algorithm to find motifs in multiple RNA secondary structures. More recently, we developed a loop model for representing RNA secondary structures. Based on this loop model we designed a dynamic programming algorithm, called RSmatch [15], for aligning two RNA secondary structures. The time complexity of RSmatch is $O(mn)$, where $m$ and $n$ are the sizes of the two compared structures, respectively. The RSmatch method is implemented in a web server, called RADAR (acronym for *RNA Data Analysis and Research*) [13], which is accessible at http://datalab.njit.edu/biodata/rna/RSmatch/server.htm. In Figure 2, the common region of two RNA secondary structures pertaining to homo sapiens sequences is portrayed using XRNA where the local matches found by RSmatch are highlighted with the green color.

Below, we describe a histogram-based scoring method to uncover novel conserved RNA stem-loops in eukaryotic UTRs using the RSmatch tool. This method is an upward extension of our previously developed histogram-based algorithm for DNA sequence classification [18]. Given a set of RNA secondary structures, the method uses RSmatch to perform pairwise alignments by comparing two RNA structures at a time in the set. Given an optimal local alignment between two structures $A$ and $B$ found by RSmatch, the set of bases in the aligned region of $A$ is denoted by $\mathcal{Q}_A = \{A_i, A_{i+1}, ..., A_j\}$ where $A_i$ ($A_j$, respectively) is the 5′-most (3′-most, respectively) nucleotide not aligned to a gap. The set of bases in the aligned region of $B$ is denoted by $\mathcal{Q}_B = \{B_m, B_{m+1}, ..., B_n\}$ where $B_m$ ($B_n$, respectively) is the 5′-most (3′-most, respectively) nucleotide not aligned to a gap. Each nucleotide $A_k \in \mathcal{Q}_A$ that is not aligned to a gap scores $|j - i + 1|$ points. All the other bases in the structure $A$ receive 0 point. Thus, the larger the aligned region between $A$ and $B$, the higher score each base in the region has. When aligning the structure $A$ with another structure $C$, some bases in $\mathcal{Q}_A$ may receive non-zero points and hence the scores of those bases are accumulated. Thus, the bases in a conserved RNA motif will have high scores.

To validate our approach, we conducted experiments to evaluate the effectiveness of this scoring method. The conserved stem-loops we considered were

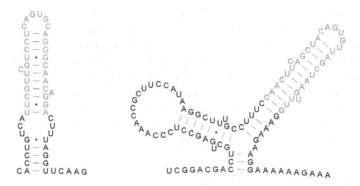

**Fig. 2.** Alignment of two RNA secondary structures where the local matches found by RSmatch are highlighted with the (light) green color

IRE motifs, which contained about 30 nucleotides, located in the 5'-UTRs or 3'-UTRs of mRNAs coding for proteins involved in cellular iron metabolism. The test dataset was prepared as follows. By searching human RefSeq mRNA sequences from the National Center for Biotechnology Information (NCBI) at http://www.ncbi.nlm.nih.gov/RefSeq/, we obtained several mRNA sequences, within each of which at least one IRE motif is known to exist. We then extracted the sequences' UTR regions as indicated by RefSeq's GenBank annotation and used PatSearch [10] to locate the IRE sequences. Each IRE sequence was then extended from both ends to obtain a 100 nt sequence. These sequences were mixed with several "noisy" sequences with the same length, where the noisy sequences are UTR regions of mRNA sequences that do not contain IRE motifs. All the resulting sequences were then folded by the Vienna RNA package [11] using the "RNAsubopt" function with setting "-e 0". This setting can yield multiple RNA structures with the same free energy for any given RNA sequence.

Figure 3 shows the score histograms for two tested RNA structures. It was observed that clusters of bases with high scores correspond to the IRE motifs in the RNA structures. Similar clusters of bases with high scores corresponding to the IRE motifs were observed in the other IRE-containing RNA structures, but not in the "noisy" structures. This result indicates that our histogram-based scoring method is able to detect biologically significant motifs in multiple RNA structures.

## 3   The RmotifDB System

RmotifDB is designed for storing the RNA structural motifs found in the UTRs of eukaryotic mRNAs. It is a web-based system which supports retrieval and access of RNA structural motifs from its database. The system allows the user to search RNA structural motifs in an effective and friendly way. RmotifDB is accessible on the Internet at http://datalab.njit.edu/bioinfo/. It is developed using Perl-CGI, Java, C and Oracle.

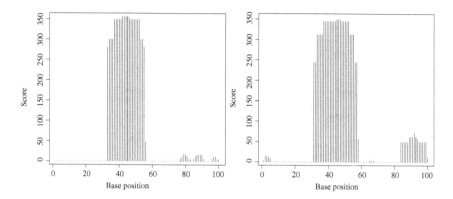

**Fig. 3.** Diagrams illustrating the effectiveness of the proposed scoring method. IRE is found around base positions 30-60 in the RNA structures corresponding to the diagrams respectively.

The RNA structural motifs stored in RmotifDB come from three sources. The primary source consists of manually collected motifs from biomedical literature. Scientists who used this database can also submit motifs to RmotifDB. The interface where scientists can submit RNA structural motifs is shown in Figure 4. Lastly, motifs are obtained from those RNA structures discovered by a wide variety of motif mining methods (such as the method described in Section 2). These motif mining methods may find new, or unknown motifs, which are also stored in RmotifDB.

Figure 5 shows the search interface of RmotifDB. The system provides two search options: query by sequence (QBS) and query by structure (QBR). With QBS, the user enters an RNA sequence in the standard FASTA format and the system matches this query sequence with motifs in the database using either RSmatch or Infernal [8]. Since RSmatch accepts, as input data, RNA secondary structures only, the system needs to invoke Vienna RNA v1.4 [11] to fold the query sequence into a structure before a match is performed. With QBR, the user enters an RNA secondary structure represented by the Vienna style Dot Bracket format [11] and the system matches this query structure with motifs in the database using RSmatch. The result is a ranked list of motifs that are approximately contained in the query sequence or the query structure. In addition, the user can search RmotifDB by choosing a Gene ID or RefSeq ID from a pre-defined list of Gene IDs and RefSeq IDs provided by the RmotifDB system where the Gene IDs and RefSeq IDs are obtained from http://www.ncbi.nlm.nih.gov/RefSeq/. This pre-defined list contains the IDs of the genes (mRNA sequences) used by our motif mining methods to discover the structural motifs stored in RmotifDB. The result of this search is a list of structural motifs containing the query gene ID (Gene ID or RefSeq ID).

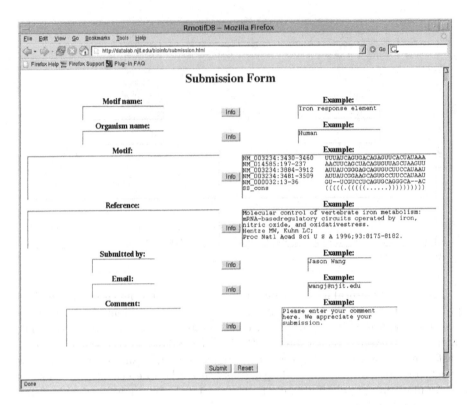

**Fig. 4.** The interface of RmotifDB where scientists can submit RNA structural motifs

## 4  Integrating RmotifDB with the Gene Ontology

In browsing the search results returned by RmotifDB, the user can click each motif to see detailed information concerning the motif. Figure 6 shows the result of
displaying a motif and its related information. Here the motif is an iron response
element (IRE) in humans shown in the Stockholm format [8]. This format is a
multiple sequence alignment output with structural annotation in the Vienna
style Dot Bracket format. The motif is depicted in the bottom right-hand corner
of the window. Also displayed are the Gene Ontology (GO) information concerning the motif and relevant articles in PubMed (not shown in the screenshot)
that publish this motif.

In general, a motif contains multiple genes (mRNA sequences) with similar
functions. The GO entries and their URLs that are highly associated with the
motif are collected and stored in RmotifDB. The GO entries in three categories,
including molecular function, biological process and cell component, are obtained
from the Gene Ontology Consortium (http://www.geneontology.org). The mapping information between the GO entries and the genes is obtained from the
LocusLink database [17]. A hypergeometric test [6] is used to measure the significance of the association between the motif and each of the GO entries. The

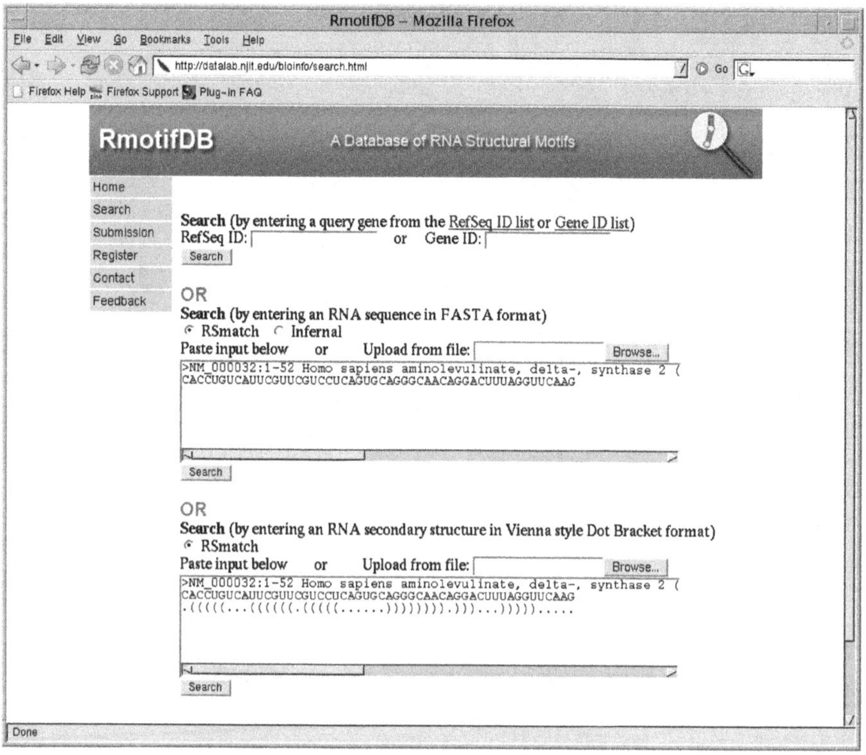

**Fig. 5.** The search interface of RmotifDB

significance is shown as the $t$-value next to each GO entry in Figure 6. The hypergeometric test is appropriate here, since it is a finite population sampling scheme with the entire population being divided into two groups—those that are associated with a particular GO entry and those that are associated with the other GO entries.

In the hypergeometric test, there are four parameters: (1) $m$, the number of white balls in an urn, (2) $n$, the number of black balls in the urn, (3) $k$, the number of balls drawn from the urn, and (4) $x$, the number of white balls drawn from the urn. The probability that $x$ out of the $k$ balls drawn are white from the urn containing $m + n$ balls is

$$f(x, m, n, k) = \frac{\binom{m}{x}\binom{n}{k-x}}{\binom{m+n}{k}} \tag{1}$$

where $x \leq \min(m, k)$.

For each RNA structural motif $M$ containing multiple genes, all GO entries are examined to evaluate their associations with $M$. Through the mapping

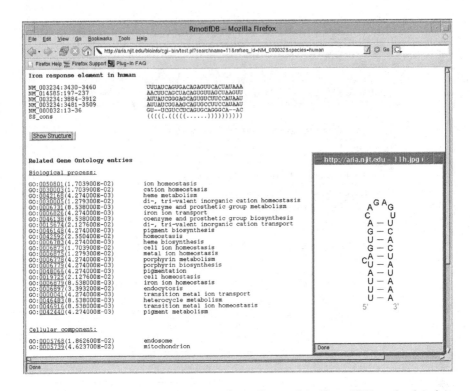

**Fig. 6.** The output showing a structural motif stored in RmotifDB and related information. The $t$-value inside the parentheses next to each GO entry indicates the significance of the association between the motif and the GO entry. The smaller the $t$-value, the more significant the association is.

information between $M$ and a GO entry $G$ in a GO category $C$ [17], we are able to calculate four numbers: (1) $N_1$, the number of genes associated with any GO entry in $C$, (2) $N_2$, the number of genes associated with $G$ in $C$, (3) $N_3$, the number of genes in $M$ associated with any GO entry in $C$, and (4) $N_4$, the number of genes in $M$ associated with $G$ in $C$, where $N_1 \geq N_2$ and $N_3 \geq N_4$. The $t$-value of the GO entry $G$ is calculated by

$$t(G) = f(N_4, N_2, N_1 - N_2, N_3) \qquad (2)$$

where the function $f$ is as defined in Equation (1). In general, the smaller the $t(G)$ value, the more significant the association between $G$ and $M$ is. RmotifDB displays $G$ together with its $t$-value if $t(G)$ is smaller than a user-adjustable parameter value (0.05 here).

## 5    Conclusion

In this paper we presented an RNA structural motif database, called RmotifDB, and described some features of RmotifDB as well as techniques used for integrating RmotifDB with the Gene Ontology. Besides integrating with the Gene Ontology, RmotifDB also supports standard formats such as accession numbers from ReqSeq as well as commonly used Gene IDs. Therefore, anyone familiar with data associated with these identification numbers can retrieve the sequences without difficulty. It is our plan in the future to integrate RmotifDB further with other popular databases creating a seamless environment for the user.

We developed a motif mining method capable of discovering structural motifs in eukaryotic mRNAs. Our system provides a search interface supporting structure-based searches in RNAs. Studying RNA data has been a popular topic in the biological database community. There are many databases and software on RNA (cf. the RNA World Site at http://www.imb-jena.de/RNA.html). Most of these databases, however, while providing structural information, do not give the user the power to query using the structural information like RmotifDB.

The system presented here is part of a long-term project [13,20] aiming to build a cyberinfrastructure for RNA data mining and data integration. This cyberinfrastructure complements existing RNA motif databases such as Rfam and UTRdb [12,16], which lack structure-based search functions. Data mining and data integration [5,7] in bioinformatics has emerged as an important field at the interface of information technology and molecular biology. Our cyberinfrastructure will contribute to this field in general, and RNA informatics in particular. In future work we plan to develop new data mining and data integration techniques for finding motifs in various organisms and for integrating the motifs with several biomedical ontologies.

## Acknowledgments

The authors thank the anonymous reviewers for their constructive suggestions, which helped to improve the presentation and quality of this paper. We also thank Drs. Jianghui Liu, Vivian Bellofatto, Yongkyu Park and Bin Tian of the New Jersey Medical School in the University of Medicine and Dentistry of New Jersey for helpful discussions while preparing this paper.

## References

1. Akmaev, V.R., Kelley, S.T., Stormo, G.D.: Phylogenetically enhanced statistical tools for RNA structure prediction. Bioinformatics 16, 501–512 (2000)
2. Bakheet, T., Frevel, M., Williams, B.R., Greer, W., Khabar, K.S.: ARED: human AU-rich element-containing mRNA database reveals an unexpectedly diverse functional repertoire of encoded proteins. Nucleic Acids Res. 29, 246–254 (2001)
3. Bindewald, E., Shapiro, B.A.: RNA secondary structure prediction from sequence alignments using a network of k-nearest neighbor classifiers. RNA 12, 342–352 (2006)
4. Bindewald, E., Schneider, T.D., Shapiro, B.A.: CorreLogo: an online server for 3D sequence logos of RNA and DNA alignments. Nucleic Acids Res. 34, 405–411 (2006)

5. Cohen.-Boulakia, S., Davidson, S.B., Froidevaux, C.: A user-centric framework for accessing biological sources and tools. In: Proc. of the 2nd International Workshop on Data Integration in the Life Sciences, pp. 3–18 (2005)

6. Dalgaard, P.: Introductory Statistics with R. Springer (2004)

7. Davidson, S.B., Crabtree, J., Brunk, B.P., Schug, J., Tannen, V., Overton, G.C., Stoeckert, Jr.C.J.: K2/Kleisli and GUS: experiments in integrated access to genomic data sources. IBM Systems Journal 40, 512–531 (2001)

8. Eddy, S.R.: A memory-efficient dynamic programming algorithm for optimal alignment of a sequence to an RNA secondary structure. BMC Bioinformatics, vol. 3(18) (2002)

9. Gorodkin, J., Stricklin, S.L., Stormo, G.D.: Discovering common stem-loop motifs in unaligned RNA sequences. Nucleic Acids Res. 29, 2135–2144 (2001)

10. Grillo, G., Licciulli, F., Liuni, S., Sbisa, E., Pesole, G.: PatSearch: a program for the detection of patterns and structural motifs in nucleotide sequences. Nucleic Acids Res. 31, 3608–3612 (2003)

11. Hofacker, I.L.: Vienna RNA secondary structure server. Nucleic Acids Res. 31, 3429–3431 (2003)

12. Jones, S.G., Moxon, S., Marshall, M., Khanna, A., Eddy, S.R., Bateman, A.: Rfam: annotating non-coding RNAs in complete genomes. Nucleic Acids Res. 33,D121–D124 (2005)

13. Khaladkar, M., Bellofatto, V., Wang, J.T.L., Tian, B., Zhang, K.: RADAR: an interactive web-based toolkit for RNA data analysis and research. In: Proc. of the 6th IEEE Symposium on Bioinformatics and Bioengineering, pp. 209–212 (2006)

14. Lewis, B.P., Shih, I.H., Jones-Rhoades, M.W., Bartel, D.P., Burge, C.B.: Prediction of mammalian microRNA targets. Cell. 115, 787–798 (2003)

15. Liu, J., Wang, J.T.L., Hu, J., Tian, B.: A method for aligning RNA secondary structures and its application to RNA motif detection. BMC Bioinformatics, vol 6(89) (2005)

16. Mignone, F., Grillo, G., Licciulli, F., Iacono, M., Liuni, S., Kersey, P.J., Duarte, J., Saccone, C., Pesole, G.: UTRdb and UTRsite: a collection of sequences and regulatory motifs of the untranslated regions of eukaryotic mRNAs. Nucleic Acids Res. 33, D141–D146 (2005)

17. Pruitt, K.D., Katz, K.S., Sicotte, H., Maglott, D.R.: Introducing RefSeq and LocusLink: curated human genome resources at the NCBI. Trends Genet. 16, 44–47 (2000)

18. Wang, J.T.L., Rozen, S., Shapiro, B.A., Shasha, D., Wang, Z., Yin, M.: New techniques for DNA sequence classification. Journal of Computational Biology 6, 209–218 (1999)

19. Wang, J.T.L., Shapiro, B.A., Shasha, D., Zhang, K., Currey, K.M.: An algorithm for finding the largest approximately common substructures of two trees. IEEE Transactions on Pattern Analysis and Machine Intelligence 20, 889–895 (1998)

20. Wang, J.T.L., Wu, X.: Kernel design for RNA classification using support vector machines. International Journal of Data Mining and Bioinformatics 1, 57–76 (2006)

# B-Fabric: A Data and Application Integration Framework for Life Sciences Research

Can Türker, Etzard Stolte, Dieter Joho, and Ralph Schlapbach

Functional Genomics Center Zurich (FGCZ), UZH / ETH Zurich
Winterthurerstrasse 190, CH–8057 Zurich, Switzerland
{tuerker,stolte,joho,schlapbach}@fgcz.ethz.ch

**Abstract.** Life sciences research in general and systems biology in particular have evolved from the simple combination of theoretical frameworks and experimental hypothesis validation to combined sciences of biology/medicine, analytical technology/chemistry, and informatics/statistics/modeling. Integrating these multiple threads of a research project at the technical and data level requires tight control and systematic workflows for data generation, data management, and data evaluation. Systems biology research emphasizes the use of multiple approaches at various molecular and functional levels, making the use of complementing technologies and the collaboration of many researchers a prerequisite. This paper presents B-Fabric, a system developed and running at the Functional Genomics Center Zurich (FGCZ), which provides a core framework for integrating different analytical technologies and data analysis tools. In addition to data capturing and management, B-Fabric emphasizes the need for quality-controlled scientific annotation of analytical data, providing the ground for integrative querying and exploitation of systems biology data. Users interact with B-Fabric through a simple Web portal making the framework flexible in terms of local infrastructure.

## 1 Introduction

Experimental research in the life sciences in general and systems biology as an emerging concept aims at characterizing complex biological organisms and functions at the systems level [3]. To achieve the required data width and depth with high accuracy and reliability, all components of a research project have to fulfill highest standards: experimental design needs to feature proper statistics and selection of experimental parameters and biological systems, experimental analysis requires to generate highly accurate and reproducible data at the highest sensitivity and specificity level for various molecular species, and data analysis has to ensure proper statistics and interpretation in order to allow for validation of the data in iterative experiment and knowledge discovery cycles.

In practical terms, scientists working in a life science research project follow a general data acquisition workflow [6] using different analytical platforms and instruments. The instruments generate raw data files which are usually stored on the instrument PC, together with the parameter files from the analysis run. In a second step, these files are copied to different PCs where analytical software tools are running. There, reformatting or processing of the raw data takes place and the analysis software generates additional data files in the form of report files that also have attached processing and visualization

S. Cohen-Boulakia and V. Tannen (Eds.): DILS 2007, LNBI 4544, pp. 37–47, 2007.
© Springer-Verlag Berlin Heidelberg 2007

parameters or bear additional links to external references. Already at this stage, analytical data is inherently distributed across multiple files in multiple locations. After going through additional cycles of data generation and analysis, data management and tracking becomes a challenge for the individual researcher. In a collaborative environment, the issues of data consistency and accessibility are even more critical. Unless all data and files generated in the analytical process are managed together with their semantic context, i.e., the lab book annotation of files and data, usage of the primary data by other researchers in the collaboration or peer-reviewing network is not possible. Thus, huge data resources will become useless for future integrated research. However, capturing the analytical raw data together with its semantic context is not easy. Usually, the instrument software is configured to place the acquired raw data on a file server. Setting up and maintaining the software environment so that the software reads and writes its data from and to a central data drive is highly software-specific. As the semantic context of an analytical data set not only includes the instrument configuration but also all relevant information from upstream workflow steps including the generation and treatment of the analytical sample, this metadata is usually not accessible since it is only written in the lab book of the researcher. Hence, mechanisms are required to enforce users to provide this information together with the experimental data.

As the complexity of the analytical systems in integrated functional genomics or systems biology research has reached a level where specific, isolated application-oriented data management and analysis has become apparently inefficient, the B-Fabric project has been initiated at the Functional Genomics Center Zurich (FGCZ) to build an infrastructure for integrated management of experimental data, scientific annotations, and metadata. B-Fabric also functions as a central LIMS which supports successful research 1) by integrating data from different sources, and 2) by ensuring that the dispersed data is available for future use. B-Fabric covers all main aspects of an integration framework, including safety, standard access protocols, transparent access, authentication, authorization, and access control. In particular, B-Fabric enforces the entry of scientifically and analytically required data such that users will then be able to browse and search for data entries across experiments, projects, and instruments.

In this paper, we introduce B-Fabric. Section 2 sketches its basic notions and goals while Section 3 presents its architecture and highlights some implementation issues. Workflow issues are discussed in Section 4. Section 5 shows how an instrument can be integrated into the B-Fabric framework. Section 6 concludes the paper.

## 2   Basic Notions and Goals of B-Fabric

Each *research project* using integrative functional genomics technologies may encompass multiple experiments. An experiment may contain a number of measurements on a number of samples and extracts, respectively. Following standards for systems biology [1], such as MIAME, MGED, or SBML, a *sample* is a biological source, from which the nucleic acids, proteins, or any other species of molecules to be investigated can be extracted from. In a controlled setting, each sample is given a unique ID and the sample is associated with a number of annotations, such as species (organism) for model systems or general terms for patient samples including sex or age. Further annotations that are relevant to a particular sample are its development stage, organism part (tissue or organ), cell type, genetic variation, individual genetic characteristics, state (disease vs. normal), treatment type (e.g., pharmacological treatment, heat shock, food deprivation), and separation technique. The notion of an *extract* refers to a sample derivate that is used in a measurement. An extract is prepared from a sample in a specific way using laboratory techniques. Each extract has a unique ID and is associated with a number of annotations, which are precisely describing how the extract was prepared. Useful extract annotations are the extraction method, extraction type (for example for nucleic acids: RNA, mRNA, genomic DNA), amplification (for nucleic acids: RNA polymerases, PCR) and purification parameters (precipitation, spin column). As a special feature, an extract may be prepared directly from a sample or can be composed of multiple extracts.

In the nomenclature for the analytical workflow, each measurement may contain more than one part, called a *workunit*. A workunit captures related measurements at a specific point in time of the experiment. It contains all the files that are necessary to restore the instrument software to the exact state when this type of snapshot was taken, including instrument settings files, software parameters, and log files. Within the workflow, the user decides when a workunit is complete, specifies the created data, adds metadata and commits it to B-Fabric central repository. The notion of a workunit is central to B-Fabric since it defines the main unit of storage and search. From a data modeling point of view, a workunit associates related file(s) with metadata, e.g., information about the extract used in the measurement. The types of files as well as the metadata may vary from instrument to instrument. Moreover, the workunit-file-extract association may vary from instrument to instrument. In some cases, each workunit is associated with one (raw) file and a corresponding extract. In other cases, a workunit comprises several files associated to one extract, or a workunit may even consist of several file-extract associations.

The basic goal of B-Fabric is to store all data generated from experiments centrally in a way that allows the subsequent data access, analysis, and visualization to be location-independent and to combine the data with all relevant metadata at the point of data generation. In detail, B-Fabric supports experimental researchers and bioinformaticians in their daily work, providing a number of advantages:

- *Data capturing and provisioning*: All experimental data is captured with its semantic context and is provided as needed.
- *Reproducibility*: Experimental data is stored together with instrument parameters and configuration files in order to being able to reproduce an experiment.
- *Uniform access*: All data (experimental, derived, metadata) is uniformly accessible from everywhere through a Web portal.
- *Federated search and integrated analysis*: Users may run queries against all public B-Fabric data (about projects, experiments, samples, extracts etc.). In this way, an integrated, inter-experiment, inter-project analysis becomes possible.
- *Transparency*: The user does not need to care about where and how the data is stored. B-Fabric functions as data fabric capturing and providing the data transparently through a Web portal.
- *Reliability*: All B-Fabric data is continuously backed up.
- *Security*: User's data is stored in a secure repository that is access controlled.

# 3   Architecture and Implementation of B-Fabric

The architecture of B-Fabric consequently follows the idea of a system gluing a set of loosely-coupled components together to satisfy the specific requirements of the overall system. Figure 1 gives an overview of the architecture of B-Fabric.

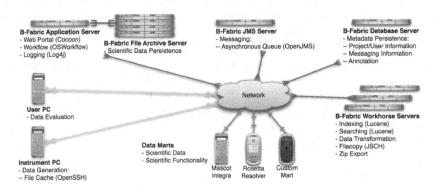

**Fig. 1.** Overview of the Architecture of B-Fabric

- *User PCs* are usual computers running a Web browser to enable access to B-Fabric. From a user PC, a scientist interacts with B-Fabric to search for and download a workunit (or parts of it) for analysis purposes. The user PCs at FGCZ provide a number of tools for data analysis and/or data visualization.
- *Instrument PCs* refer to computers that generate or hold scientific data to be imported into B-Fabric. From an instrument PC, a user interacts with B-Fabric through a Web browser to create and commit a workunit.
- *Data Marts* refer to external (autonomous) systems that supports specialized scientific data management, analysis, and/or visualization. Users use the marts to investigate the results of their experiments in detail. In its current setting, B-Fabric has one mart (Rosetta Resolver) for the detailed management and analysis of transcriptomics experiments and one mart (Mascot Integra) for the detailed management and analysis of proteomics experiments. For some instrument PCs, the associated workunits cannot be assigned to one of these two marts since the marts are not able to capture all specific information of the workunit. For these cases, B-Fabric implements a custom mart which is able to manage the corresponding information.
- *B-Fabric Application Server* acts as Web portal providing users a consistent and access controlled front-end. It is responsible for all user interaction with B-Fabric, and consequently implements the presentation logic. Among others, this portal includes forms for the scientific annotation of workunits and interfaces for browsing and searching. The portal can be accessed by a usual Web browser from everywhere, e.g. from an instrument PC to import data into B-Fabric or from a user PC to download data from B-Fabric. The portal dynamically adapts its layout and underlying workflows according to the computer from which the user accesses B-Fabric, e.g., in case of an access via an instrument PC, the workflow for creating and storing a workunit is adapted to the specifics of the corresponding instrument.

- *B-Fabric File Archive Server* stores and manages the workunits. All associated data files are stored in the file archive that is only accessible via the B-Fabric portal.
- *B-Fabric Database Server* stores and manages all metadata of B-Fabric, including the base administrative data, such as project or user data, as well as the annotations of the workunits, samples etc. Above that, it manages all messaging information that is used to communicate between the B-Fabric components.
- *B-Fabric JMS Server* provides persistent message queues for asynchronous communication between B-Fabric components. The latter place their messages (tasks) in the corresponding queues which are monitored by the responsible components.
- *B-Fabric Workhorse Server(s)* waits for messages arriving on the queues configured for a process component doing a single task. Typical tasks of a workhorse are the copying of data from an instrument PC to the file archive, the metadata extraction from the experimental data, and the indexing of the data for the search engine, to name a few. After having performed a task, a workhorse places a message on the corresponding queue to inform about the state of the task.

In principle, all B-Fabric components could be deployed on one computer. However, in a typical B-Fabric setting, the components will be distributed over several computers due to performance and availability reasons. Especially the workhorses will run on more than one machine to distribute the heavy load of data copying and processing. In this way, B-Fabric is able to scale up with increasing workloads.

Currently, B-Fabric provides an integration framework with basic functionality on top to store and search for all data generated from experiments. In more detail, following features are supported currently: creation and storage of workunits, export of workunits (e.g. to the instrument or analysis computer), zip-download of workunits, browsing through B-Fabric data, basic and advanced search, creation and management of collections, sample/extract registration, and workflow and error management. Besides, B-Fabric is linked together with our user and project management tool.

The current implementation of B-Fabric consists of 166 Java classes and 212 XML files. In total, it comprises more than 65000 lines of code without Java scripts, plain text configuration files etc. The implementation synthesizes a number of well-known technologies, which are open source, mainly written in Java, and heavily exploit XML. In the following, we sketch the key technologies of the B-Fabric architecture.

**Apache Cocoon - Web Application Portal and Integration Framework.** All B-Fabric components for the web-based user interaction are built on Apache Cocoon [2], which is an open source web application development framework. Cocoon has been chosen because it represents a mature application development framework and has an extremely advanced form handling model which eases the implementation of the different instrument-specific workflows. Cocoon strictly follows the idea of separation of concerns. It separates the content, application logic, and layout concerns. Thus, application developers may work on the presentation of an application without affecting the content or the application logic, for example. This reduces the long-term overhead associated with maintaining a complex code base and allows the different concerns in maintaining the application to be isolated and distributed to different development teams. Cocoon is based on the pipeline and sitemap model. Abstractly spoken, a pipeline processes a (Web) request and provides a response to it. In detail, a pipeline consists of a sequence of steps, starting with a data generation (retrieving) step, followed by arbitrary many data transformation steps, and finished by a data serialization step. Each step is dynamically configurable. The result of the data generation step is provided as an XML file, which is then transformed using XSLT, and finally serialized to a specified format for presentation. For each request, Cocoon allows to specify one or more pipelines. A sitemap is an XML file which among others configures matchers to choose a specific pipeline for an incoming request. The processing of a request is thus determined by a sitemap. This very flexible mechanism allows easy configuring of any type of data source as data generator, arbitrary transformations of the data, and finally presentation of the data in any form (e.g. XML or PDF). This feature plays an important role in B-Fabric where new data sources have to be plugged into the system dynamically.

**PostgreSQL - Data Management.** PostgreSQL [11] is a powerful and mature open source SQL database system which provides native programming interfaces for several languages, including Java which was a prerequisite for being a component of B-Fabric. PostgreSQL was chosen for B-Fabric since it is currently the most sophisticated open source database system which satisfies the demands of B-Fabric w.r.t. stability, performance, and scalability. In principle, B-Fabric could run with any other SQL database system, too. B-Fabric uses PostgreSQL for managing project and user data, instrument-specific metadata, scientific annotations, and B-Fabric messaging information.

**OJB - Object-Relational Mapping.** Apache ObJectRelationalBridge (OJB) [7] is an object-relational mapping tool that supports transparent persistence for Java Objects against relational databases. OJB has been chosen for B-Fabric since it is extremely flexible and smoothly fits into the Cocoon framework. Besides, OJB has a number of advanced features that are useful for B-Fabric, e.g. object caching, lazy materialization and distributed lock-management with configurable transaction-isolation levels. The object-relational mapping is specified using XML files and resides in a dynamic layer, which can be manipulated at runtime to change the behavior of the persistence kernel. B-Fabric uses OJB to transparently map between the B-Fabric Java application objects and the corresponding relational PostgreSQL database entries.

**OSWorkflow - Workflow Management.** B-Fabric uses OSWorkflow [10] from the open source project OpenSymphony as workflow engine. OSworkflow was chosen because it is extremely flexible and can seamlessly be integrated into the Cocoon framework. OSWorkflow is different from most other workflow systems available since it does not require a graphical tool for defining workflows. It is a low-level, Java-based workflow engine that processes workflows described in XML. Beside sequences of steps, it supports split and join steps, and thus allows parallel execution of steps. This important feature is required by some central B-Fabric workflows, e.g. to detach the workunit annotation (requires user interaction) from the copying of the workunit files (which usually tend to be very large in size). Using parallel steps, the user can start (and finish) annotating while the copying takes place.

**Apache Lucene - Indexing and Search.** B-Fabric uses the open source tool Apache Lucene [5] to support full-text search on all B-Fabric (meta)data. Lucene was chosen because it is a high-performance, cross-platform text search engine that can seamlessly be integrated into the Cocoon framework. Lucene offers several powerful information retrieval features through a simple API. It supports incremental indexing as well as batch indexing. Its search algorithms among others support ranked results, different query types (phrase queries, wildcard queries, proximity queries etc.), fielded searching (e.g., title, author, contents), date-range searching, complex queries based on boolean operators, sorting by any field, and allows simultaneous update and searching. Especially the fielded searching is a feature of Lucene that is of high value for B-Fabric, e.g., to reduce the search space to certain workunits, sample, extracts etc. B-Fabric uses Lucene to provide a simple as well as an advanced search on different granularities.

**OpenJMS - Messaging.** OpenJMS [8] is an open source implementation of Sun Microsystems's Java Message Service (JMS), which supports asynchronous (and synchronous) message delivery, guaranteed delivery of messages, persistent messaging, and especially a point-to-point and publish-subscribe messaging model. Above that, it integrates with Servlet containers and supports in-memory and database garbage collection together with many communication protocols like TCP, RMI, HTTP and SSL. Besides all these characteristics, OpenJMS was chosen because it is scalable and can cope with the level of services of B-Fabric. B-Fabric uses OpenJMS to manage the various message queues, together with the JMS component of Cocoon which is responsible for retrieving new workflow steps to be executed. Services without a Cocoon-based web interface like the workhorses communicate directly with OpenJMS.

**Apache log4j - Logging.** The open source logging utility Apache log4j [4] allows for enabling logging at runtime without modifying the application binary. The logging behavior can be controlled by editing a configuration file, without touching the application binary. B-Fabric uses log4j to insert log statements into the code. In this way, it provides a low-level, but always applicable method for application debugging. Besides, log4j is designed so that log statements can remain in shipped code without incurring a heavy performance cost. Both features mentioned above are essential especially for complex distributed systems, such as B-Fabric.

**OpenSSH - Connectivity.** OpenSSH [9] is a free SSH tool, which encrypts all data traffic to effectively eliminate eavesdropping, connection hijacking, and other attacks. Besides, OpenSSH provides secure tunneling and several authentication methods, and supports all SSH protocol versions. B-Fabric uses OpenSSH to securely transmit data from the instrument PCs to the B-Fabric file archive server. Note that in principle any data copy library can be used instead of OpenSSH, as long as it is Java library.

## 4   User Versus B-Fabric Workflows

Scientists performing experiments using analytical technologies run through a (manual) user workflow. They enter physically the analytical laboratory with their samples and/or extracts, prepare the instrument to run the measurements, copy the generated results to a place where they can access them using analysis software, start the software to analyze the results, run several analysis steps, and finally save all relevant results (i.e. the corresponding data files) on a server or a CD. Some of these workflow steps are repeated several times depending on the results and the type of experiment.

A common workflow template ensures that all scientists take the same steps through B-Fabric, enforcing a rigorous control over the flow of data. As an example, scientists will need to annotate samples and files before getting access to the produced raw data and before further analysis and publication of the data can take place. In this context it is important to distinguish between user workflows and B-Fabric workflows. Both types of workflows differ in multiple ways: A user workflow describes the actual steps a scientist takes while he performs his research. A B-Fabric workflow on the other hand specifies the steps that have to be taken by B-Fabric to support the scientist. Some steps, e.g. the copying of data into the B-Fabric file archive, do not require user intervention, while others, e.g. the annotation step, involve a user interaction.

Figure 2 illustrates the two different views on workflows. From the user point of view, the workflow consists of the steps: 1) sample/extract preparation, 2) running the experiment, 3) creating a workunit to store the experimental results, and 4) providing scientific annotations for the workunit. From B-Fabric point of view, the workflow has the following steps: 1) creating a workunit to store experimental results, 2a) providing scientific annotations for the workunit and 2b) copying the experimental data to the data storage, and 3) indexing all the data. As can be seen, the user interacts with B-Fabric only in two steps: workunit creation and annotation input. According to this workflow, the user is free to create workunits based on files of any type.

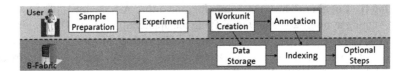

**Fig. 2.** Example User vs. B-Fabric Workflow

## 5   Instrument Integration

Before an instrument can be integrated into the B-Fabric framework, the following instrument-specific questions must be answered:

- What is the input and output of the instrument? For instance, in most cases, the input is one extract, while the output is one or more raw data files together with some log files which describe the generation of the raw data files.
- Where does the instrument writes its output? Is this place configurable? The information about the location of the raw data files is essential for transparently reading and importing the files into B-Fabric. For each instrument to be integrated into B-Fabric, the corresponding location must be specified.
- Are there any instrument-specific annotations that should be managed by B-Fabric (in addition to the generic sample and extract annotations)? Depending on the technology associated with an instrument, the set of meaningful annotations may vary. These annotations are needed to better understand and investigate the research results, especially in comparison to other research results.

- Can a B-Fabric data mart be used for the management of the data of the instrument in order to exploit its analytic functionality for data analysis. If there exists such a data mart, does it provides an interface to import data into mart? If not, the B-Fabric custom mart needs to be extended to capture the instrument-specific annotations.
- Which workflows are affected by the instrument? There is at least one workflow that applies to all instruments. It is the workunit creation workflow. Of course, this workflow may vary from instrument to instrument. The instrument integrator has not to specify only the corresponding B-Fabric workflow but also has to define the interaction between B-Fabric and the user.

Technically, the integration of an instrument into the B-Fabric framework requires the execution of the following tasks:

1. An OpenSSH server has to be installed and run on the instrument PC. OpenSSH allows a secure transfer of experimental data between B-Fabric and the instrument and analysis PCs.
2. A data provider must be implemented. The instrument-specific data provider component tells B-Fabric from which location the experimental data has to be imported into B-Fabric.
3. A data mart has to be determined where the experimental data generated with this instrument shall be managed and analyzed in detail. In case there is no mart for that instrument, the B-Fabric custom mart has to be extended appropriately.
4. One or more instrument-specific workflows have to be specified using the XML notation of OSWorkflow. These workflows define the flow of control w.r.t. the generation and usage of data of that instrument.

In the following, we sketch how to integrate an instrument into the B-Fabric framework. We use the Affymetrix instrument as example. Assume the OpenSSH server is running the corresponding instrument computer. The following lines depict a part of the XML file which implements the corresponding data provider:

```
<?xml version="1.0"?> <xconf xpath="/cocoon">
  <component>
  <!-- Data provider for Affymetrix instrument -->
  <component-instance class="bfabric.dataprovider.impl.SshBasedDataProvider"
                      logger="bfabric.data-provider.affymetrixgenechip"
                      name="affymetrixgenechip">
    <name>Affymetrix GeneChip System</name>
    <research-area>Transcriptomics</research-area>
    <data-mart>Resolver</data-mart>
    <workflow>affymetrixgenechip</workflow>
    <optional-resources>
      <base-path>$dataprovider.affymetrixgenechip.base-path</base-path>
      <include-pattern>.*\.CEL|.*\.cel</include-pattern>
    </optional-resources>
  </component-instance>
  </component>
  ...
</xconf>
```

Note that the data provider is configured such that it knows which data mart is target for metadata digestion and which research area is associated with this instrument.

Furthermore, the data provider defines a list of resource location names to choose from when creating a workunit. Finally, the data provider knows which workflow is target for the workunit creation.

Figure 3 shows a snapshot of the OSWorkflow designer tool which models the workunit creation workflow for the Affymetrix instrument.The details of the workflow is usually specified using the XML notation. For instance, since the copying is realized by sending a JMS message to the import queue, the data copying step determines the corresponding queue (import) and action (jms-import). This is done by setting the corresponding XML-tags of that step to the appropriate values. Due to space restrictions, we however omit the presentation of the lengthy XML specification of a workflow.

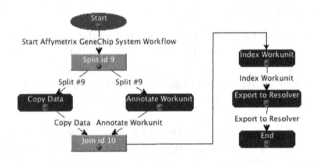

**Fig. 3.** B-Fabric workflow for the Affymetrix workunit creation

Altogether, we may state that a new component can be plugged into B-Fabric by creating a few XML configuration files which provide the necessary information about the component and the workflows it is involved in. After having integrated some instruments into the framework this task get easier since the existing XML configuration files can be copied and adapted to the specifics of the new instrument to be installed.

# 6    Conclusions and Outlook

B-Fabric provides the foundation for integrative biological research over multiple analytical technologies, workflows, and users, by effectively storing all relevant analytical raw data and experimental parameter information in conjunction with the corresponding biological and workflow annotation. B-Fabric also supports a new way for research departments and technology centers to maximize their innovation, by making all data accessible to the researchers and allowing integrated searches on this data. The researchers benefit from the secure, long-term storage combined with easy access to and download of the research results. Interpretability and reproducibility of research results is facilitated by capturing all semantically related data together with annotations that are conform to standards. The rigorous workflow-driven "one-time" annotation and the transparent federated search leads to significant overall time savings for the scientists.

From the implementation of B-Fabric we learned some not really new but often neglected important lessons: 1) As data generators such as instruments or applications are constantly in flux, the integration infrastructure must provide templates to quickly adapt to changes. 2) Since data access and provisioning as well as authentification and access control are difficult to add to existing systems, a data fabric approach with a global namespace for all file-based data is essential. 3) A simple combined data browsing and querying interface is the basis for successful usage of the data because life sciences researchers are not used to formulate complex database queries. 4) A workflow-driven approach that demands the entering of appropriate metadata and annotations from the researcher is crucial since otherwise the generated huge raw data files will be of no use for anybody else. 5) Data quality is a huge problem that requires advanced strategies.

Building on the modular architecture, future releases of B-Fabric will also provide the integration platform for queries across projects, users, organisms, molecule species, and analytical technologies. By these means, B-Fabric is the essential first step in achieving quality-controlled and reproducible integration of biological research data necessary for integrative research and systems biology. Modules to be implemented will include technology-dependent scaling functions, which allow for estimations of correlations between the different molecular, spatial and temporal data sets.

# References

1. Brazma, A., Krestyaninova, M., Sarkans, U.: Standards for Systems Biology. Nature Reviews 7, 593–605 (2006)
2. Apache Cocoon. http://cocoon.apache.org/
3. Hack, C.J.: Integrated transcriptome and proteome data: the challenges ahead. Briefings in Functional Genomics & Proteomics 3(3), 212–219 (2004)
4. Apache log4j. http://logging.apache.org/log4j/docs/
5. Apache Lucene. http://lucene.apache.org/
6. Ludäscher, B., et al.: Scientific Workflow Management and the Kepler System. Concurrency and Computation: Practice and Experience, vol. 18(10) (August 2006)
7. Apache ObJectRelationalBridge - OJB. http://db.apache.org/ojb/
8. OpenJMS. http://openjms.sourceforge.net/
9. OpenSSH. http://www.openssh.com/
10. OpenSymphony - OSWorkflow. http://opensymphony.com/osworkflow/
11. PostgreSQL. http://www.postgresql.org/
12. Yu, J., Buyya, R.: A Taxonomy of Scientific Workflow Systems for Grid Computing. In: SIGMOD Record, vol. 34(3) (September 2005)

# SWAMI: Integrating Biological Databases and Analysis Tools Within User Friendly Environment

Rami Rifaieh, Roger Unwin, Jeremy Carver, and Mark A. Miller

The San Diego Supercomputer Center, University of California San Diego,
9500 Gilman Drive La Jolla, CA 92093-0505
{rrifaieh, unwin, jjcarver, mmiller}@sdsc.edu

**Abstract.** In the last decade, many projects have tried to deal with the integration of biological resources. Web portals have flourished online, each providing data from a public provider. Although these online resources are available with a set of manipulation tools, scientists, researchers, and students often have to shift from one resource to another to accomplish a particular task. Making a rich tool set available along with a variety of databases, data formats, and computational capabilities is a complex task. It requires building a versatile environment for data integration, data manipulation, and data storage. In this paper, we study the requirements and report the architectural design of a web application, code named SWAMI, which aims at integrating a rich tool set and a variety of biological databases. The suggested architecture is highly scalable in terms of adding databases and new manipulation tools.

**Keywords:** Biology Workbench, Biological data integration, Biological tool integration.

## 1 Introduction

Modern molecular biology research is increasingly reliant upon the production and analysis of very large amounts of digital data (gene sequencing, microarrays etc.). The assembly and analysis of these data provides the basis for posing and verifying hypotheses on a very large scale. Data analyses typically require individual scientists to assemble enormous amounts of data locally and to integrate this data with data gathered from remote public or private providers. The aggregated data is then queried or analyzed using search or algorithmic tools. Often there is a significant time investment in collecting, organizing, and preparing the proper formats for algorithmic analysis. In fact, scientists may spend ~80% of their time in assembling data (e.g. data manipulation, extracting subset of data from external files, reformatting data, moving data, etc.) and preparing it for analysis [1].

It would be of great benefit to provide researchers with tools that can speed this process. There are three stages where help is needed. First, the investigator must be able to manage data created locally. This means producing data of a known format. This step is unique to individual laboratories, and must be managed to some extent within that context. Second, the data must be compared with data from other sources.

S. Cohen-Boulakia and V. Tannen (Eds.): DILS 2007, LNBI 4544, pp. 48–58, 2007.

In general, public biological resources are freely accessible, and provide tools for navigation, querying, and searching through websites. For instance, EMBL, DDBJ, and NCBI provide the tools SRS, getentry and Entrez, respectively, for locating required sequence data. In addition to data resources, many sites provide online tools for data analysis as well. However, when local and remote data must be aggregated for analysis, online tools at a single site are seldom adequate. Moreover, extracting, integrating, and cataloging data from several sources requires unification of independent heterogeneous data sources, subsequent selection of the needed sources, and the transformation of the content to fit with users' needs and perspectives. This includes understanding data format compatibilities and data conversion possibilities. All of these issues impede scientific data analysis (c.f. [2]). Since discovery (and not data manipulation) is the ultimate goal of end users, the ability to aggregate data in the presence of appropriate analytical tools would be extremely helpful, particularly when manipulation of data formats is required prior to analysis.

While the problem is fairly well defined, and the requirements are known or can be gathered, no environment that addresses these three needs in a robust and scalable way is currently available. The SWAMI project was undertaken to create such an environment. SWAMI is an outgrowth of the Biology Workbench [3] (BW) which was the first web application to provide an integrated data management and analysis environment. Our goal is to re-engineer the BW while preserving its spirit, so as to reduce user overhead for collecting, storing, manipulating, and analyzing virtually any biological data using any specified tool, in a way could, in time, be scaled meet the needs of the entire community.

The paper is organized as follows: Section 2 describes the prior art in Biological data and tool integration, Section 3 studies the requirements gathered from the community, Section 4 describes how these requirements and prior art can be assembled into a versatile web application, and Section 5 summarizes these experiences, and forecasts future directions for SWAMI.

## 2   Background and Related Works

Despite much effort, available solutions for meaningful integration, transformation, and manipulation of biological data have not achieved the full functionality required by domain researchers. Indeed, biological data are diverse and generated independently in many "omics" subfields. Public data sources rely upon existing data management technologies and involve multiple transformations between different levels of data granularity and reconcile the difference in data structure and format. The complex biological data space is matched by an equally complex set of algorithmic tools used in data analysis. Tasks such as sequence assembly, similarity searches, structure visualization and phylogenetic inference require a variety of analytical bioinformatics applications.

### 2.1   Integrating Biological Data

Many efforts have been made to provide solutions for biological data integration; the basic approaches reported are as navigational, warehousing, and mediator based:

**Navigational:** The most widely used data integration tool in life sciences relies on flat file storage. Data are extracted from public data sources and assembled into one or more flat files. These flat file(s) are indexed and/or cross-referenced according to table and/or specific field locations. This kind of solution has been implemented in public resources, including SRS [4, 5], BioNavigator [6], Entrez [7], SeqHound [8], Lucegene [9], and the BW [3, 10]. Although, this solution is simple to maintain and easy to use, it does not provide the rich queries and query optimization benefits of relational modeling.

**Warehousing (data translation)** is accomplished by materializing data from multiple resources into a local warehouse and executing all queries on the warehoused data. Data warehousing requires the use of Extraction Transformation and Load (ETL) tools to load data, and map it to a materialized global schema. The overhead of creating the ETL tools to import data into a common schema is offset by efficient query execution and query optimization, and grater control over data consistency/redundancy. Local storage of the integrated data allows local curation: users can add annotations in tables created for this purpose. A more difficult problem in warehousing solutions is scalability: if many frequently-updated databases are to be warehoused, the combined overhead of loading and re-indexing the databases can be problematic. Implementations in this category include GUS [11], Atlas [12], BioSQL [13], BioMart [14], BioWarehouse [15], and Chado[16].

**Federated/mediator based approaches (query translation)** leave data in their native schema, typically in their home production environment. At runtime, a mediator translates user queries, optimizes them, and maps them to available remote or local RDBMS. The mediated schema is a virtual one: data are accessed in their original physical schema. The overhead in this approach comes from constructing the tools to register the remote data resources, and to map and optimize queries across them. In exchange for this investment, all responsibility for data management, curation, and updating is left to the data provider. This solution has specific drawbacks, including lack of control of data, lack of permission to directly query a remote resource, and vulnerability to sudden loss of service due to physical or logical changes at the provider site. Implementations in this category include Tambis [17], K2/Kleiski [11], and DiscoveryLink [18].

### 2.2  Integrating Biological Tools

Several attempts have also been made to create integrated software environments for staging and analyzing biological data. Some examples of integrated software environments are:

**The Biology Workbench [3]:** BW was the first web application to present users with an integrated environment for tools and data. The BW provides web-based interfaces to search 33 databases, store the search results, and route stored sequences to 66 sequence analysis/ presentation tools [3]. This approach remains the most common for heterogeneous biological databases [19].

The **Pasteur Institute Software Environment** (PISE) [20] provides a web application framework that is highly scalable, where each analytical tool is described by a PISE-XML document. Using the PISE DTD, the document provides all the information required to spawn a static web page, together with the information to assemble a PERL/CGI command line to run the program. Currently, PISE supports 200+ tools, freely available through a web application at the Pasteur Institute, however, it provides no data resources or data storage area for users.

**NC Bioportal Project:** The NC Bioportal [21] was created as an extensible bioinformatics portal. Bioportal links static PISE interfaces [20] to grid computing resources. Its design combines rapid, scalable tool deployment with the ability to map jobs over distributed computing resources. Access to integrated data resources is currently provided via web application to North Carolina students and researchers only.

**MIGenAS:** (**M**ax Planck **I**ntegrated **Gen**e **A**nalysis **S**ystem): MiGenAS [22] is a web application that integrates bioinformatics software tools with locally generated microbial sequence data. Its capabilities cover the complete processing chain for microbial genome sequencing. The MiGenAS portal is available only to researchers within the Max Planck Institute, and their colleagues. We are not aware of any planned public release of the underlying software.

**Anabench (Biology Workbench):** AnaBench [23] is a Web/CORBA-based workbench environment integrating bioinformatics sequence analysis software in a flexible manner. This web application provides tool access, including the EMBOSS suite, and a workflow pipelining system [24] but does not provide access to data resources from external public providers. Like the BW, access and user data storage is provided free of charge to registered users.

**Thick Client Solutions:** Smart Client software (packages that must be downloaded and installed locally) represents another design approach. These tools offer the ability to construct workflows linking tools in sequence, and pipelining data through the workflow. Packages that are currently available include BioSteer [25], Pegasys [26], Kepler [27], Taverna [28] and Gemstone [29]. While these tools offer significant promise as an alternative to portals, the user must have the ability and the administrative privileges to install the client locally.

# 3  Requirements

The SWAMI project was created to extend the functionality of the BW by providing 1) a dynamic set of data resources, 2) improved data search tools, 3) a dynamic set of analytical tools with static *and* interactive interfaces, 4) improved data/task management capabilities so data can be annotated, modified, and assembled by the users, and 5) provisions for growth of the resource and expansion of its services.

## 3.1  Non Functional Requirements

The first step in gathering user requirements for the SWAMI project was to examine the current user population of the BW, which will be the initial population served by SWAMI. User surveys revealed that BW is essentially serving educational community. Through interviews with instructors, it became clear that non-functional requirements of the educational community include minute by minute reliability (uptime), low latency, and stability of content. In contrast to the education community, researchers rely less on minute by minute access, and slow responses to larger data submissions or more computationally intensive tasks are tolerated. Based on the assumption that both researchers and educators can benefit from new features and capabilities deployed in the most stable and reliable environment possible, we have elected to tailor SWAMI development and architecture choices to give priority to requirements for stability and predictability. This means new capabilities will be deployed more slowly, but with more stability. Additional non-functional requirements were gathered, and are posted at http://www.ngbw.org.

## 3.2 Functional Requirements

**Integration of biological databases:** SWAMI should integrate a large set of publicly available data sources. This data will be stored in a public area that is accessible by all users for query and search. The initial list of data sources to be supported (based on initial user requests) includes most Genbank DBs, Uniprot, and PDB. Further addition of DBs will be made based on user feedback, so the data model must be extensible to include new data types. Users can import data of any format and data type known to the application, and the application must provide format checking for any uploaded data. Uploaded data, user experiments, and user results will be stored in a dedicated private user DB, which will store data created, metadata stored, organized, and annotated by the individual users.

**Flexible management of analysis tools:** Both public and user data resources should be integrated with a wide variety of analytical and modeling tools, with a point-and-click interface, and with no file format compatibility problems. A main challenge is to permit facile addition of new manipulation tools (open source tools) and to provide easy and user-friendly access. The application should also allow data flow between compatible programs with essentially no user effort. Consequently, the application should enable pipelining data between analytical tasks and routing data from one tool to the next. The implementation should allow users to modify, save, and download results in printable/editable/publishable formats.

**High-end visualization tools:** SWAMI should provide enhanced visualization capabilities, including interactive tools for visualizing molecular structures and phylogenetic trees. Users should be able to interactively import, construct, modify, and save structure models and tree models in real time. The tools provided must be lightweight and run effectively from the server side without requiring software downloads or plug-in installations. We will not develop such tools, but rather integrate the numerous existing tools using applet/servlet technologies.

**User configurable interface:** The applications should provide role-based user applications. User interfaces should differ depending on role, for example, teacher/student or beginner/advanced. Users should ultimately be given the capability to create their own, "custom" user interfaces without being constrained by the preconceptions of the design team. A "smart" or "thick" client interface that increases the power and functionality of the toolkit, and permits local storage will be added to a subsequent release for users with administrative control of their computers. This will help them managing locally their data and interact with the system for data search and tools usage provided sufficiently robust compute resources.

# 4  System Description

The main architectural challenge is to create a reliable, stable, well defined system while at the same time enabling the extensibility and scalability both of data resources and tools.

## 4.1  Core Architecture Design

The SWAMI architecture design evolved in the context of user requirements and following best practices in terms of separation of concerns among components.

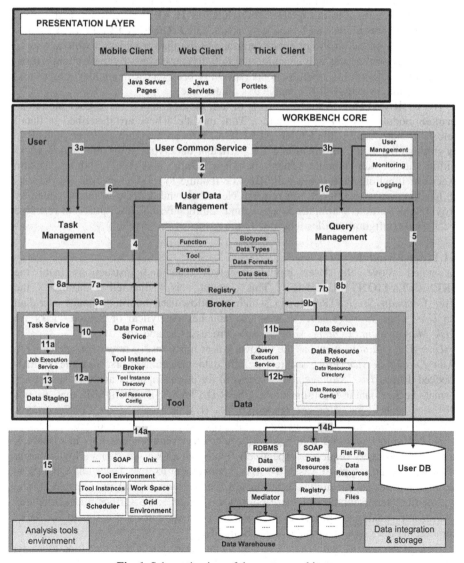

**Fig. 1.** Schematic view of the system architecture

System capabilities are created as a set of independent services. This makes it easier to think of functionality of each component rather than of the system as whole. Consequently, the separation of functionalities means that addition of a new data type, database, or tool requires as little change as possible in other components.

SWAMI architecture features three modules, which are coordinated through a fourth module, an "intelligent" Broker that contains all the specific information needed by individual components. The Broker contains this information in a logical registry that describes properties of data types, data sources, and tools supported by the application. A schematic representation of the architecture designed to fulfill this design goal is shown in Figure 1.

The architecture design includes a **PRESENTATION LAYER** that receives user requests, passes them to the **CORE WORKBENCH APPLICATION**, and returns application results to the user by the same route. The **PRESENTATION LAYER** has a published Application Programmer Interface (API) that provides flexible access to SWAMI. Browser access will be supported initially, and access by thick clients (user-downloaded software) or mobile clients (specialty browsers with smaller screens and lower resource systems) will be enabled at a later date.

The **CORE APPLICATION** consists of four major components: a **User** module, a **Broker** module, a **Data** module and a **Tool** module. These are described in detail below. As noted above, the **CORE APPLICATION** architecture achieves scalability by storing all specific information about particular data types, tools, etc, in a *Registry* within the **Broker** module. Thus, the **User, Data,** and **Tool** modules contain executive functions, and contact the **Broker** module to receive specific instructions on handling and executing a user request. New tools and data types can be added by providing definitions within the *Registry*, and adding physical descriptions of resource location in configuration files that specify the names of servers.

### 4.1.1 User Module
As noted above, the **User** module receives data and instructions from the **PRESENTATION LAYER**. This input is directed through the *User_Common_Service* (Process 1), which can allow user input to: manage user data via *User_Data_Management* (Process 2), launches Analytical or Search Tasks via *Task_Management* or *Query_Management* (Processes 3a and 3b).

*User_Data_Management* has the following user functions: (i) refers uploaded data to *Data_Format_Service* (Process 4) to check for data formatting errors. (ii) routes format-checked data to *Task_Management* (Process 6) for use as input for analysis. (iii) routes format-checked data/user settings to the **USER DATABASE** (an RDBMS that sustains all user data) for storage (Process 5). (iv) accesses *User_Configurations*, *User_Monitoring*, and *User_Logging* modules (Process 16), which modify user settings.

*Task_Management* and *Query_Management* serve parallel functions in managing user tasks and query requests, respectively. Note that a Query is, from a logical point of view, a specific type of Task. We find it convenient to treat Tasks and Queries as separate entities in our architecture, but emphasize their logical relationship by the symmetrical representation of these entities in Figure 1. To further underscore this relationship, analogous processes in **Task** or **Data** modules are assigned the same number, but are given an alphabetical sub-qualifier.

*Task_Management* and *Query_Management* have the following functions: (i) interrogate the *Registry* in **Broker** to identify the resources available and the formats required by these resources (Process 7). (ii) combine information from the *Registry* with instructions from *Task_Management* and *Query_Management* and pass it to an executive service: *Task_Service* or *Data_Service* (Process 8).

### 4.1.2 Broker Module

The **Broker** module interacts with **User**, **Tool**, and **Data** modules via APIs. The **Broker** module contains a *Registry* that provides logical information about each tool and data resource known to the application.

The *Registry* is the "brain" of the application; it contains all of the logical information about the application and data sets/types known. The *Registry* defines semantics using ontological concepts and relationships (axioms). Information included in the *Registry* is gathered using an ontology language (OWL [30]). By using an ontology language, and adding a reasoning engine (Pellet [31]) the **Broker** can respond more intelligently to requests from user queries or task requests.

The information about tools in the *Registry* includes tool functions, tool names, input parameters required, output parameters delivered, and results presentation formats (.pdf, .ps, .txt). The *Registry* also contains all information known to the application about data resources. This includes ***Biotypes*** (which describe the biological nature of the entity: i.e. protein, DNA, lipid, etc); ***Data Types*** (which describe whether the data is a sequence, structure, genome, etc.), ***Data Format*** (which describes the presentation format: e.g. if a sequence is in fasta, asn.1, etc.), and ***Data Sets*** (which describes where the data came from EBI, GenBank, PDB, etc.).

The isolation of all logical information about application resources, new tools, data resources, and data types in the *Registry* means new Tools and Data types can be added by editing the registry without extensive modification of other modules.

### 4.1.3 Tool and Data Modules

The **Tool** and **Data** modules are conceptually identical. Both modules interact with the **Broker** and **User** modules, but they do not interact directly with each other. The principal function of the **Tool/Data** modules is to receive **user** requests as input combined with instructions from the **Broker** (Processes 8 and 9), decompose those requests into atomic executable jobs or queries, and return those results to the **User** module with instructions obtained from the **Broker.**

*Task/Data_Service* receives ***Task/Query*** information from *Task/Query_Management* (Process 8). If the ***Task*** is a workflow, for example, *Task_Service* decomposes the ***Task*** into a sequence of ***Jobs***, and manages their execution. This is accomplished by submitting the first ***Job*** for execution, then passing the output of the first ***Job*** as input to the second ***Job***, and so forth. Similarly, *Data_Service* assimilates queries and composes individual queries, and passes them to the executive service (Process 11).

*Job/Query_Execution_Service* (Process 11) receives ***Jobs/Queries*** from *Task/Query_Preparation_Service* and prepares command line arguments or formulates SQL queries. In both modules, a *Broker* (Process 12) receives command line arguments for jobs, and locates the specific hardware, the specific tool implementation (executable), and arranges scheduling used to execute the job. *Tool_Instance_Broker* contains the physical addresses of available hardware, knowledge of binary locations on those machines, information on available web services, etc. Similarly, *Data_Resource_Broker* contains the physical addresses of available databases, flat file locations, information on available web services, etc.

## 4.2  Tools and DB Integration Design

The SWAMI architecture can access and manage new tools and data sources once they are registered within the *Broker*. Registration occurs at both logical and physical levels. Logical integration occurs in *Registry*, where relevant logical information for each new tool or data type is stored. For each new tool, the accepted data formats, parameters required for the analysis, command line flags to modify the binary function, etc. must be registered. For new database resources, the data sets, data types, biotypes and any other parameters necessary to specify the types of information located in the database must be registered. Physical integration occurs within the **Tool** and **Data** modules, in configuration files.

### 4.2.1  Scalable Data Integration

There are two ways to proceed for handling data integration, building an ad-hoc data integration solution or using existing solutions (Section 2). We explored these two alternatives. The first is a simple, incremental improvement on the current WB system (code named CherryPicker). Cherry Picker parses flat file DB into records and tokenizes them into DB lookup tables. It allows arbitrary fields to be identified and indexed (as character or numeric), facilitating more targeted searches. Our long term goal is to integrate the databases by semantically matching fields in different databases. We tested existing biological warehousing databases (Atlas [12], GUS [11], and Biowarehouse [15]), which come with embedded ETL tools used to populate DB schema with data and record coming from available resources.

To permit users to interact with the DB using domain language (e.g. sequences, genes, proteins) biological data types (instead of string, integer, float, etc.) will be used in the query formulation [5]. In this respect, SWAMI *ontology* serves as a global schema for all DB known to the SWAMI application. It includes all the concepts and terms to be queried by the user and their representations in the physical storage. The association between the ontology and the data storage schema is done through the registration process. A similar work on using ontology concepts for registering and querying multiple DB schemas is being carried out in [32].

### 4.2.2  Scalable Tool Integration

To achieve tool scalability, we must address the problem of how to create a new interface and register each tool. This can be a significant problem if more than a handful of tools are to be implemented. We chose to take advantage of tool XML documents available through PISE [20] to solve this problem. We can implement all the 200+ tools of the PISE by using PISE-XML files and (i) extracting the XML interface information, and creating a .jsp page in the PRESENTATION LAYER. (ii) extracting all information about tool parameters and writing this information to the *Registry* in **Broker.** (iii) reading PISE XML files and writing a SWAMI-XML document. All of the PISE XML documents can be converted by a single mapping step, making this process highly scalable. For future tools, a single XML file will be required in order to integrate the new tool into the system.

In addition, the user requirement for specific tools to view and modify protein structure and phylogenetic trees can easily be met by taking advantage of applet and servlet technologies. We have already implemented the Sirius protein structure viewer [33] in this way. We plan to incorporate other interactive tools based on user requests.

# 5 Conclusion

Advances in computational and biological methods during the last decade have remarkably changed the scale of biomedical and genomic research. Current research on biomedicine and genomics is characterized by immense volume of data, accompanied by a tremendous increase in the number of databases and tools. This wealth of information and data presents a major challenge for users. Although, many of the available resources are public and accessible, users are constrained by the number of analysis tools provided by individual site. A cross-sites manipulation is only available through user tedious management (conversion format, saving intermediate data, etc.). The goal of this work is to bring together contributions from publicly available resources and analysis tools to the end user. This New Generation of Biology Workbench (SWAMI) is designed to tackle the issues of growing need of data access, data storage, and availability of computational resources. The design proposed emphasizes modularity and separation of functionalities. The goal of this architecture design is to minimize dependence of individual components on other components, allowing versatility in changing and adapting the software to new functionalities, just as the requirements for the SWAMI dictate. Our original design guarantees to incrementally add more resources (databases and tools) on demand of the users. The SWAMI architecture manages new tools and data sources once they are registered within the application through logical and physical registration process. Future work will include incremental implementation of the system; a first public release is expected shortly; please visit http://www.ngbw.org for more details.

# References

1. Scientific Data Management Center (2005) http://sdm.lbl.gov/sdmcenter/
2. Abiteboul, S., Agrawal, R., Bernstein, P., Carey, M., Ceri, S., Croft, B., DeWitt, D., et al.: The Lowell database research self-assessment. Commun. ACM 48(5), 111–118 (2005)
3. Subramaniam, S.: The Biology Workbench (1998) http://workbench.sdsc.edu/
4. Etzold, T., Argos, P.: SRS – An Indexing And Retrieval Tool For Flat File Data Libraries. CABIOS 9, 49–57 (1993)
5. Documentation, S.: SRS at the European Bioinformatics Institute (2006) http://srs.ebi.ac.uk/srs/doc/index.html
6. Entigen, Bionavigator - BioNode & BioNodeSA: Overview (2001) http://www.entigen.com/ library
7. National Library of Medicine (2005) http://www.ncbi.nlm.nih.gov/entrez/query.fcgi?DB=pubmed
8. Michalickova, K., Bader, G.D., Dumontier, M., Lieu, H., Betel, D., Isserlin, R., Hogue, C.W.: Seqhound: biological sequence and structure database as a platform for bioinformatics research. BMC Bioinformatics 3(1), 32 (2002)
9. GMOD, LuceGene: Document/Object Search and Retrieval (2007) http://www.gmod.org/?q=node/83
10. Subramaniam, S.: The biology workbench - A seamless database and analysis environment for the biologist. Proteins-Structure Function and Genetics 32(1), 1–2 (1998)
11. Davidson, S.B., Crabtree, J., Brunk, B.P., Schug, J., Tannen, V., Overton, G.C., Stoeckert, C.J.: K2/Kleisli and GUS: Experiments in integrated access to genomic data sources. IBM Systems Journal 40(2), 512–531 (2001)
12. Shah, S.P., Huang, Y., Xu, T., Yuen, M.M.S., Ling, J., Ouellette, B.F.F.: Atlas - a data warehouse for integrative bioinformatics. BMC Bioinformatics 6 (2005)

13. Lapp, H.: BioSQL (2006) http://www.biosql.org/wiki/Main_Page
14. Haider, S., Holland, R., Smedley, D., Kasprzyk, A.: BioMART Project (2007) http://www.biomart.org/index.html
15. Lee, T.J., Pouliot, Y., Wagner, V., Gupta, P., Stringer-Calvert, D.W.-J., Tenenbaum, J.D., Karp, P.D.: BioWarehouse: a bioinformatics database warehouse toolkit. BMC Bioinformatics 7, 170–184 (2006)
16. GMOD, Getting Started with Chado and GMOD (2007) http://www.gmod.org/getting_started
17. Stevens, R., Baker, P., Bechhofer, S., Ng, G., Jacoby, A., Paton, N.W., Goble, C.A., Brass, A.: TAMBIS: Transparent access to multiple bioinformatics information sources. Bioinformatics 16(2), 184–185 (2000)
18. Haas, L., Schwartz, P., Kodali, P., Kotlar, E., Rice, J., Swope, W.: DiscoveryLink: A System for Integrated Access to Life Sciences Data Sources. IBM Systems Journal 40(2), 489–511 (2001)
19. Kohler, J.: Integration of Life Sciences Databases. BIOSILICO 2, 61–69 (2004)
20. Letondal, C.: A Web interface generator for molecular biology programs in Unix. Bioinformatics 17(1), 73–82 (2001)
21. RENCI, The NC BioPortal Project (2007) http://www.ncbioportal.org/
22. Rampp, M., Soddemann, T., Lederer, H.: The MIGenAS integrated bioinformatics toolkit for web-based sequence analysis. Nucleic Acids Research 34(Web Server issue), W15–W19 (2006)
23. Badidi, E., De Sousa, C., Lang, B.F., Burger, G.: AnaBench: a Web/CORBA-based workbench for biomolecular sequence analysis. BMC Bioinformatics 4, 63–72 (2003)
24. Badidi, E., De Sousa, C., Lang, B.F., Burger, G.: FLOSYS–a web-accessible workflow system for protocol-driven biomolecular sequence analysis. Cell. Mol. Biol (Noisy-le-grand) 50(7), 785–793 (2004)
25. Lee, S., Wang, T.D., Hashmi, N., Cummings, M.P.: Bio-STEER: a Semantic Web workflow tool for Grid computing in the life sciences. Future Generation Computer Systems 23, 497–509 (2007)
26. Shah, S.P., He, D.Y.M., Sawkins, J.N., Druce, J.C., Quon, G., Lett, D., Zheng, G.X.Y., Xu, T., Ouellette, B.F.: Pegasys: software for executing and integrating analyses of biological sequences. BMC Bioinformatics 5, 40–48 (2004)
27. Altintas, I., Berkley, C., Jaeger, E., Jones, M., Ludäscher, B., Mock, S.: Kepler: An Extensible System for Design and Execution of Scientific Workflows. In: 16th Intl. Conf. on Scientific and Statistical Database Management (SSDBM'04), Santorini Island, Greece (2004)
28. Oinn, T., Addis, M., Ferris, J., Marvin, D., Senger, M., Greenwood, M., Carver, T., Glover, K., Pocock, M.R., Wipat, A., Li, P.: Taverna: a tool for the composition and enactment of bioinformatics workflows. Bioinformatics 20(17), 3045–3054 (2003)
29. Baldridge, K., Bhatia, K., Greenberg, J.P., Stearn, B., Mock, S., Sudholt, W., Krishnan, S., Bowen, A., Amoreira, C., Potier, Y.: GEMSTONE: Grid Enabled Molecular Science Through Online Networked Environments. In: Life Sciences Grid 2005, LSGrid, Singapore (2005)
30. McGuinness, D.L., van Harmelen, F.: OWL Web Ontology Language (2007) http://www.w3.org/TR/owl-features/
31. Parsia, B., Sirin, E.: Pellet: An OWL DL Reasoner. Poster. In: Third International Semantic Web Conference (ISWC2004), Hiroshima, Japan (2004)
32. Nambiar, U., Ludaescher, B., Lin, K., Baru, C.: The GEON portal: accelerating knowledge discovery in the geosciences. In: Eighth ACM international Workshop on Web information and Data Management (WIDM '06). Arlington, Virginia, USA: ACM Press, New York, NY, pp. 83–90 (2006)
33. Buzko, O.: SIRIUS: An Extensible Molecular Graphics and Analysis Environment (2007) http://sirius.sdsc.edu

# $^{my}$Grid and UTOPIA: An Integrated Approach to Enacting and Visualising in Silico Experiments in the Life Sciences

Steve Pettifer*, Katy Wolstencroft, Pinar Alper, Teresa Attwood,
Alain Coletta, Carole Goble, Peter Li, Philip McDermott, James Marsh,
Tom Oinn, James Sinnott, and David Thorne

The University of Manchester, School of Computer Science, Kilburn Building
Oxford Road, Manchester, M13 9PL, United Kingdom
steve.pettifer@manchester.ac.uk

**Abstract.** In silico experiments have hitherto required ad hoc collections of scripts and programs to process and visualise biological data, consuming substantial amounts of time and effort to build, and leading to tools that are difficult to use, are architecturally fragile and scale poorly. With examples of the systems applied to real biological problems, we describe two complimentary software frameworks that address this problem in a principled manner; $^{my}$Grid/Taverna, a workflow design and enactment environment enabling coherent experiments to be built, and UTOPIA, a flexible visualisation system to aid in examining experimental results.

**Keywords:** Workflows, visualisation, web services, in silico experimentation.

## 1 Introduction

The life science community has relied on information technology to store and process data since at least the late 70s. Advances in laboratory techniques and technologies have led to the exponential growth of machine-readable biological data, the management and analysis of which was only made possible with the increase of raw compute power, network performance and digital storage capacity. The in silico life science community is large and encompasses a broad area of research, at one extreme examining the chemical properties of small molecules, and at the other, modeling and understanding how complete biological systems function. Increasingly, the community has moved from the isolated study of individual molecules and data objects to studying whole genomes, proteomes and metabolomes.

In many cases in silico experimentation involves chaining together a series of analysis and data resources from a number of different locations. Tools initially designed to work in isolation on comparatively small localised data sets

---

* Corresponding author.

S. Cohen-Boulakia and V. Tannen (Eds.): DILS 2007, LNBI 4544, pp. 59–70, 2007.

are now being required to work together in a much broader distributed environment. With the subtle and complex concepts involved in biology, this integration presents significant technological challenges. Improvised assemblies of scripts and programs working on fluctuating file formats were once a viable means of performing these analyses, however such assemblies require substantial effort to develop and maintain, and lack the robustness, reproducability and auditability required by today's scientific community. Consequently, alternatives are needed that are able to deal with both the distributed nature of tools and resources and the heterogeneity of data and data formats.

In the post-genomic era, in silico and in vitro experiments are strongly interlinked. Increasingly, biological research involves the generation and analysis of large amounts of complex data using high throughput techniques which require both wet laboratory and dry computational techniques. To be successful, both types of experiment must adhere to the classical scientific experimental process with all its associated rigour. This experimental 'life-cycle' is essentially the same for both in vitro and in silico research: an experiment is designed to support a particular hypothesis; the experiment is executed; the results are recorded and analysed; and the findings are published to the scientific community for discussion, verification and for the generation of new hypotheses.

In this paper we describe the integration of two complimentary systems, <sup>my</sup>Grid[11] and UTOPIA[12], designed to support a more streamlined and rigorous in silico experimentation process. <sup>my</sup>Grid is a collection of components for building, enacting and managing workflow experiments in the life sciences. UTOPIA is a suite of interactive visualisation and analysis tools for examining and analysing the results of such experiments. Each system targets different problems from the life science domain: <sup>my</sup>Grid enables the automated integration of distributed resources and UTOPIA allows scientists to visualise and interact with experimental results. The functionality of one system compliments the other, providing the scientist with a platform that can support the whole in silico experiment life-cycle.

In the following sections we describe the individual systems' functionality before describing their integration and application.

## 2    <sup>my</sup>Grid

<sup>my</sup>Grid (http://www.mygrid.org.uk/) is a collection of loosely coupled components built on the technologies of web services, workflows and the semantic web. The design philosophy is one of openness and ease of extension. Consequently the architecture is an extensible set of components that can be adopted independently but which are designed to work together in a distributed service-oriented environment.

At the heart of <sup>my</sup>Grid is the Taverna workbench workflow environment. Taverna provides a framework for designing, executing and sharing workflow experiments using distributed web services. At present, there are over 3000 distributed services that Taverna can access from across the world, predominantly owned by third parties. These include resources from major service providers, such as the NCBI, DDBJ and the EBI, as well as specialist 'boutique' service providers offering resources and analysis tools that they themselves have developed for particular fields of research which they wish to share with the community. The reason that Taverna can access so many resources is that it does not impose any data models upon the service providers: for a web service to be integrated into Taverna, all that is required is the URL of the WSDL file. In this way, $^{my}$Grid can maximise the number and variety of resources accessible to scientists, and minimise the complexity for scientists adding their own services. Taverna can also access other types of service, for example, local java services, Biomart databases, bespoke beanshell scripts and also Biomoby services [7], further maximising coverage of the in silico biology domain.

As well as the ability to access distributed resources, Taverna allows the automation of experiments. The workflow itself defines how and when during an experiment a service should be invoked and the workflow can iterate over multiple data objects, enabling repetitive tasks to proceed without the scientist's intervention. By combining services together in Taverna workflows, scientists can automatically access and analyse large amounts of data from a large number of distributed resources from their own desktops. Accessing these resources at their source means that individual scientists do not require local supercomputing power and do not have the overhead associated with the maintenance of data resources.

## 2.1 Service Discovery

Taverna addresses the problem of accessing and interoperating between distributed services, but their distribution creates the requirement for manageable service discovery. The Feta Semantic Discovery component [9] enables services to be discovered by the biological functions they perform and the descriptions of the resources they can consume and produce.

Scientists generally know the methods or analyses they wish to use in an experiment, but they do not necessarily know what individual services are called, or where those services might be hosted. In order to address this problem services need to be annotated with a common set of terms that describe the various attributes necessary for their discovery. These will include, for instance, their input, output and biological task. These descriptions are delivered by annotating services with terms from the $^{my}$Grid ontology, over which Feta can browse and query.

## 2.2 The $^{my}$Grid Ontology

The $^{my}$Grid ontology describes the biological in silico research domain and the dimensions with which a service can be characterised from the perspective of the scientist [15,14]. Consequently the ontology is logically separated into two distinct components: the service ontology and the domain ontology. The domain ontology describes the types of algorithms and data resources used in silico, and the types of data that may be derived from, or used by, these resources. It effectively acts as an annotation vocabulary including descriptions of core bioinformatics data types and their relationships to one another. The service ontology describes the physical and operational features of web services, for example, where they are located, how many inputs and outputs they have and the type of service they are.

The <sup>my</sup>Grid ontology is written in OWL, the Web Ontology Language [5]. Using a formal description logic language means that we can support service discovery mediated both by scientists browsing for a service and computational service discovery using OWL-based reasoning, for example, when detecting semantic mismatches (incompatibilities) between services in workflows [3] and when automatically finding intermediaries to connect two incompatible services [6].

The Feta semantic discovery component of <sup>my</sup>Grid is essential for the integration of <sup>my</sup>Grid and UTOPIA. The provision of semantic descriptions of services also implicitly provides semantic descriptions of any data objects consumed or produced by the workflow. UTOPIA can exploit these descriptions to determine the types of data it is being presented with.

<sup>my</sup>Grid has been used in many areas of biological research, including genotype/phenotype correlations, functional genomics, systems biology, proteomics, microarray analyses and data integration. To demonstrate the added value of integrating <sup>my</sup>Grid and UTOPIA, we will use an example of previous research using <sup>my</sup>Grid, a study into the genetics of Graves Disease [8].

### 2.3   A <sup>my</sup>Grid Use-Case

Graves Disease is an autoimmune disease that causes hyperthyroidism. A project using <sup>my</sup>Grid workflows investigated the complex genetic basis of the disease by identifying and characterising genes located in regions of the human chromosome that showed linkage to Graves Disease. The experiment involved analysing microarray data to determine genes differentially expressed in Graves Disease patients and healthy controls, and then characterising these genes (and any proteins encoded by them) in an annotation pipeline. Gathering and examining all available information about potential candidates allowed researchers to determine which genes were the most likely candidates for further laboratory investigation. The outcome of the research was the identification of the gene I kappa B-epsilon as being involved in Graves Disease.

In this paper, we concentrate on the annotation pipeline workflow from the Graves Disease experiment. This workflow begins with the results of the microarray analysis with an Affymetrix probeset identifier and extracts information about genes encoded in this region. For each gene, the objective is to extract evidence from other data sources to potentially support it as a good candidate for disease involvement. The workflow (figure 1) was designed for the Graves Disease experiment, but it is equally applicable to similar candidate gene studies. It collects information on Single Nucleotide Polymorphisms in coding and non-coding regions, protein products, protein structure and functional features, metabolic pathways and Gene Ontology terms.

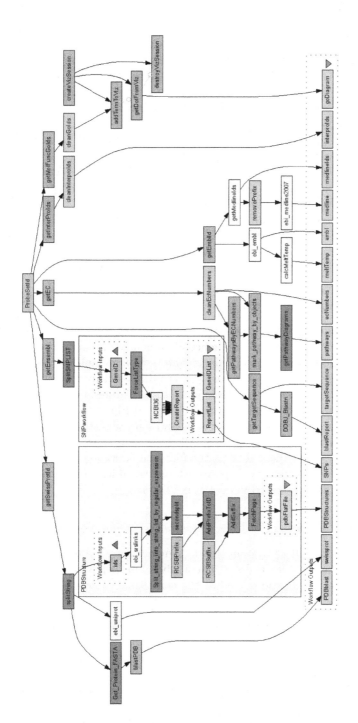

**Fig. 1.** Graves Disease workflow in Taverna

# 3   UTOPIA

Having designed and executed an experiment using Taverna, the UTOPIA system (http://utopia.cs.manchester.ac.uk/) addresses the next stage in the scientific cycle – that of analysis and interpretation. The emphasis here is on providing the scientist with intuitive ways of viewing their results in a variety of different forms without the need to write extra code or perform unnecessary tasks to convert between file formats or move between views.

Though the design of the initial in silico experiment benefits from access to the widest possible variety of tools and algorithms, the visualisation and analysis of data typically revolves around a comparatively small set of representations such as molecular structure, protein or genetic sequence, graph-based representations and various statistical and multi-dimensional data plots. Tools for visualising these effectively, however, are difficult to build well. UTOPIA provides a small set of high-quality and intuitive display tools, imbued with well-founded interaction metaphors, and made seamlessly interoperable and able to access a large number of data formats via an underlying semantic model shared with Taverna.

## 3.1   The Visualisation Tools

The UTOPIA suite currently has three released front-end applications: CINEMA is a fully-featured multiple sequence alignment tool, supporting many of the common desktop direction manipulation interaction metaphors such as 'drag-and-drop' and 'cut-and-paste'; Ambrosia[13] is a 3D structure viewer, exploiting modern Graphical Processing Unit techniques to accelerate high quality rendering of very large molecular models in real time; and Find-O-Matic provides an iTunes-like interface for discovering services and data objects. Examples of the tools in operation are shown in figure 3. All these tools share data in real-time via the underlying model described in the following section and are able to visually annotate the objects they display with biological features generated from services and workflows accessed via Feta and Taverna.

## 3.2   The Data Model

At the core of the UTOPIA system is a data model designed to be rich enough to capture the semantics of the data to be analysed in such a way that it can be exchanged between applications, and at the same time sufficiently light-weight such that it can be interrogated in real time to extract the data required by the interactive visualisation tools [10]. To achieve this balance between richness and efficiency – and also for conceptual elegance – the model is split in to a number of orthogonal spaces.

First, a distinction is made between *structure* and *annotation*: concepts that are accepted as 'fundamental facts' within a domain, and concepts that annotate or enrich the knowledge of the structure in some way but are in themselves either 'received wisdom', fuzzy, or refer to a process or collection of structural concepts.

Unlike in the physical and mathematical sciences where discoveries are axiom based, very few of the concepts in the biological domain can be thought of as absolute truths: beyond such things as atoms, bonds, residues and sequences the majority of biological features contain degrees of uncertainty or ambiguity that must be somehow represented within our model in order that it can be rendered as a visual object. UTOPIA's **structure space** is therefore quite small, and consists of four types of node: bonds, atoms, residues and sequences. All other concepts are mapped as *annotations* that project onto this structure space, and comprise **annotation space**. Each annotation may map to a single node in *structure space*, or to a set of nodes. An annotation may also have associated provenance. Optionally, a set of annotations may have an ontological structure projected onto it from **semantic space** to give it meaning in a particular domain or context and so that annotations can be classified and grouped in a hierarchy if appropriate. Finally, **variant space** represents uncertainty, conflict, and alternatives within a data set. A *variant* node maps onto a set of structural nodes that all maintain to represent the same data, and provides a mechanism for making any identifiable ambiguity or conflict explicit in the model.

What may initially appear as an overly philosophical way of storing data in fact yields a simple universal model that can represent actual data, uncertainty, conflicts and ambiguities, and be annotated in an arbitrary way using extensible ontological structures. Certain areas of the life sciences – for example, sequence or structure analysis – will heavily populate the structural space, with some annotations and semantics. Other areas – such as systems biology which deals in 'higher level' biological concepts – will concentrate on generating hierarchies and networks of annotations. Ambiguity and conflict exist in most areas: the 'same' protein from UniProt and PDB may have differing residues or atoms; the signalling network representing what is nominally the 'same' organism may have apparently different metabolites and reactions depending on its source.

The separation of these spaces allows their implementation to be tailored for their most common use in visualisation: a certain amount of 'heavyweight' computational reasoning may be required to infer that an Enzyme is-a-kind-of Protein so that it can be viewed in a sequence viewing tool; however the data structures and algorithms to support this must not interfere with the need to rapidly extract 10s of 1000s of annotations that form a systems biology graph, or the 100,000 or so atoms 30 times a second in order to be able to render a ribosomal complex as an interactive 3D structure.

This underlying model allows UTOPIA to gather and integrate data from a wide variety of heterogeneous sources and to generate a canonical internal representation that can be visualised by any of the front-end tools. Tools negotiate with the model using terms from the semantic space, e.g. 'can render sequences of residues with regional annotations', 'can show a fingerprint motif' or 'can display a structure of atoms with regional annotations', and thus do not have to be aware of file formats or the means of accessing remote sources of data. The richness of the model has two additional important features:

- Multiple UTOPIA tools are inherently aware that they are viewing the same biological concept, albeit potentially in radically different forms (e.g. as a residue sequence, as a molecular structure, and as a frequency plot). Thus modifications made to the data in real time by one tool and injected in to the underlying model are immediately reflected in any other.
- Biological concepts are exposed as 'first class citizens' in the interface itself, thus the tools are aware that the user has selected 'a sequence', 'an alignment of sequences', 'a signaling pathway', a 'cell compartment' and so on. This is especially important in terms of UTOPIA's integration with <sup>my</sup>Grid, and is explained in more detail in section 4.

### 3.3   Access to Remote Resources

UTOPIA provides tools for retrieval, visualisation and interactive manipulation of biological data; it does not of itself provide any algorithms or mechanisms for performing computational analysis of the data it manages: all such features are accessed via third party software. Plugin components called conduits connect UTOPIA to other sources of data and computation such as scripts, executable programs, web services and workflows. Terms from the <sup>my</sup>Grid ontology are used to annotate these, allowing UTOPIA to expose the functionality in appropriate parts of its tools' interfaces and to inject concepts into its model from the input and output formats used by the 3rd party software.

## 4   The Integrated System

UTOPIA and <sup>my</sup>Grid support complimentary aspects of the in silico experimentation process. The strength of <sup>my</sup>Grid lies in its ability to discover services

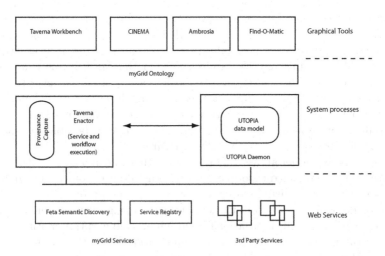

**Fig. 2.** Combined architecture

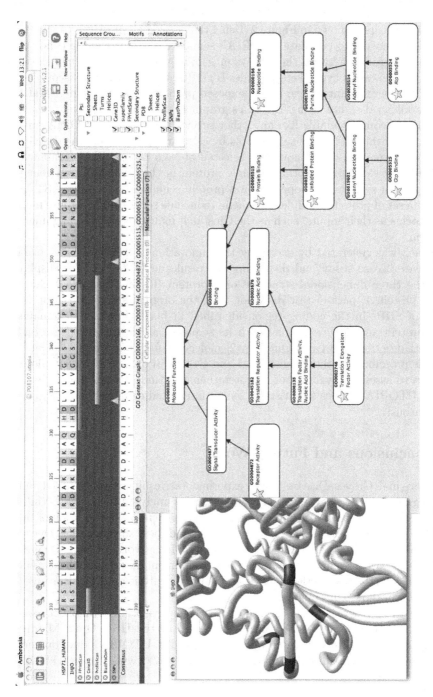

**Fig. 3.** Using UTOPIA to examine the output of a Taverna workflow

and to compose and enact workflows in order to generate auditable results. UTOPIA provides a flexible means of visually analysing such results. Their combined architecture is shown in Figure 2. Using the ontology that drives its semantic service discovery to populate UTOPIA's semantic space provides a coherent means of integration between the two systems, giving coverage of much of the in silico experimentation cycle. Feta provides UTOPIA with a means of discovering services by their semantic tags, and is exposed primarily via the Find-O-Matic tool. Workflows designed using Taverna's graphical interface, once semantically annotated, become available like any other service from within UTOPIA, and are exposed within the visualisation tools by comparing their semantic description to the semantics of the currently selected objects in UTOPIA's graphical interfaces. For example, selecting a single protein in the CINEMA tool generates a menu of services that operate on a protein sequence, such as BLAST[1]; selecting a set of proteins modifies this menu to include services that require multiple proteins as their inputs, such as the Clustal[4] multiple-sequence alignment algorithm.

Figure 3 was generated by executing the annotation pipeline workflow from the Graves disease study and displaying the results in UTOPIA. The display shows the three dimensional structure of a protein (in this case, Heat shock 70 kDa protein 1, part of test data not from the original study) with the locations of SNPs in the coding region identified in black. Functional domains of the protein are displayed along with its sequence above the structure, and the insert describes the Gene Ontology[2] molecular function terms associated with this protein. Combining functional and structural data in this way allows scientists to assess the validity of a protein being a candidate for disease association. UTOPIA enables scientists to see an integrated display of all this data simultaneously.

## 5   Conclusions and Future Work

In the original Graves Disease work, experimental results were gathered but not integrated, so for interpretation each result was analysed separately. Whilst this was effective, it was also time-consuming and inefficient. The addition of the UTOPIA tools to the experimental process addressed these shortcomings. Due to the extensible architectures and common semantic frameworks of both Taverna and UTOPIA, we were able to combine their complimentary functionalities. The outcome is a method for performing comprehensive analysis and visualization of data from workflow experiments.

At the time of writing, the Taverna workbench has been downloaded over 30,000 times, and development continues as part of the UK's Open Middleware Infrastructure Institute (http://www.omii.ac.uk). UTOPIA's development continues under the auspices of the EMBRACE (http://www.embracegrid.info) Network of Excellence.

# References

1. Altschul, S.F., Gish, W., Miller, W., Myers, E.W., Lipman, D.J.: Basic local alignment search tool. Mol. Biol. 215, 403–410 (1990)

2. Ashburner, M., Ball, C.A., Blake, J.A., Botstein, D., Butler, H., Cherry, J.M., Davis, A.P., Dolinski, K., Dwight, S.S., Eppig, J.T., Harris, M.A., Hill, D.P., Issel-Tarver, L., Kasarskis, A., Lewis, S., Matese, J.C., Richardson, J.E., Ringwald, M., Rubin, G.M., Sherlock, G.: Gene ontology: tool for the unification of biology. the gene ontology consortium. Nat. Genet. 25(1), 25–29 (2000)

3. Belhajjame, K., Embury, S.M., Paton, N.W.: On characterising and identifying mismatches in scientific workflows. In: Data Integration in the Life Sciences, pp. 240–247 (2006)

4. Higgins, D., Thompson, J., Gibson, T., Thompson, J.D., Higgins, D.G., Gibson, T.J.: CLUSTAL W: Improving the sensitivity of progressive multiple sequence alignment through sequence weighting, position-specific gap penalties and weight matrix choice. Nucleic Acids Research 22, 4673–4680 (1994)

5. Horrocks, I., Patel-Schneider, P.F., van Harmelen, F.: OWL: the making of a web ontology language. Web Semantics 1(1), 7–26 (2003)

6. Hull, D., Zolin, E., Bovykin, A., Horrocks, I., Sattler, U., Stevens, R.: Deciding matching of stateless services. In: 21st Nat. Conf. on Artificial Intelligence (AAAI06) Boston, MA, USA (2006)

7. Kawas, E., Senger, M., Wilkinson, M.D.: Biomoby extensions to the taverna workflow management and enactment software. BMC Bioinformatics, (Electronic) Journal Article Research Support, Non-U.S. Gov't, vol. 7(523), pp. 1471–2105 (2006)

8. Li, P., Hayward, K., Jennings, C., Owen, K., Oinn, T., Stevens, R., Pearce, S., Wipat, A.: Association of variations on i kappa b-epsilon with graves' disease using classical and mygrid methodologies. In: 3rd UK e-Science All Hands Meeting, Nottingham, UK (2004)

9. Lord, P., Alper, P., Wroe, C., Goble, C.: Feta: A light-weight architecture for user oriented semantic service discovery. In: Gómez-Pérez, A., Euzenat, J. (eds.) ESWC 2005. LNCS, vol. 3532, pp. 17–31. Springer, Heidelberg (2005)

10. McDermott, P., Sinnott, J., Thorne, D., Pettifer, S., Attwood, T.: An architecture for visualisation and interactive analysis of proteins. In: Proc. of 4th Int. Conf. on Coordinated and Multiple Views in Exploratory Visualization, London, UK, pp. 55–65 (July 2006)

11. Oinn, T., Greenwood, M., Addis, M., Alpdemir, M.N., Ferris, J., Glover, K., Goble, C., Goderis, A., Hull, D., Marvin, D., Li, P., Lord, P., Pocock, M., Senger, M., Stevens, R., Wipat, A., Wroe, C.: Taverna: lessons in creating a workflow environment for the life sciences. Concurrency and Computation: Practice and Experience 18, 1067–1100 (2005)

12. Pettifer, S., Sinnott, J.R., Attwood, T.K.: UTOPIA: user friendly tools for operating informatics applications. Comparative and Functional Genomics 5, 56–60 (2004)

13. Thorne, D., Pettifer, S., Sinnott, J., Attwood, T.: Unifying abstract and physical molecular model interaction. In: Proc. of EGUK Theory and Practice of Computer Graphics Conf. (2005)

14. Wolstencroft, K., Alper, P., Hull, D., Wroe, C., Lord, P., Stevens, R., Goble, C.: The mygrid ontology: Bioinformatics service discovery. Int. Journ. of Bioinformatics Research and Applications (IJBRA) (2007)
15. Wroe, C., Stevens, R., Goble, C.A., Roberts, A., Greenwood, M.: A suite of daml+oil ontologies to describe bioinformatics web services and data. Int. Journ. of Cooperative Information Systems 2(2), 197–224 (2003)

# A High-Throughput Bioinformatics Platform for Mass Spectrometry-Based Proteomics[*]

Thodoros Topaloglou[1], Moyez Dharsee[2], Rob M. Ewing[2], and Yury Bukhman[3]

[1] Information Engineering, Dept of Mechanical & Industrial Eng., University of Toronto
[2] Infochromics, MaRS Discovery District, Toronto
[3] Campbell Family Institute for Breast Cancer Research, University Health Network, Toronto

**Abstract.** The success of mass spectrometry-based proteomics in emerging applications such as biomarker discovery and clinical diagnostics, is predicated substantially on its ability to achieve growing demands for throughput. Support for high throughput implies sophisticated tracking of experiments and the experimental steps, larger amounts of data to be organized and summarized, more complex algorithms for inferring and tracking protein expression across multiple experiments, statistical methods to access data quality, and a streamlined proteomics-centric bioinformatics environment to establish the biological context and relevance of the experimental measurements. This paper presents a bioinformatics platform that was built for an industrial mass spectrometry-based proteomics laboratory focusing on biomarker discovery. The basis of the platform is a robust and scalable information management environment supported by database and workflow management technology that is employed for the integration of heterogeneous data, applications and processes across the entire laboratory workflow. This paper focuses on selected features of the platform which include: (a) a method for improving the accuracy of protein assignment, (b) novel software tools for protein expression analysis that combine differential MS quantitation with tandem MS for peptide identification, and (c) integration of methods to aid the biological relevance and statistical significance of differentially expressed proteins.

## 1 Introduction

Proteomics studies yield volumes of data characterized by heterogeneity and diversity in data formats, processing methods, and software tools and databases that are involved in order to transform spectral data into relevant and actionable information for scientists. As the size of, and the demand for, proteomics studies increases, laboratories are compelled to introduce more automation and increase throughput. Support for high throughput means more sophisticated tracking of experiments and the experimental steps, larger amounts of data to be organized and summarized, more complex algorithms for inferring and tracking protein expression across multiple experiments, statistical methods to access data quality, and a streamlined proteomics-centric bioinformatics environment to establish the biological context and relevance

---

[*] The work reported in this paper was carried out by the authors at MDS Proteomics / Protana.

S. Cohen-Boulakia and V. Tannen (Eds.): DILS 2007, LNBI 4544, pp. 71–88, 2007.

of the experimental measurements. In addition to the involvedness of data management, high throughput relates to the comprehensiveness of the analysis. Automation of computational steps enables fast processing of large datasets based on set parameters that reflect the common case. As a result, data processing can miss correct answers or admit false ones. The alternative is manual inspection of the data which lowers throughput. One solution is to incorporate appropriate methodologies within the data management strategy that help to detect and filter out poor data. Integration of such methodologies requires solid data management foundations that support queries to integrated data across the laboratory processes.

In this paper we present a proteomics data management and bioinformatics platform that our team was tasked to develop, for an industrial mass spectrometry-based proteomics laboratory focusing on biomarker discovery. Key requirements were to tackle the entire laboratory lifecycle, ensure high throughput, i.e., hundreds of MS acquisitions per week, and support both protein identification and protein quantitation (expression) analysis workflows. The landscape of MS data analysis, prior to the work described here, included a variety of specialized stand-alone applications for MS data acquisition, spectral analysis and protein identification (e.g., Mascot, Sequest, X!Tandem), manual means for protein quantitation, and no commercial-of-the-self (COTS) proteomics data management solutions and close to zero interoperability between existing applications.

We opted for a solution that is based on a systems integration approach, i.e., built upon a modular framework where each *task* can be fulfilled by a class of *components* that satisfy the task's interface, and provide a framework where data can flow from one component to another; components are software packages, custom programs, databases, or interfaces through which users access and change data. We committed to maximize the use of industrial strength, commercial technology and tools such as database and workflow management systems and visualization tools where applicable. For the most part, we adopted the "hub-and-spoke" model for application interoperation where a common database system plays the role of the hub. Data processing workflows were modeled and implemented in a commercial workflow management engine. The backbone of the platform therefore comprised a robust and scalable information management environment supported by database and workflow management technology that is employed for the integration of heterogeneous data, applications and processes across the entire laboratory process.

The novel aspects of our work are divided into two categories. First is the definition of a reference architecture of a system for acquisition, management, analysis and interpretation of MS proteomics data that was implemented and used in production. The second is the development of novel methods for protein inference, peptide quantitation and differential protein expression analysis. This paper focuses on the methods, algorithms and their implementation. We also discuss the structure and the main features of a protein sequence and annotation database and associated environment that was developed to provide timely access to data and tools and to enable data sharing and collaboration among scientists.

The rest of this paper is organized as follows. Section 2 introduces some concepts of mass spectrometry and proteomics and a case study used to demonstrate the functionality of the informatics platform. Section 3 presents the main functions of the computational analysis workflows supported by the platform, with emphasis on the algorithmic aspects and the platform components that were introduced to address these functions. Section 4 describes the overall information management and bioinformatics platform. Section 5 presents related work, and Section 6 concludes with a summary of our contributions and discussion on continuing challenges and future work directions.

## 2  Background

**Mass Spectrometry**

The goal of proteomics is to identify and quantify all the expressed proteins expressed in a biological system under specific conditions. Mass spectrometry (MS) has developed into the method of choice for achieving this goal (Aebersold and Mann 2003). The success of MS as a powerful analytical technology for biomolecule analysis is due to several technological accomplishments including the development of efficient protein ionization methods and advances in mass analyzer hardware. In addition, protein separation methods, including 1-D and 2-D gel electrophoresis and gel-free methods, make a variety of complex sample types amenable to MS analysis, including biofluids such as urine or blood plasma, cell lines and tissue extracts.

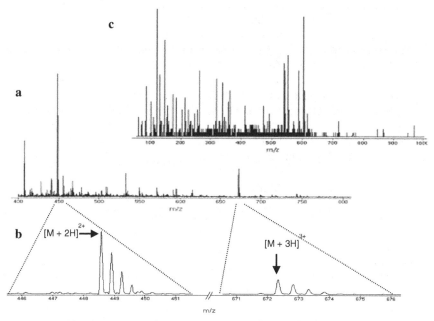

**Fig. 1. (a)** MS spectrum from rat urine sample scanned at elution time 55.2 min of a 210 minute LC gradient. **(b)** Isotopic clusters for a molecular species with monoisotopic mass (M) 1342.67 amu; both $[M + 2H]^{2+}$ and $[M + 3H]^{3+}$ ions were detected for this species. **(c)** MS/MS fragmentation spectrum for 2+ precursor ion observed 672.34 m/z, recorded at elution time 54.95 min. This spectrum was identified by Mascot as peptide LKAALSENEQLK with observed mass 1342.66 Da, corresponding to the mass of the molecular species shown in b.

A vast and multi-disciplinary scientific basis underlies the application of MS in proteomics, a detailed description of which is well beyond the scope of this paper. Here, we provide a brief outline of concepts particular to the analysis of complex samples with the MS platform that was employed in the case study. A protein sample is typically subjected to denaturation and enzymatic digestion (e.g. with Trypsin) to produce a mixture consisting primarily of peptide fragments. The mixture is injected into a liquid chromatography (LC) column for separation, and pumped at a low flow rate towards a capillary terminus at which a high electric field transforms the liquid material into charged ion droplets. The latter process is known as electrospray ionization (ESI) (see (Smith, Loo et al. 1990) for a review). The ion current emerging from the ESI tip is introduced into a mass spectrometry device where ions are separated on the basis of their mass to charge ratios (m/z). Ions at different m/z values are counted, resulting in a mass spectrum (MS). In tandem MS, or MS/MS mode, ions are typically subjected to high-energy collision with neutral gas molecules, leading to peptide fragmentation; the recorded tandem mass spectrum captures m/z and intensities values of the fragments (Chernushevich, Loboda et al. 2001). Continuous flow of the sample mixture is subjected to periodic scanning and detection of ions spanning the duration of the LC elution gradient.

An MS spectrum is an ion intensity signal recorded over an m/z range (Figure 1a). A charge state distribution series results for each molecular species detected, with adjacent peaks in a series assumed to be separated by exactly one charge (carried by a proton in the positive ion mode), thus enabling the calculation of accurate molecular weight using any two peaks within a series. Other reliable charge state determination algorithms have been proposed that are based on the detection of isotopic clusters (Horn, Zubarev et al. 2000), (Zhang and Marshall 1998), (Senko, Beu et al. 1995) (Figure 1b). A mass accuracy as low as 1-2 ppm was reported recently using a linear quadrupole ion trap coupled with an FTICR analyzer (Syka, Marto et al. 2004). A fragmentation pattern provided by an MS/MS spectrum (Figure 1c) can be used to deduce the amino acid sequence for the parent peptide (or precursor ion). Software for *de novo* and database search-based sequencing has made possible the automated identification of hundreds of proteins per sample.

## A Biomarker Discovery Case Study

The objective of this study was to identify proteins in rat urine that are indicators of drug-induced kidney damage. This study was conducted by MDS Proreomics / Protana Inc. in order to demonstrate its discovery proteomics technology in clinically relevant fluids such as urine. Here we discuss the aspects of the study pertaining to data management and data analysis and describe the key bioinformatics methods and systems involved in carrying out the study.

The study involved 15 rats, 10 of which were treated with a drug known to cause kidney damage and the remaining 5 used as controls. Urine samples were taken at 3 time points after treatment. The drug causes proteinuria, a condition in which total protein concentration in urine is significantly increased. The urine samples were taken before the onset of proteinuria, near its peak, and after it had mostly subsided. Both treated and control rats were young and continued to grow through the course of the experiment. Auxiliary study data including specimen parameters such as weight over time, gender, clinical tests, histopathology report, sample amount and concentration, and sample preparation measurements, were recorded in the platform's sample tracking database (SATS).

LC-MS acquisitions were performed in batches containing 1 control and 2 treated rats, producing a total of 5 batches. The samples from a given time point and batch were always analyzed back-to-back in order to facilitate an accurate control/treated comparison. The order of batches and the order of samples within each batch were randomized. The study setup and acquisition information were also captured in SATS. LC-MS/MS acquisitions were run on a QStar Hybrid XL mass spectrometer in an intensity-dependent sequencing mode, allowing for the determination of fragmentation spectra of the most intense ions.

## 3   Bioinformatics Methods for MS Proteomics

The discrete steps of the biomarker discovery study workflow are outlined in Table 1. Each step involves a sequence of tasks or subsequent workflows that generate or transform data. In the rest of this section, we use the case study to demonstrate capabilities of the platform, focusing on those that are technically challenging and involve novel work.

The main objective of a biomarker discovery study is the identification of differentially expressed proteins that are biologically relevant and statistically significant. To achieve this, raw MS data are processed via two parallel analysis workflows: protein identification and peptide quantitation (Figure 3).

**Table 1.** Overview of bioinformatics in biomarker discovery lifecycle

| |
| --- |
| **Sample Processing** |
| 1.   Register samples and sample attributes in the sample tracking system |
| 2.   Create acquisition plan applying principles of statistical experimental design |
| **MS Data Acquisition** |
| 3.   Collect data files from MS instruments |
| 4.   Copy MS data to centralized file storage for analysis |
| **MS Data Processing** |
| 5.   Derive quantitative profiles of peptides based on MS |
| 6.   Search MSMS data against a protein sequence db and derive peptide sequences |
| 7.   Integrate peptide profile data in a matrix based on the study structure |
| **Bioinformatics  Analysis and Interpretation** |
| 8.   Assign peptides to proteins by clustering peptide hits across multiple acquisitions |
| 9.   Report statistically significant differential peptides and proteins |
| 10. Evaluate biological relevance of significant hits using available annotation & iterature |
| 11. Mine over-represented annotation themes in the study |

Quantitative differential analysis can be summarized as the measurement and interpretation of protein expression in biological samples aimed at detecting differences that are due to conditions present in the sample (e.g. disease, treatment). The analysis is *quantitative* because abundances are derived for each molecular species (e.g., peptides) detected in the samples; it is *differential* because we compare abundances across samples. In the case study, for example, expression profiles of peptides and proteins are calculated in samples taken from animals at different time points, in an effort to identify proteins that may be related to the early onset of nephrotoxicity. Methods for computing such profiles from complex biological samples using mass spectra are described in (Lilien, Farid et al. 2003), (Petricoin, Ardekani et al. 2002), (MacCoss, Wu et al. 2003), (Johnson, Mason et al. 2004) .

**Protein Assignment**

Peptide sequences are assigned based on interpretation of their fragmentation (MSMS) spectra. Several methods are available for this, including de novo sequencing and database searching (for a recent review see (Baldwin 2004)). The standard practice in our platform is through database-searching using the Mascot search engine (Mascot; www.matrixscience.com) and a protein sequence database. A known limitation of the database searching approach is that search engines make incorrect assignments (Nesvizhskii, Keller et al. 2003). Considerable effort within the proteomics community is being directed at refining peptide scoring and distinguishing high-quality spectra. The solution that we implemented, at the time of this study, was to accept peptide identifications above a fixed Mascot score threshold. The score threshold was originally defined empirically. Each MS/MS acquisition data file (all 45 of the case study) was searched using Mascot. The resulting data-files were parsed into a relational database of mass spectrometry experiments and analyses (MSdb), including also the Mascot search-parameters (such as modifications, enzyme information, etc). We later extended the search engine score with a second score derived using an empirical statistical model. The integration of a second score is a straightforward process in MSdb, as it is also the case with results from additional search engines.

The choice and size of the search database plays an important role in peptide matching (Fenyo and Beavis 2003), (Cargile, Bundy et al. 2004). Intuitively, a large, comprehensive, multi-species database increases the chances of a match and of false positives, where a smaller species specific database reduces false positive matches. An important factor in peptide matching is the redundancy of the database. Our approach is to generate non-redundant sequence search databases that are appropriate to the study in hand. A central component of our bioinformatics infrastructure is therefore the construction of a non-redundant protein sequence database. In contrast to existing public-domain efforts (e.g. IPI (Kersey, Duarte et al. 2004)), we define redundant proteins as being proteins from the same species that are identical in sequence and length. We built and maintained an in-house protein sequence and annotation database, called AIDA (for details see Section 4), based on the eukaryotic proteins from all major sources of protein sequence information (SwissProt, TrEMBL, GenBank etc). Each distinct entry in AIDA, i.e., a unique sequence and species pair, has a unique and stable protein identifier, termed PI. Using AIDA as the source, we generate appropriate, versioned sequence databases for MS/MS searches. A combined human, rat and mouse protein sequence database was used in the current study.

The next step is the protein inference. The objective in this step is to identify a minimal set of proteins consistent with the observed peptides. Protein inference is a non-trivial process that is complicated by the 'many-to-many' relationships between MS/MS spectra and candidate peptides and between peptides and protein sequences (Yang, Dondeti et al. 2004), (Tabb, McDonald et al. 2002), (Kristensen, Brond et al.

2004). Biomarker discovery studies, however, like our case study, commonly possess a property that helps to improve the sensitivity of the protein inference process. This property is replication. As multiple biological and/or technical replicate samples are analyzed, the same population of peptides is assumed to be present in these samples. However, the mass spectrometer focuses at selected ions due to constraints like length of duty cycle or sample complexity. Merging therefore multiple search results leads to a richer pull of peptides that can be exploited to infer proteins. A similar parsimonious approach is outlined by Yang et al (Yang, Dondeti et al. 2004). The protein inference method is based on a clustering operation that minimizes the set of proteins that best explains the set of observed peptides. The equivalent operation of the search engine is no longer applicable as we consider multiple acquisitions, although Mascot has lately introduced this function.

Briefly, our clustering operation works as a two-step process. In the first step, proteins are grouped according to shared sets of matching peptides (for two proteins to be grouped one of the sets of peptides must either be equal to, or a proper subset of the other). Second, redundant clusters are filtered out by ranking clusters by peptide complement and then iteratively removing clusters whose peptide complement is redundant with respect to two or more other clusters. The output of this clustering process is stored in MSdb. The resulting set of clustered proteins and peptides is further refined by selecting a representative 'anchor' protein from each cluster. The anchor protein of a given cluster is identified by taking the top ranked protein by number of peptides. In some cases, there may be multiple top-ranked proteins (insufficient peptide evidence to distinguish the proteins). In these cases, other selection criteria are applied such as choosing an anchor by species (i.e. if for a given cluster in our rat urine study, a human and rat protein with equivalent peptide evidence were identified – the rat protein would be selected as the anchor), or by annotation (i.e. selecting the best-annotated protein).

**Peptide Profile Matrix Construction**

A major task in the quantitation workflow is the generation of the profile matrix. Intuitively, a peptide matrix integrates quantitative peptide data from multiple samples. The structure of the profile matrix is illustrated in Figure 2. A row in this matrix represents a molecular species (a putative tryptic peptide) measured in a set of samples, while a column represents the molecular profile of a distinct sample or state. Each cell therefore represents the presence (or absence) of a putative peptide in a given state, characterized by monoisotopic mass, intensity, elution time, and sequence, among other attributes. This profile matrix is the starting point for subsequent bioinformatics and statistical analyses.

| | $S_1$ | ... | $S_i$ | ... | $S_n$ |
|---|---|---|---|---|---|
| $M_1$ | | | | | |
| | | | | | |
| $\dot{M}_i$ | | | $C_{ij} = \{...,Mass_{ij}, Time_{ij}, Intensity_{ij}, Sequence_{ij, ...}\}$ | | |
| | | | | | |
| $\dot{M}_m$ | | | | | |

Fig. 2. Profile Matrix

The derivation of a profile matrix involves a series of steps that are summarized as follows. First, raw spectral peaks are extracted from the MS data file, isotopic clusters are resolved, and each isotopic cluster is assigned a charge state and molecular mass. Second, ions for the same molecular species observed at different charge states and different chromatographic elution time points are combined, producing a set of distinct molecular species detected in a given sample. Finally, samples are compared and molecular species are brought together to produce the rows of the profile matrix. Next, we discuss each step in detail with the help of the case study.

The first step is a pre-processing and signal analysis step that takes as input the raw MS signal and produces a list of monoisotopic peaks. Raw MS signal peaks are extracted from the the acquisition file with the help of the instrument vendor's supplied software. In the case study, raw signal peaks are extracted from each of the 45 data files using in an extractor application that utilizes the Analyst QS API (www.sciex.com), resulting in an *extracted peak* list for each MS spectrum. The data extractor program provides the automation that is needed for high-throughput processing. An extracted peak consists of an ion mass-to-charge ratio (m/z), intensity, and elution time. The extracted intensities are channelled through a second order, 5-point Savitsky-Golay smoothing routine and then processed by the THRASH algorithm (Horn, Zubarev et al. 2000). THRASH was modified slightly for processing TOF data from the QStar Hybrid XL instrument. THRASH resolves isotopic clusters in an MS spectrum and assigns charge state and monoisotopic mass to each cluster, thus producing a set of monoisotopic peaks. The THRASH parameters and the output data are saved in MSdb. In the case study, nearly 18 million isotopic clusters were generated, averaging 404,161 per sample.

The next steps involve two grouping operations applied to monoisotopic peaks in order to obtain a set of molecular species detected for each sample:

(1) In a given MS spectrum, monoisotopic peaks with matching mass (within a specified mass tolerance window) are grouped and their intensities summed, bringing together ions having acquired different charges but representing a single molecular species (such as the peaks for the 2+ and 3+ ions shown in Figure 1c).

(2) Molecular species appearing in multiple adjacent MS scans are grouped to yield a chromatographic peak. Mass and elution time occurring at the intensity apex are recorded for each peak. A chromatographic peak corresponds to the intensity profile of an eluting putative peptide. Peptide sequences assigned by Mascot, as described in the Protein Assignment section, are linked to chromatographic peaks on the basis of peptide neutral mass and precursor ion elution time. Intuitively this corresponds to a fuzzy join operation on mass and time attributes as some tolerance mass and time intervals are applied. This join connects the abundance information (intensity) of a putative peptide with the peptide sequence information (identity), if it is available. In total, 161,939 chromatographic peaks were constructed in the study, with an average of 3,680 per sample.

Operations (1) and (2) produce a list of chromatographic peaks per sample.

The next step involves two alignment operations that complete the construction of the profile matrix. Because the experimental design involved batches of samples, we first align by batch and then across all batches. This step is varied depending on the experimental design and in particular on the replication or batching factors.

(3) A pairwise chromatographic peak alignment algorithm is applied to the chromatographic peaks within a batch. In the case study, each batch consists of 1 control and 2 treated samples. In each batch, a reference sample is randomly selected, and each non-reference sample is aligned with the reference, resulting in shifted chromatographic peak elution times. Alignment is based on a piecewise polynomial function computed using elution times of control points common between samples. After alignment, chromatographic peaks are compared and grouped across the set of samples in each batch on the basis of mass, elution time, charge composition, and sequence where available. This results in a set of unique putative peptides for each batch. The case study produced a total of 100,623 unique putative peptides across the 15 batches. Each batch contained on average 6,708 unique putative peptides.

(4) Average monoisotopic mass and elution time are calculated for each putative peptide in a batch. Peptides consisting entirely of singly-charged ions are assumed to be products of electronic and chemical noise and are filtered out. Mass, elution time and sequence, if assigned, are compared between peptides from different batches. The case study was finally condensed in profile matrix containing a final count of 22,349 distinct putative peptides (matrix rows).

Table 2 show the significant reduction in data points that is achieved during each step of this process. An important feature of this process is that all data generated after raw peak extraction, from the monoisotopic ion peaks generated by THRASH, to MS/MS precursor ions and sequences assigned by Mascot, to the cells of the profile matrix itself are stored in a relational database, MSdb, This provides an efficient way of querying and comparing data points during the various grouping operations, and for carrying the profile matrix forward for further analysis.

**Table 2.** Data points generated in the case study after MS peak extraction

|  | Total | Mean per batch | Mean per sample |
|---|---|---|---|
| Isotopic Clusters (Monoisotopic peaks) | 17,783,082 | 1,185,539 | 404,161 |
| Molecular Species (2) | 161,939 | 10,796 | 3,680 |
| Putative Peptides grouped within Batches (3) | 100,623 | 6,708 | N/A |
| Putative Peptides grouped across Batches (4) | 22,349 | N/A | N/A |

## Identification of Differentially Expressed Proteins

The goal of the case study is to identify proteins that are differentially expressed, e.g., low in time point 1, high in time point 2, and low again in time point 3. The information needed to make this decision is made available in the profile matrix and in the protein assignment process. The method for identifying differentially expressed proteins is outlined in Table 3. First, the profile matrix is filtered such that only those peaks detected in at least 75% of the samples and at least one of the treatment groups are retained. After filtering, the matrix retained a total of 2,035 rows. The filtered matrix is analogous to data matrices produced by microarray studies, enabling the use of many of the analysis techniques developed for microarrays, as reviewed in (Simon, Korn et al. 2003).

Statistical analysis here aims to answer two questions: (a) which peptides (rows of the matrix) are differentially expressed with statistical significance, and (b) which peptides link to the same protein cluster according to the protein inference algorithm described earlier. The details of the statistical analysis are beyond the scope of this paper. The results of the case study are summarized in Table 3. Specifically, the peptide clustering algorithm produced 345 protein clusters, 216 of which were linked to data rows that passed the reproducibility filter. Altogether, 32% of all reproducible data rows were linked to a protein sequence. This quantitative assessment concludes that about 200 urine proteins were affected by drug treatment. These proteins were supported by more that 500 peptides, for which both high-confidence sequence assignment and sufficient intensity data were calculated.

**Table 3.** Quantitation and identification stream statistics

| | |
|---|---|
| **Quantitation** | |
| Profile matrix rows, total | 22,349 |
| Rows passing reproducibility filter | 2,035 |
| Reproducible rows linked to high-confidence sequences | 661 |
| **Identification** | |
| High-confidence sequences, total | 1,264 |
| High-confidence sequences linked to reproducible rows | 550 |
| Clusters, total | 345 |
| Clusters linked to reproducible rows | 216 |

## Biological Interpretation

A significant challenge with high-throughput "-omic" technologies is the interpretation of the resulting data. In the current study, using the proteomic data processing workflows and tools supported by the informatics platform we identified around 200 rat urine proteins that are differentially expressed in drug-treated and control animals. Just the identity of these proteins does not reveal much about the effects of drug treatment. Assessment however of the available annotation for the proteins in the list – such as sequence-based annotations (e.g. functional domains, signal peptides) or function-based annotations (e.g. biological process, pathways, etc), may provide some insights of the effects of drug treatment. In addition, these annotations can be grouped by type and using annotation mining tools to identify functions, motifs, etc which are over- or under-represented in the overall set (Hosack, Dennis et al. 2003), (Zeeberg, Feng et al. 2003), (Boyle, Weng et al. 2004). In the current study, for example, biological processes such as renal cell repair and glomerular damage were identified which are consistent with known biological processes induced by the action of the drug in the rat kidney.

To facilitate these kinds of annotation-based analyses, we rely on AIDA. AIDA which stands for Automated Integration of Datasets and Applications, in addition to being a protein index as it was described in Section 3, is also a protein annotation database. AIDA collates protein annotation from many major public domain sources. A home developed extract-transform-load (ETL) application, called the data feeder, is responsible for assembling and keeping the data in AIDA up to date. The data feeder maintains information on the schema of certain subsets of the external data sources and its correspondence to the AIDA schema, such that it selectively extracts information from these sources, it transforms it to the local schema, performs the necessary entity matching operations, and loads this information into AIDA. In addition, each protein entry in AIDA has several fields that hold pre-computed annotations that are calculated using a number of computational annotation tools, such as sequence feature prediction programs to pre-compute domains, signal peptides etc. These pre-computed predictions are stored in the database for rapid query access. A central feature in AIDA, as previously described, is the assignment of a unique PI (protein identifier). Each PI represents a unique combination of protein sequence and species, so that identical sequences from a given species will be grouped as a single PI (identical sequences are required to be identical in both sequence and length).

The PI has the following properties: (a) The PI is persistent, i.e., the identity of a unique sequence, in a given species, will never change. Oftentimes, a unique sequence changes identity or has multiple identities (e.g. GenBank identifiers or GIs) in the public sources. Using a PI, a sequence is immune from public ID changes, yet all these corresponding GIs are linked to the PI. (b) PI-based, AIDA generated sequence databases used for MS/MS searches are free of the problem of reporting the same sequence under different public identifiers (e.g. GI) that may change beyond our control. (c) Sequence-based information such as annotations attached to different versions of a sequence are brought together under the same PI. (d) The PI maintains provenance of protein sequences, since any ID or name changes of the sequence in the public sources is recorded in the PI record. (e) The PI functions as the cross-database link between AIDA and MSdb.

Understanding as much as possible about the biological meaning of an observed differential protein is an important part of the process of the identification of viable biomarker or drug target candidates. To this end, AIDA provides a uniform view of many of the known associations (e.g. sequence features, biological processes, pathways) for protein sequences.

## 4   Bioinformatics Platform Implementation

Figure 3 provides a schematic overview of the bioinformatics platform which comprises systems that deliver the functionality of the methods described above, and a backbone information management environment intended for high-throughput, scalable and proteomics analysis. The system is structured as a series of workflows that load data into a family of integrated databases which are made available to data analysts and scientific users via a range of front-end analysis and visualization applications. In what follows we briefly discuss the information and workflow management backbone of the system, and a representative front-end application.

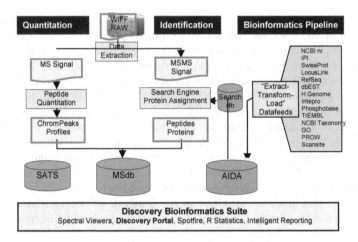

**Fig. 3.** Bioinformatics Platform Overview

## Data and Workflow Management

Even for a relatively small study such as the case study described in this paper, there is a need to keep track of acquisitions, analysis operations, study structure, sample-specific factors and program parameters for interpretation of the results. This information may not be stored in a single *place* as it is generated at different times and by different systems. In order to facilitate tracking of data and processes throughout the laboratory workflow, we organized the data stores back-ending the various processing steps as a federation of cross-referencing databases. The significant members of the federation include a sample tracking database (SATS), a repository for mass spectrometry experiments analysis results (MSdb), and a protein index and annotations database (AIDA). The common reference entity between SATS and MSdb is a uniquely assigned MS acquisition identifier. Similarly, MSdb and AIDA share a common reference for protein entries (PI). Maintenance of referential integrity between these independently maintained databases is performed by custom programs. For convenience and easier cross-database queries, the individual databases are federated using DB2 Information Integrator (www.ibm.com).

The systems that support the proteomics laboratory workflows are both distributed and heterogeneous. For example, MS data acquisitions are performed using vendor-supplied applications running on the workstations that control the instruments; data files stored on these workstations are automatically copied to a network file system; finally MSdb is updated with acquisition-relevant records, at which point the identification workflow is triggered to extract MS/MS data, perform Mascot searches, and parse the results into MSdb, etc. Organizing into a collection of integrated databases is only part of the overall solution. Improving application interoperability and automating data transfers between applications represents another significant simplification and source of productivity gains. We achieve this through the deployment of workflow management solutions. In addition to simplifying data analysis by automating processing steps, the introduction of workflow management also improves utilization of computing resources. The data processing workflows were automated using TurboWorx (www.turboworx.com), a commercial solution for workflow management, while the bioinformatics pipeline that maintains AIDA is supported by an in-house workflow management solution.

## Discovery Portal

The Discovery Portal (DP) is a web-based collaboration environment that allows scientists to access information on proteins, maintain their projects, access proteomics analysis applications, map proteins to scientific literature and explore external databases. The Discovery Portal was designed to be the single point of access for all internal and external applications and data in the bioinformatics area. It creates a hub of information based on the family of databases mentioned earlier and provides an access interface to them.

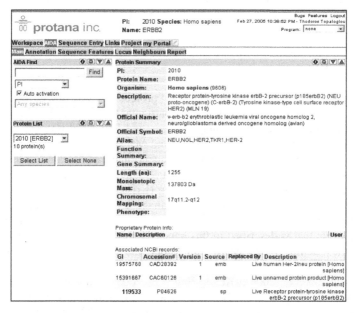

**Fig. 4.** The Discovery Portal showing the AIDA find, Protein list and Protein summary portlets

The Discovery Portal is built using the Jetspeed portal application builder with the goal to enable aggregation of multiple Web applications is a single page as portlets, and allow users to customize and personalize the content of the page. Important portlets include the AIDA query, AIDA annotation (Fig. 4), custom reports, and a literature mining portlet, among others. The Discovery Portal is easily extendable and enables economic and rapid development of interfaces for bioinformatics applications and data sources.

The portal supports a variety of ways to integrate applications. Specifically it implements a container framework with standardized navigation features and style sheets, and a specification of a general method for deploying applications inside the framework. It also provides a common user authentication component for controlling data access and ownership. The portal maintains a runtime data object holding the data/information to be rendered, and a publisher/subscriber pattern to facilitate acting on modifications to the data object. This allows for results (proteins) of an operation, to be available for use in another application, making it really easy to run multiple applications on the same data without copying, pasting and opening of new applications.

The Workspace is another key portlet that further aids user collaboration and application interoperability. The motivating principle behind the Workspace is that experiments and bioinformatics analyses produce lists of proteins that scientists may want to store, share, modify or compare. Within the Portal, the scientific user can create protein lists as workspace objects and open them in some other application or promote them to the Portal's runtime object. The Workspace facilitates both persistent storage of user selected data and data exchange between portal applications. The latter reduces the proliferation of disparate spreadsheets and promotes interoperability. Workspace protein lists may also be associated with an open-ended set of attributes, both user-defined and derived from underlying databases. An XML database is employed to tackle the representation and storage of Workspace objects.

# 5   Related Work

Until recently, the focus of informatics in proteomics was the development of methods for interpreting spectral data and in particular tools for identification (Mascot (Perkins, Pappin et al. 1999), X!Tandem (Craig and Beavis 2004)) and quantitation (Xpress (Li, Zhang et al. 2003)) of MS/MS and MS spectra. As the volume of data produced by LC-MS experiments and subsequent analyses, became an issue, stand alone tools were extended with some data management functionality (Mascot Integra (www.matrixscience.com), EPICenter (Kristensen, Brond et al. 2004)), or integrated in laboratory information management systems (Proteus LIMS (www.genelogics.com)). Still, these systems address only a part of the discovery process lifecycle that is described in this paper. High-throughput laboratories such as EMSL at the Pacific Northwestern Laboratory and the Seattle Proteomics Center (SPC) at the Institute of Systems Biology have developed significant platforms (Kiebel, Anderson et al. 2004), (Keller, Eng et al. 2005) with similar functionality but different architecture than ours. The SPC system, called trans-proteomics pipeline, is of particular interest because it promotes open standards and XML formats in order to enhance interoperability and assist integration of third party tools. Instrument vendors also market solutions for management and analysis of proteomics data (Protein Expression System (www.waters.com)) that combine protein identification and quantitation analyses. Several proteomics companies have also developed their own custom platforms that are specific to their needs (ProteomicIQ (www.proteomesystems.com), CellCarta (www.caprion.com)). Although these systems share common goals, they differ in their focus, functions and throughput support. Comprehensive comparison of those would be very hard due to limited or no access to their methods, architecture and technical specifications. To the best of our knowledge, this paper is the first to provide details on the algorithms and implementation of such platform. Recently, bioinformatics software vendors have launched products with support for the entire LC-MS proteomics workflows and functionality tailored to biomarker discovery that share features with our work including the Elucidator system (www.rosettabio.com) and Proteomarker (www.infochromics.com). Proteomarker has evolved from the work reported here and is designed to provide rapid and accurate quantitative analysis of LC-MS datasets.

Our work connects with, and benefits from, many disciplines including algorithms for protein identification (Baldwin 2004) and differential proteomics analysis (Listgarten and Emili 2005), integration of protein sequence and annotation databases (Kersey, Duarte et al. 2004), scientific workflow management (Ludascher and Goble 2005), implementation of data provenance in scientific databases (Simmhan, Plale et al. 2005), and proteomics data representation (Taylor, Paton et al. 2003) and analysis (Pedrioli, Eng et al. 2004).

# 6  Discussion

We have presented a bioinformatics platform for a mass spectrometry based proteomics. We have also provided an account of specific challenges and solutions pertaining to selected aspects of the platform. The design and implementation of the platform and of its components are based on the requirement of an industrial proteomics laboratory focusing in novel biomarker discovery.

The contributions of the platform are methods for accurate assignment of proteins to observed peptides, construction of a peptide profile matrix from raw mass spectra that enables statistical analysis for establishing significant expression profiles. A distinguishing feature of the platform is the integration of these components through an underlying infrastructure for data and workflow management, enabling us to connect otherwise disparate data and processes, from sample preparation, mass spectrometry, protein identification, spectral analysis, and biological interpretation.

The benefits of integrating data and processes in a proteomics laboratory are especially important for biomarker discovery. First, it improves the efficiency and throughput of data analysis. For example, an earlier study conducted in the same laboratory, of the same biological model consisting of 18 acquisitions, using manual methods for grouping spectral peaks and spreadsheets for collating data, required 2 months for data analysis and produced a smaller yield of peptide and protein hits. In the case study described here, involving differential quantitative analysis of 45 samples, construction of the profile matrix was rendered a routine task completed in a matter of days, amounting to a ten-fold reduction in data processing time. Second, results and data from intermediate processing steps are accessible through database queries. For example, for a differential protein selected by statistical analysis, supporting data can be readily reported at all levels, from protein cluster, to peptide, to individual raw MS peaks across all samples. Third, an integrated platform helps the scientist to focus on data interpretation and enhancement of discovery methods, rather than on manually running software and locating and connecting diverse data.

There are several continuing challenges and directions for future work. In protein identification, a significant drawback is the inability to assign sequences to all statistically meaningful peptide expression profiles. This reflects technological limitations of MS instrumentation, as well as shortcomings of existing protein identification algorithms. The development of an empirical statistical scoring method for peptide hits, based on physical and chemical properties of spectra, along the lines of (Nesvizhskii, Keller et al. 2003) can give rise to good identifications and improve the sensitivity of database searches. In the construction of the profile matrix, we make important assumptions pertaining to digestion efficiency, chromatography, ion detection, and fragmentation. Currently, algorithm parameters settings such as elution time tolerance windows for peak construction and comparison, or relative abundance thresholds for establishing significant differentials, are set based on the results of various reproducibility studies conducted to support those assumptions. However, a continuing challenge is to fully characterize effects of instrument configuration on resulting data. The storage of comprehensive process and experimental data in an integrated database is a step forward to this direction.

Biomarker discovery studies generate large and complex datasets, the analysis of which necessitates advanced data management and analysis tools. As shown here, such tools broaden the understanding of the data and offer insights for technical improvements.

**Acknowledgements.** We would like to thank our colleagues in former MDS Proteomics / Protana: Panos Economopoulos, Anne-Marie Simmie, Michael Li, Peter Chu, Huicheng Wang, Adrian Pasculescu and Derek Lee for contributing development effort; former colleagues Soeren Schandorff (AIDA), Søren Larsen (Discovery Portal), Adel Shrufi, Mark Robinson, Zhe Wang, Chris Orsi (DAS), John Sulja and Kevin Mok for software contributions. Special thanks to Nancy Ng, Jian Chen, Henry Duewel for providing the case study data, Ian Stewart for software requirements specs, and Shane Climie for educating us in biomarker discovery.

# References

Aebersold, R., Mann, M.: Mass spectrometry-based proteomics. Nature 422(6928), 198–207 (2003)

Baldwin, M.A.: Protein identification by mass spectrometry: issues to be considered. Mol. Cell Proteomics 3(1), 1–9 (2004)

Boyle, E.I., Weng, S., et al.: GO::TermFinder–open source software for accessing Gene Ontology information and finding significantly enriched Gene Ontology terms associated with a list of genes. Bioinformatics 20(18), 3710–3715 (2004)

Cargile, B.J., Bundy, J.L., et al.: Potential for false positive identifications from large databases through tandem mass spectrometry. J Proteome Res. 3(5), 1082–1085 (2004)

Chernushevich, I., Loboda, A., et al.: An introduction to quadrupole-time-of-flight mass spectrometry. Journal of Mass Spectrometry 26, 859–865 (2001)

Craig, R., Beavis, R.C.: TANDEM: matching proteins with tandem mass spectra. Bioinformatics 20(9), 1466–1467 (2004)

Fenyo, D., Beavis, R.C.: A method for assessing the statistical significance of mass spectrometry-based protein identifications using general scoring schemes. Anal. Chem. 75(4), 768–774 (2003)

Horn, D.M., Zubarev, R.A., et al.: Automated Reduction and Interpretation of High Resolution Electrospray Mass Spectra of Large Molecules. Journal of American Society for Mass Spectrometry 11, 320–322 (2000)

Hosack, D.A., Dennis, Jr., G., et al.: Identifying biological themes within lists of genes with EASE. Genome Biol, vol. 4(10) (2003)

Johnson, K.L., Mason, C.J., et al.: Analysis of the Low Molecular Weight Fraction of Serum by LC-Dual ESI-FT-ICR Mass Spectrometry: Precision of Retention Time, Mass, and Ion Abundance. Analytical Chemistry 76, 5097–5103 (2004)

Keller, A., Eng, J., et al.: A uniform proteomics MS/MS analysis platform utilizing open XML file formats. Molecular Systems Biology (2005)

Kersey, P.J., Duarte, J., et al.: The International Protein Index: an integrated database for proteomics experiments. Proteomics 4(7), 1985–1988 (2004)

Kiebel, G.R., Anderson, G.A., et al.: Proteomics Research Information Storage and Management (PRISM) System, Pacific Northwest National Laboratory (2004)

Kristensen, D.B., Brond, J.C., et al.: Experimental Peptide Identification Repository (EPIR): an integrated peptide-centric platform for validation and mining of tandem mass spectrometry data. Mol. Cell Proteomics 3(10), 1023–1038 (2004)

Li, X.-J., Zhang, H., et al.: Automated Statistical Analysis of Protein Abundance Ratios from Data Generated by Stable-Isotope Dilution and Tandem Mass Spectrometry. Analytical Chemistry 75(23), 6648–6657 (2003)

Lilien, R., Farid, H., et al.: Probabilistic Disease Classification of Expression-Dependent Proteomic Data from Mass Spectrometry of Human Serum. Journal of Computational Biology 10(6), 925–946 (2003)

Listgarten, J., Emili, A.: Statistical and computational methods for comparative proteomic profiling using liquid chromatography-tandem mass spectrometry. Mol. Cell Proteomics 4(4), 419–434 (2005)

Ludascher, B., Goble, C.: Guest Editors' Introduction to the Special Section on Scientific Workflows. SIGMOD Rec. 34(3), 4–5 (2005)

MacCoss, M.J., Wu, C.C., et al.: A Correlation Algorithm for the Automated Quantitative Analysis of Shothun Proteomics. Analytical Chemistry 75(24), 6912–6921 (2003)

Nesvizhskii, A.I., Keller, A., et al.: A statistical model for identifying proteins by tandem mass spectrometry. Anal. Chem. 75(17), 4646–4658 (2003)

Pedrioli, P.G., Eng, J.K., et al.: A common open representation of mass spectrometry data and its application to proteomics research. Nat. Biotechnol 22(11), 1459–1466 (2004)

Perkins, D.N., Pappin, D.J., et al.: Probability-based protein identification by searching sequence databases using mass spectrometry data. Electrophoresis 20(18), 3551–3567 (1999)

Petricoin, E., Ardekani, A., et al.: Use of proteomic patterns in serum to identify ovarian cancer. Lancet 7(9306), 572–577 (2002)

Senko, M., Beu, S., et al.: Automated Assignment of Charge States from Resolved Isotopic Peaks for Multiply Charged Ions. Journal of American Society for Mass Spectrometry 6, 52–56 (1995)

Simmhan, Y., Plale, B., et al.: A Survey of Data Provenance in e-Science. SIGMOD Rec. 34(3), 31–36 (2005)

Simon, R.M., Korn, E.L., et al.: Design and Analysis of DNA Microarray Investigations. Springer, Heidelberg (2003)

Smith, R., Loo, J., et al.: New Developments in Biochemical Mass Spectrometry: Electrospray Ionization. Analytical Chemistry 62, 882–899 (1990)

Syka, J., Marto, J., et al.: Novel Linear Quadrupole Ion Trap/FT Mass Spectrometer: Performance Characterization and Use in the Comparative Analysis of Histone H3 Post-translational Modifications. Journal of Proteomics Research 3, 621–626 (2004)

Tabb, D.L., McDonald, W.H., et al.: DTASelect and Contrast: tools for assembling and comparing protein identifications from shotgun proteomics. J Proteome Res. 1(1), 21–26 (2002)

Taylor, C.F., Paton, N.W., et al.: A systematic approach to modeling, capturing, and disseminating proteomics experimental data. Nat. Biotech 21(3), 247–254 (2003)

Yang, X., Dondeti, V., et al.: DBParser: web-based software for shotgun proteomic data analyses. J Proteome Res. 3(5), 1002–1008 (2004)

Zeeberg, B.R., Feng, W., et al.: GoMiner: a resource for biological interpretation of genomic and proteomic data. Genome Biol, vol. 4(4) (2003)

Zhang, Z., Marshall, A.: A Universal Algorithm for Fast and Automated Charge State Deconvolution of Electrospray Mass-to-Charge Ratio Spectra. Journal of American Society for Mass Spectrometry 9, 320–332 (1998)

# Bioinformatics Service Reconciliation by Heterogeneous Schema Transformation

Lucas Zamboulis[1,2], Nigel Martin[1], and Alexandra Poulovassilis[1]

[1] School of Computer Science and Information Systems, Birkbeck, Univ. of London
[2] Department of Biochemistry and Molecular Biology, University College London

**Abstract.** This paper focuses on the problem of bioinformatics service reconciliation in a generic and scalable manner so as to enhance interoperability in a highly evolving field. Using XML as a common representation format, but also supporting existing flat-file representation formats, we propose an approach for the scalable semi-automatic reconciliation of services, possibly invoked from within a scientific workflows tool. Service reconciliation may use the AutoMed heterogeneous data integration system as an intermediary service, or may use AutoMed to produce services that mediate between services. We discuss the application of our approach for the reconciliation of services in an example bioinformatics workflow. The main contribution of this research is an architecture for the scalable reconciliation of bioinformatics services.

## 1 Introduction

In recent years, the bioinformatics field has seen an explosion in the number of services offered to the community. These platform-independent software components have consequently been used for the development of complex tasks through service composition within workflows, thereby promoting reusability of services. However, the large number of services available impedes service composition and so developing techniques for semantic service discovery that would significantly reduce the search space is of great importance [12].

After discovering services that are relevant to one's interests, the next step is to identify whether these services are functionally compatible. Bioinformatics services are being independently created by many parties worldwide, using different technologies and data types, hindering integration and reusability [21]. In particular, after discovering two such services, the researcher needs to first identify whether the output of the first is compatible with the input of the second based on a number of factors, such as the technology employed by each service, the representation format and the data type used.

In practice, compatible services are rare. Within Taverna (see `http://taverna.sourceforge.net`), service technology reconciliation is addressed by using Freefluo [19], an extensible workflow enactment environment that bridges the gap between web services and other service types, such as web-based REST services (stateless services that support caching). However, the researcher still needs to reconcile the outputs and inputs of services in terms of content, data

S. Cohen-Boulakia and V. Tannen (Eds.): DILS 2007, LNBI 4544, pp. 89–104, 2007.

type and representation format, spending time and effort in developing functionality that, even though essential for the services to interoperate, is irrelevant to the experiment.

The primary cause of this problem is the existence of multiple different data types and representation formats used even for basic concepts, such as DNA sequences. These data types and representation formats, used for the same or overlapping concepts, have been developed over the years by collaborative work between researchers and/or industry and so even though standardisation efforts are important and encouraged by the community, non-standardised efforts are likely to persist and new ones are bound to appear in this constantly evolving field. For this reason, service composition solutions that take into consideration this factor are essential. Unfortunately, most current tools concentrate on a specific data type and representation format (or combinations of pairs of types and formats, when translation is needed) to accomplish a highly specific task, rather than being generic [13]. As a result, reusability of existing tools is low.

Another common practice in bioinformatics is the use of flat-file representation formats for the overwhelming majority of data types, while the adoption rate of XML is low. This practice does not allow the application of Semantic Web technologies and solutions to their full extent, such as semantically annotating fields within a bioinformatics data type. For example, even though it is possible to annotate a service as having FASTA output, it is not possible to annotate the different fields within the non-tagged FASTA data type. But, even if a data type *is* tagged, e.g. UniProt, annotation cannot be performed in a generic way, as it would require data type-specific annotation tools.

We also observe that, even though the use of semantic annotations is key to service discovery and composition, service providers are disinclined to supply comprehensive annotations for their services. Relying on a centralised approach for such a task is clearly not scalable, and so any proposed solution for the reconciliation of bioinformatics services must ensure that the amount of required annotations is kept to a minimum and that it is reused as much as possible.

We argue that (a) the use of XML and (b) allowing the annotation and manipulation of service inputs and outputs at a fine-grained level, can boost service interoperability in a scalable manner. We therefore propose and exemplify an architecture for the reconciliation of services by exploiting the (manual) semantic annotation of service inputs and outputs using one or more interconnected ontologies, and the subsequent automatic restructuring of the XML output of one service to the required XML input of another. Although our approach uses XML as the common representation format, non-XML services are also supported by the use of converters to and from XML. Our schema and data transformation approach is supported by the AutoMed heterogeneous data integration system (see http://www.doc.ic.ac.uk/automed) and can accommodate two types of service reconciliation: either using AutoMed as a service itself, e.g. from within a workflow tool, or using AutoMed to generate mediating services.

In the remainder of this paper, Section 2 first reviews current approaches related to service interoperability. Section 3 then provides an overview of the

AutoMed system, to the level of detail necessary for this paper. Section 4 introduces our proposed approach for a scalable solution to the problem of bioinformatics service reconciliation. Section 5 presents our ongoing work in applying our approach to the reconciliation of bioinformatics services. Section 6 provides an overall discussion of our approach and gives our plans for future work.

## 2   Related Work

In the context of service composition, research such as [20,18,2] has mainly focused on service technology reconciliation, matchmaking and routing, assuming that service inputs and outputs are a priori compatible. This assumption is restrictive, as it is often the case that two services are semantically compatible, but cannot interoperate due to data type and/or representation format mismatches.

This problem has forced service consumers to handle such mismatches with custom code from within the calling services. In an effort to minimise this issue and promote service reusability, $^{my}$Grid (see http://www.mygrid.org.uk) has fostered the notion of *shims* [7], i.e. services that act as intermediaries between services and reconcile their inputs and outputs. However, a new shim needs to be manually created for each pair of services that need to interoperate. [8] states that, even though in theory the number of shims that $^{my}$Grid needs to provide is quadratic in the number of services it contains, the actual number of shims should be much smaller. However, this manual approach is not scalable, as in 2005 $^{my}$Grid gave access to 1,000 services [12] and this number is now over 3,000.

[3] describes a scalable framework that uses mappings to one or more ontologies, possibly containing subtyping information, for reconciling the output of a service with the input of another. The sample implementation of this framework is able to use mappings to a single ontology in order to generate an XQuery query as the transformation program.

We observe that [3] only provides for shim generation, whereas our approach, by using the AutoMed data integration system, provides a uniform approach to workflow and data integration, both of which are key aspects of in silico biological experiments. Furthermore, the work presented here differs from [3] in a number of aspects and provides a more generic solution to the problem of bioinformatics service reconciliation. First, we also consider services that produce or consume non-XML data and also allow primitive data type reconciliation, whereas [3] does not. Moreover, we allow 1-$n$ GLAV correspondences, compared to the 1-1 LAV correspondences of [3] and we also define a methodology for reconciling services that correspond to more than one ontology. We also note that our XML restructuring algorithm is able to avoid loss of information during data transformation, by analysing the hierarchical nature of the source and target schemas and by using subtype information provided by the ontologies.

[22] also uses a mediator system for service composition. However, the focus is either to provide a service over the global schema of the mediator whose data sources are services, or to generate a new service that acts as an interface over other services. In contrast, we use the AutoMed toolkit to reconcile a sequence

of semantically compatible services that need to form a pipeline: there is no need for a single 'global schema' or a single new service to be created.

Concerning the use of ontologies for data integration, a number of approaches have been proposed. For example, [1] uses an ontology as a virtual global schema for heterogeneous XML data sources using LAV mapping rules, while [4] undertakes data integration using mappings between XML data sources and ontologies, transforming the source data into a common RDF format. In contrast, we use XML as the common representation format and focus on restructuring the source data into a target XML format, rather than on integration.

## 3   Overview of AutoMed

AutoMed is a heterogeneous data transformation and integration system which offers the capability to handle virtual, materialised and hybrid data transformation/integration across multiple data models. It supports a low-level **hypergraph-based data model (HDM)** and provides facilities for specifying higher-level modelling languages in terms of this HDM. An HDM schema consists of a set of nodes, edges and constraints, and each modelling construct of a higher-level modelling language is specified as some combination of HDM nodes, edges and constraints (the constraints are expressed in the IQL query language — see below).

For any modelling language $\mathcal{M}$ specified in this way (via the API of AutoMed's Model Definitions Repository) AutoMed provides a set of primitive schema transformations that can be applied to schema constructs expressed in $\mathcal{M}$. In particular, for every construct of $\mathcal{M}$ there is an add and a delete primitive transformation which add to/delete from a schema an instance of that construct. For those constructs of $\mathcal{M}$ which have textual names, there is also a rename primitive transformation.

Instances of modelling constructs within a particular schema are identified by means of their *scheme* enclosed within double chevrons $\langle\langle \ldots \rangle\rangle$. AutoMed schemas can be incrementally transformed by applying to them a sequence of primitive transformations, each adding, deleting or renaming just one schema construct (thus, in general, AutoMed schemas may contain constructs of more than one modelling language). A sequence of primitive transformations from one schema $X_1$ to another schema $X_2$ is termed a *pathway* from $X_1$ to $X_2$ and denoted by $X_1 \to X_2$. All source, intermediate, and integrated schemas, and the pathways between them, are stored in AutoMed's Schemas & Transformations Repository.

Each add and delete transformation is accompanied by a query specifying the extent of the added or deleted construct in terms of the rest of the constructs in the schema. This query is expressed in a functional query language, IQL [9]. Also available are extend and contract primitive transformations which behave in the same way as add and delete except that they state that the extent of the new/removed construct cannot be precisely derived from the rest of the constructs. Each extend and contract transformation takes a pair of queries that specify a lower and an upper bound on the extent of the construct. These bounds may be Void or Any, which respectively indicate no known information about the lower or upper bound of the extent of the new construct.

The queries supplied with primitive transformations can be used to translate queries or data along a transformation pathway $X_1 \to X_2$ (see [15,16] for details). For translating data from $X_1$ to data on $X_2$ the add, extend and rename steps are used. The queries supplied with primitive transformations also provide the necessary information for these transformations to be automatically *reversible*, in that each add/extend transformation is reversed by a delete/contract transformation with the same arguments (including the same query arguments), while each rename is reversed by a rename with the two arguments swapped. As discussed in [15], this means that AutoMed is a **both-as-view (BAV)** data integration system: the add/extend steps in a transformation pathway correspond to Global-As-View (GAV) rules while the delete and contract steps correspond to Local-As-View (LAV) rules. If a GAV view is derived from solely add steps it will be *exact* in the terminology of [11]. If, in addition, it is derived from one or more extend steps using their lower-bound (upper-bound) queries, then the GAV view will be *sound* (*complete*) in the terminology of [11]. Similarly for LAV views. An in-depth comparison of BAV with the GAV and LAV approaches to data integration can be found in [15], while [16,17] discusses the use of BAV in a peer-to-peer data integration setting. [10] discusses how Global-Local-As-View (GLAV) rules [5,14] can also be derived from BAV pathways. We note that AutoMed and BAV transform both schema and data together, and thus do not suffer from any data/schema divide.

## 4    Bioinformatics Service Reconciliation

In this section, we present the problems encountered during service reconciliation and describe our proposed approach for overcoming them, including a brief discussion of how our approach could be incorporated within a workflow tool. We then provide details of XML DataSource Schema (XMLDSS), the XML schema type used in our approach, and of our own earlier work on schema transformation using ontologies that has been extended to enable service reconciliation.

### 4.1    Proposed Approach

Consider a service $S_1$ that produces data that need to be consumed by another service $S_2$. In general, the following issues need to be resolved when trying to handle data exchange between $S_1$ and $S_2$:

1. **Data model heterogeneity:** different data models (e.g. legacy flat files and XML) or different schema types (e.g. DTD and XML Schema) may be used. It may also be the case that a service producing or consuming XML data does not have an accompanying XML schema.
2. **Semantic heterogeneity:** schematic differences caused by the use of different terminology, or describing the same information at different levels of granularity.

3. **Schematic heterogeneity:** schematic differences caused by modelling the same information in different ways. This heterogeneity is common to all data modelling languages, but is amplified in XML due to its hierarchical nature, as well as the possibility of using elements with a single text node and attributes interchangeably.
4. **Primitive data type heterogeneity:** differences caused by the use of different primitive data types, e.g. `int` and `varchar`, for the same concept.

To resolve these issues, we propose the following 4-step approach, illustrated in Figure 1:

**Step 1: XML as the common representation format.** We handle differences in the representation format by using XML as the common representation format. If the output/input of a service is not in XML, then a format converter is needed to convert to/from XML.

**Step 2: XMLDSS as the schema type.** We use our own XMLDSS schema type for the XML documents input to and output by services. An XMLDSS schema can be automatically extracted from an XML document or automatically derived from an accompanying DTD/XML Schema, if one is available.

**Step 3: Correspondences to typed ontologies.** We use one or more ontologies as a 'semantic bridge' between services. Providers or users of services semantically annotate the inputs and outputs of services by defining correspondences between an XMLDSS schema and an ontology. Ontologies in our approach are typed, i.e. each concept is associated with a data type, and so defining correspondences resolves issues 2 and 4 discussed above.

**Step 4: Schema and data transformation.** We use the AutoMed toolkit to automatically transform the XMLDSS schema of the output of service $S_1$ to the XMLDSS schema of the input of service $S_2$. This is achieved using the two automatic algorithms discussed in Section 4.4.

If service $S_1$ does not have an accompanying DTD or XML Schema for its output, sample XML output documents for $S_1$ must be provided, and these must represent all valid formats that $S_1$ is able to produce, so as to create an XMLDSS schema that represents all possible instances of the output of $S_1$. If this is not possible, then an XMLDSS can be extracted at run-time for every new instance XML document output by $S_1$. The same applies for the input of $S_2$.

### 4.2 Integration of Approach With Workflow Tools

Our architecture for service reconciliation supports two different approaches identified below, depending on the preferred form of interoperability between AutoMed and the workflow tool.

**Mediation service.** With this approach, the workflow tool invokes service $S_1$, receives its output, and submits this output and a handle on service $S_2$

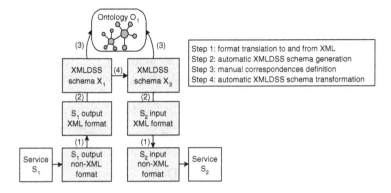

**Fig. 1.** Reconciliation of services $S_1$ and $S_2$ using ontology $O_1$

to a service provided by the AutoMed system. This uses our approach to transform the output of $S_1$ to a suitable input for consumption by $S_2$.

**Shim generation.** With this approach, the AutoMed system is used to generate shims, i.e. tools or services for the reconciliation of services, by generating transformation scripts which are then incorporated within the workflow tool.

In the following, we provide an overview of the shim generation architecture. The mediation service architecture is described in more detail in Section 5.

With the shim generation approach, AutoMed is not part of the architecture, and so it is necessary to export AutoMed's mediation functionality described and exemplified in Section 5. This functionality consists of the format converters, the algorithms for generating an XMLDSS schema from an XML document, DTD or XML Schema, and the XMLDSS schema transformation algorithms.

Format converters are not a part of the AutoMed toolkit and so can be used from within a workflow tool, without exporting any AutoMed functionality. The converters can be either incorporated within the workflow tool, or their functionality can be imported using services. As an example, a number of shims in $^{my}$Grid are format converters.

The XMLDSS schema type is currently used only within the AutoMed system, but it does not require AutoMed functionality. As a result, the XMLDSS schema generation algorithms can be used from within a workflow tool in the same way as format converters.

The two XMLDSS schema transformation algorithms described in Section 4.4 are currently tightly coupled with the AutoMed system, since they use the BAV approach, which is currently supported only by AutoMed. To use our approach without dynamically integrating AutoMed with a workflow tool, we need to export the functionality of the schema transformation algorithms, in order for this AutoMed-dependent functionality to be used statically by a workflow tool. To this effect, we have designed an XQuery query generation algorithm, as detailed in [25], that derives a single XQuery query $Q$, able to materialise an XMLDSS schema $X_2$ using data from the data source of an XMLDSS schema $X_1$, and a transformation pathway $X_1 \rightarrow X_2$. In summary, to derive query $Q$, the

algorithm first uses AutoMed's Query Processor to create the IQL view definition $V$ of each construct $c$ of $X_2$ in terms of constructs of $X_1$, and to translate
each $V$ into an equivalent XQuery query, $V_{XQuery}$. The algorithm then creates a
single XQuery query $Q$, for materialising $X_2$ by following a bottom-up approach
as follows. The algorithm first creates the XQuery queries for materialising the
leaf elements of $X_2$, together with their attributes and child text nodes. These
queries are then used to create the queries that materialise the parent elements
of the leaf elements, together with their attributes and text nodes. This process
is repeated until the root of $X_2$ is reached and the overall query $Q$ is formulated.

### 4.3   XML DataSource Schema (XMLDSS)

The standard schema definition languages for XML are DTD and XML Schema.
However, both of these provide grammars to which conforming documents adhere, and they do not explicitly summarise the tree structure of the data sources.
In our schema transformation setting, tree-structured schemas are preferable as
they facilitate schema traversal, structural comparison between a source and a
target schema, and restructuring of the source schema. Moreover, such a schema
type means that the queries supplied with AutoMed primitive transformations
are essentially path queries, which are easily generated.

The AutoMed toolkit therefore supports a modelling language called *XML
DataSource Schema* (XMLDSS), which summarises the tree structure of XML
documents, much like DataGuides [6]. XMLDSS schemas consist of four kinds
of constructs: Element, Attribute, Text and NestList (see [23] for details of their
specification in terms of the HDM). The last of these defines parent-child relationships either between two elements $e_p$ and $e_c$ or between an element $e_p$ and
the Text node. These are respectively identified by schemes of the form $\langle\langle i, e_p, e_c \rangle\rangle$
and $\langle\langle i, e_p, \text{Text} \rangle\rangle$, where $i$ is the position of $e_c$ or Text within the list of children
of $e_p$ in the XMLDSS schema.

In an XMLDSS schema there may be elements with the same name occurring
at different positions in the tree. To avoid ambiguity, the identifier element-
Name\$count is used for each element, where count is incremented every time the
same elementName is encountered in a depth-first traversal of the schema.

### 4.4   XML Schema and Data Transformation Using Ontologies

We now describe the two algorithms, the schema conformance algorithm (SCA)
and the schema restructuring algorithm (SRA), used in our approach to transform a source XMLDSS schema $X_1$ and its data to the structure of a target
XMLDSS schema $X_2$. In this setting, these are the XMLDSS schemas of the
outputs and inputs of services. Our own previous work in [23,26,27] addressed
the issue of XML schema and data transformation. This section describes an
extended version of the approach of [27], in that the expressiveness of the correspondences used in our approach has been enriched, and the SCA algorithm
has been extended to support this.

The SCA uses manually defined correspondences between XMLDSS schemas
$X_1$ and $X_2$ and an ontology $O$, in order to automatically transform $X_1$ and

$X_2$ into equivalent schemas $X_1'$ and $X_2'$ that use the same terms as $O$. As a result, transformation pathways $X_1 \rightarrow X_1'$ and $X_2 \rightarrow X_2'$ are created. By the bidirectionality of BAV, a pathway $X_2' \rightarrow X_2$ can be automatically derived from the pathway $X_2 \rightarrow X_2'$.

In [27], a *correspondence* defines an Element, Attribute or NestList of an XMLDSS schema by means of an IQL query over a typed ontology.[1] In particular, an Element $e$ may map either to a Class $c$; or to a path ending with a class-valued property of the form $\langle\!\langle \mathsf{p}, \mathsf{c1}, \mathsf{c2} \rangle\!\rangle$, where $p$ is the property name and $c_1$ and $c_2$ are source and target classes; or to a path ending with a literal-valued property $\langle\!\langle \mathsf{p}, \mathsf{c}, \mathsf{Literal} \rangle\!\rangle$, where $p$ is the property name and $c$ the source class; additionally, the correspondence may state that the instances of a class are constrained by membership in some subclass. An Attribute may map either to a literal-valued property or to a path ending with a literal-valued property.

We now extend the correspondences of [27] as follows. An XMLDSS scheme of the form $\langle\!\langle \mathsf{i}, \mathsf{e}, \mathsf{Text} \rangle\!\rangle$ (where $i$ denotes the order of $\langle\!\langle \mathsf{Text} \rangle\!\rangle$ in the list of children of Element $\langle\!\langle \mathsf{e} \rangle\!\rangle$) may map to a literal-valued property of the form $\langle\!\langle \mathsf{p}, \mathsf{c}, \mathsf{Literal} \rangle\!\rangle$. In addition to 1-1 correspondences, we now also allow 1-$n$ correspondences as follows. An Element/Attribute may map to more than one path over the ontology. In this case, $n$ correspondences are required, each associating the same XMLDSS Element/Attribute to a different path over the ontology, and specifying an expression that determines the part of the extent of the Element/Attribute to which the correspondence applies (an example of this is given in Section 5). This expression is in general a select-project IQL query. We note that these extended correspondences are GLAV, in contrast with the LAV correspondences defined in our own earlier work [27], as an expression over an XMLDSS construct (rather than just an XMLDSS construct) maps to a path in the ontology.[2]

The SCA uses correspondences from an Element or Attribute to a single path over the ontology to rename that construct, ensuring consistency with the terminology of the ontology. In the case of a 1-$n$ correspondence relating to an Element $e$ with parent $p$, the algorithm first retrieves all relevant correspondences, then inserts $n$ Elements under $p$ (in the position previously held by $e$), named after the paths specified by the correspondences, and finally deletes $e$ and its underlying structure. When inserting the $n$ Elements under $p$, the algorithm also replicates the underlying structure of the old Element $e$ under each one of the newly inserted Elements. A 1-$n$ correspondence relating to an Attribute is handled similarly: the owner Element is replaced by $n$ Elements with the same name, each containing a different Attribute named after the paths specified by the correspondences. A correspondence mapping an Attribute or a scheme of the form $\langle\!\langle \mathsf{i}, \mathsf{e}, \mathsf{Text} \rangle\!\rangle$ in the XMLDSS to a literal-valued property in the ontology is used to perform

---

[1] In principle, it would be possible to use more high-level query languages such as XQuery to specify correspondences in our setting. Currently, AutoMed provides an XQuery-to-IQL translator component, capable of translating (possibly nested) FLWR XQuery queries to (possibly nested) select-project-join IQL queries.

[2] Even though BAV pathways could have been used to express these GLAV mappings, we specify the mappings directly as GLAV rules for compactness.

primitive data type reconciliation: if the data type of the **Attribute** or scheme in the XMLDSS schema is not the same as in the ontology, the algorithm replaces the **Attribute** or scheme by performing a type-casting operation.

After the transformation of schemas $X_1$ and $X_2$ into schemas $X_1'$ and $X_2'$ that use the same terms as $O$, our second algorithm, the SRA presented in [27], automatically transforms $X_1'$ to the structure of $X_2'$, producing a transformation pathway $X_1' \rightarrow X_2'$. To do so, the SRA first inserts into $X_1'$ those constructs present in $X_2'$ but not in $X_1'$. After this *growing phase*, a *shrinking phase* follows, in which the SRA removes from $X_1'$ those constructs present in $X_1'$ but not in $X_2'$. The SRA is able to generate synthetic structure to avoid loss of data caused by structural incompatibilities between $X_1'$ and $X_2'$. The SRA is also able to use information that identifies an element/attribute in $X_1'$ to be either equivalent to, or a superclass of, or a subclass of an element/attribute in $X_2'$. This information may be produced by, e.g. a schema matching tool or, in our context here, via correspondences to an ontology.

Consequently, an overall transformation pathway from $X_1$ to $X_2$ can now be obtained by composing the pathways $X_1 \rightarrow X_1'$, $X_1' \rightarrow X_2'$ and $X_2' \rightarrow X_2$. This pathway can be used to automatically transform data that is structured according to $X_1$ to be structured according to $X_2$, and an XML document structured according to $X_2$ can finally be materialised (the pathway $X_1 \rightarrow X_2$ could also be used to translate queries expressed on $X_2$ to operate on $X_1$).

Note that we do not assume the existence of a single ontology. As discussed in [27], it is possible for XMLDSS schema $X_1$ to have a set of correspondences $C_1$ to an ontology $O_1$, and for XMLDSS schema $X_2$ to have a set of correspondences $C_2$ to another ontology $O_2$. Provided there is an AutoMed transformation pathway between $O_1$ and $O_2$, either directly or through one or more intermediate ontologies, we can use $C_1$ and the transformation pathway between $O_1$ and $O_2$ to automatically produce a new set of correspondences $C_1'$ between $X_1$ and $O_2$. As a result, this setting is now identical to a setting with a single ontology. There is a proviso here that the new set of correspondences $C_1'$ must conform syntactically to the correspondences accepted as input by the schema conformance process. Determining necessary conditions for this to hold is an area of future work.

## 5   Case Study

We now describe our approach in more detail and demonstrate the use of AutoMed as a mediation service by specifying a sample bioinformatics workflow. Note that listings of all service inputs, outputs and XMLDSS, XML Schema and DTD schemas discussed in this section are given in [25].

Figure 2 illustrates a sample workflow with three services that will be used to demonstrate our approach. The first service takes as input an IPI (http://www.ebi.ac.uk/IPI) accession number, e.g. IPI00015171, and outputs the corresponding IPI entry as a flat file using the UniProt (http://www.ebi.uniprot.org) format. The second service receives an InterPro (http://www.ebi.ac.uk/interpro) accession number and returns the

**Fig. 2.** Sample Workflow

corresponding InterPro entry. The third service receives a Pfam(http://
www.sanger.ac.uk/Software/Pfam accession number and returns the corresponding Pfam entry. In this workflow, two transformations are needed: $T_1$
extracts the InterPro accession number from an IPI entry using the UniProt
format, while $T_2$ extracts the Pfam accession number from an InterPro entry.

We now apply the mediation service approach described in Section 4.2, for
the reconciliation of the services of the workflow of Figure 2.

**Step 1: XML as a common representation format.** Service *getIPIEntry*
outputs a flat file that follows the UniProt representation format and contains
a single entry consisting of multiple lines. Each line consists of two parts, the
first being a two-character line code, indicating the type of data contained in
the line, while the second contains the actual data, consisting of multiple fields.

Since UniProt also has an XML representation format specified by an XML
Schema, we created a format converter that, given an IPI flat file *f* that follows
the UniProt format, converts *f* to an XML file conforming to that XML Schema.

Service *getInterProEntry* outputs an XML file and so there is no need for
a format converter. Concerning the input of the second and the third service,
they each take as input a single string, representing an InterPro/Pfam accession
number, respectively. The input XML documents for these contain a single XML
element, ip_acc and pf_acc, respectively, with a PCData node as a single child,
as shown below. For these, the format converters implement the functionality of
the XPath expressions /ip_acc/text() and /pf_acc/text(), respectively.

```
<ip_acc>InterPro_accession_string</ip_acc>
<pf_acc>Pfam_accession_string</pf_acc>
```

**Step 2: XMLDSS schema generation.** After resolving representation format
issues, we now give details on the generation of XMLDSS schemas for our setting.
As discussed above, service *getIPIEntry* outputs a flat file which is converted to
an XML file that conforms to the UniProt XML Schema. An XMLDSS schema
for the output of this service is automatically derived from that XML Schema.
Similarly, an XMLDSS schema for the output of service *getInterProEntry* is
automatically derived using the InterPro DTD schema.

Concerning the input of the second and the third service, the corresponding
XMLDSS schemas are automatically extracted by using a single sample XML
document for each, such as the ones given earlier.

**Step 3: Correspondences.** After generating the XMLDSS schemas for our
workflow, we need to specify the correspondences between these schemas and an
ontology. In this case, we have used the typed $^{my}$Grid OWL domain ontology.

In general, all XMLDSS elements and attributes should be mapped to the ontology. However, if an element or attribute cannot be mapped to the ontology, this construct is not affected by our SCA and SRA algorithms that use the correspondences to transform $X_1$ to the structure of $X_2$. An advantage of this is that data transformation is still possible with only a partial set of correspondences from an XMLDSS schema to the ontology. This property is particularly significant in terms of the applicability and scalability of our approach, as it allows for incrementally defining the full set of correspondences between an XMLDSS schema and an ontology: one can define only those correspondences relevant to the specific problem at hand, instead of the full set of correspondences.

In our example, this means that we only need to specify correspondences for those constructs of the XMLDSS schema of the output of *getIPIEntry* that contribute to the input of service *getInterProEntry*. Consequently, we need to specify correspondences for only two constructs, ⟨⟨dbReference\$9⟩⟩ and ⟨⟨dbReference\$9, id⟩⟩ (see Table 1). The first models an entry in a bioinformatics data resource, whose type is specified by ⟨⟨dbReference\$9, type⟩⟩. The type of a resource is modelled in IPI using data values, whereas in the ontology it is modelled as classes, and so $n$ correspondences are required for this construct, where $n$ is the number of types of resources that IPI supports and that also exist in the ontology. Each of these correspondences maps ⟨⟨dbReference\$9⟩⟩ to a class in the ontology representing a bioinformatics data resource record and specifies the part of the extent of ⟨⟨dbReference\$9⟩⟩ to which the correspondence applies. For example, the second correspondence states that those instances of ⟨⟨dbReference\$9⟩⟩ whose ⟨⟨dbReference\$9, type⟩⟩ Attribute has a data value of 'Pfam', map to the ⟨⟨Pfam_record⟩⟩ ontology class. Due to space limitations, but without loss of generality, we only provide the two correspondences related to InterPro and Pfam.

The XMLDSS schema of the input of service *getInterProEntry* consists of a single Element construct, ⟨⟨ip_acc⟩⟩, which corresponds to class ⟨⟨InterPro_accession⟩⟩ in the ontology, and of a NestList construct, ⟨⟨1, ip_acc, Text⟩⟩. The correspondences are given in Table 2. The correspondences for the XMLDSS schema of the input of the third service, *getPfamEntry*, are not listed as they are

**Step 4: Schema transformation.** After manually specifying correspondences, the SCA and SRA algorithms can automatically transform the outputs of services *getIPIEntry* and *getInterProEntry* to the required inputs for services *getInterProEntry* and *getPfamEntry* respectively.

Concerning the output of service *getIPIEntry*, the schema conformance algorithm (SCA) first retrieves all correspondences related to ⟨⟨dbReference\$9⟩⟩ (in this case 2 correspondences) and inserts ⟨⟨InterPro_record\$1⟩⟩ and ⟨⟨Pfam_record\$1⟩⟩, using the correspondences' expressions to select the appropriate ⟨⟨dbReference\$9⟩⟩ instances, i.e. those that have a type Attribute with value 'InterPro' and 'Pfam' respectively. As discussed in Section 4.4, the SCA then replicates under the newly inserted Elements the structure located under ⟨⟨dbReference\$9⟩⟩ (again using the correspondences' expressions to select the appropriate structure), and then removes ⟨⟨dbReference\$9⟩⟩. Note that this removal is postponed until after any other insertions are performed, as other insertions

**Table 1.** Correspondences between the XMLDSS schema of the output of $getIPIEntry$ and the $^{my}$Grid ontology

| | |
|---|---|
| **Construct:** | $\langle\!\langle$dbReference\$9$\rangle\!\rangle$ |
| **Extent:** | $[\{d\}|\{d,t\} \leftarrow \langle\!\langle$dbReference\$9, type$\rangle\!\rangle; t =' InterPro']$ |
| **Path:** | $\langle\!\langle$InterPro_record$\rangle\!\rangle$ |
| **Construct:** | $\langle\!\langle$dbReference\$9$\rangle\!\rangle$ |
| **Extent:** | $[\{d\}|\{d,t\} \leftarrow \langle\!\langle$dbReference\$9, type$\rangle\!\rangle; t =' Pfam']$ |
| **Path:** | $\langle\!\langle$Pfam_record$\rangle\!\rangle$ |
| **Construct:** | $\langle\!\langle$dbReference\$9, id$\rangle\!\rangle$ |
| **Extent:** | $[\{d,i\}|\{d,i\} \leftarrow \langle\!\langle$dbReference\$9, id$\rangle\!\rangle;$ |
| | $\{d,t\} \leftarrow \langle\!\langle$dbReference\$9, type$\rangle\!\rangle; t =' InterPro']$ |
| **Path:** | $[\{ir,l\}|\{ia,ir\} \leftarrow \langle\!\langle$part_of, InterPro_accession, InterPro_record$\rangle\!\rangle;$ |
| | $\{ia,l\} \leftarrow \langle\!\langle$datatype, InterPro_accession, Literal$\rangle\!\rangle]$ |
| **Construct:** | $\langle\!\langle$dbReference\$9, id$\rangle\!\rangle$ |
| **Extent:** | $[\{d,i\}|\{d,i\} \leftarrow \langle\!\langle$dbReference\$9, id$\rangle\!\rangle;$ |
| | $\{d,t\} \leftarrow \langle\!\langle$dbReference\$9, type$\rangle\!\rangle; t =' Pfam']$ |
| **Path:** | $[\{pr,l\}|\{pa,pr\} \leftarrow \langle\!\langle$part_of, Pfam_accession, Pfam_record$\rangle\!\rangle;$ |
| | $\{pa,l\} \leftarrow \langle\!\langle$datatype, Pfam_accession, Literal$\rangle\!\rangle]$ |

**Table 2.** Correspondences between the XMLDSS schema of the input of $getInterPro$ and the $^{my}$Grid ontology

| | |
|---|---|
| **Construct:** | $\langle\!\langle$ip_acc\$1$\rangle\!\rangle$ |
| **Extent:** | $\langle\!\langle$ip_acc\$1$\rangle\!\rangle$ |
| **Path:** | $[\{ia\}|\{ia,ir\} \leftarrow \langle\!\langle$part_of, InterPro_accession, InterPro_record$\rangle\!\rangle]$ |
| **Construct:** | $\langle\!\langle 1, $ip_acc\$1, Text$\rangle\!\rangle$ |
| **Extent:** | $\langle\!\langle 1, $ip_acc\$1, Text$\rangle\!\rangle$ |
| **Path:** | $[\{ia,l\}|\{ia,ir\} \leftarrow \langle\!\langle$part_of, InterPro_accession, InterPro_record$\rangle\!\rangle;$ |
| | $\{ia,l\} \leftarrow \langle\!\langle$datatype, InterPro_accession, Literal$\rangle\!\rangle]$ |

may need to use the extent of $\langle\!\langle$dbReference\$9$\rangle\!\rangle$ in the queries supplied with the AutoMed transformations.

The SCA then retrieves all correspondences related to $\langle\!\langle$dbReference\$9, id$\rangle\!\rangle$ (in this case 2 correspondences) and inserts Attributes $\langle\!\langle$InterPro_record\$1,InterPro _record.part_of.InterPro_accession$\rangle\!\rangle$ and $\langle\!\langle$Pfam_record\$1,InterPro_record.part_of.Pf am_accession$\rangle\!\rangle$, using the correspondences' expressions to select the appropriate $\langle\!\langle$dbReference\$9, id$\rangle\!\rangle$ instances (as discussed earlier, $\langle\!\langle$dbReference\$9$\rangle\!\rangle$ has not yet been removed). Concerning primitive data types, $\langle\!\langle$dbReference\$9, id$\rangle\!\rangle$ is of type **string**, and the same applies for all accession numbers in the $^{my}$Grid domain ontology, so there is no need for any type-casting operations.

Concerning the input of $getInterProEntry$, the SCA uses the first correspondence to rename $\langle\!\langle$ip_acc\$1$\rangle\!\rangle$ to $\langle\!\langle$InterPro_record.part_of.InterPro_accession\$1$\rangle\!\rangle$, while the second correspondence, which is a primitive data type reconciliation correspondence, is of no consequence as both the input of the service and the ontology model InterPro accession numbers using the **string** data type.

After the application of the SCA, the XMLDSS schema $X_2$ of the input of service $getInterProEntry$ contains three constructs, $\langle\!\langle$InterPro_record.part_of.Int erPro_accession$1$\rangle\rangle$, $\langle\langle$Text$\rangle\rangle$ and a NestList linking these two constructs. The XMLDSS schema of the output of service *getIPIEntry*, $X_1$, contains a number of constructs, but the only ones relevant to those of $X_2$ are $\langle\langle$InterPro_record$1\rangle\rangle$ and $\langle\langle$InterPro_record$1$,InterPro_record.part_of. InterPro_accession$\rangle\rangle$. The schema restructuring algorithm (SRA) therefore applies a number of contract transformations supplied with the queries Void and Any, so as to remove non-relevant constructs. The only non-trivial transformation is the attribute-to-element transformation: first Element $\langle\langle$InterPro_record.part_of.InterPro_accession$1\rangle\rangle$ is added to $X_1$ using the extent of Attribute $\langle\langle$InterPro_record$1$,InterPro_record.part_of.Inter Pro_accession$\rangle\rangle$, then NestList $\langle\langle$InterPro_record.part_of.InterPro_accession$1$, Text$\rangle\rangle$ is added, again using the Attribute extent, and finally the Attribute is deleted.

After applying the SRA, we finally employ the XMLDSS schema materialisation algorithm defined in [26] to materialise $X_2$, i.e. the input of service *getInterProEntry*, using data from the data source of $X_1$, i.e. the output of service *getIPIEntry*, using the transformation pathway $X_1 \to X_1' \to X_2' \to X_2$.

The application of Step 4 for the second part of our workflow is similar.

## 6   Conclusions and Future Work

In this paper we have presented a generic and scalable architecture for bioinformatics service reconciliation within a wider data transformation framework. Our approach makes no assumptions about representation format, primitive data type usage or the number of ontologies used. Moreover, this approach can be used either dynamically or statically from within a workflow tool.

The architecture exploits format converters to establish a common XML format for all service inputs and outputs, thus reducing the overall complexity of service reconciliation by establishing a common representation format. Service inputs and outputs are then abstracted using the XMLDSS schema type which can be automatically generated either from XML documents, or from accompanying DTD or XML Schema specifications using our algorithms.

Our approach is able to use correspondences to multiple ontologies for defining the semantics of services. This 'semantic bridge' is utilised by two automatic algorithms that use the correspondences to allow data transformation between services. The schema conformance algorithm is able to use 1-1 and 1-$n$ GLAV correspondences to ontologies, in order to produce schemas with no semantic heterogeneity. The schema restructuring algorithm then restructures the source schema to the target schema. This algorithm is able to avoid loss of information that may be caused due to structural incompatibilities of the data sources.

While the correspondences to ontologies must be produced manually or semi-automatically, an advantage of our approach is that correspondence reusability is promoted by allowing the use of multiple ontologies. Moreover, our approach does not require a full set of correspondences to be defined, but instead allows the definition of only those correspondences between the XMLDSS schema and the ontology that are relevant to the problem at hand - we therefore allow an incremental approach for the definition of correspondences.

The architecture has been illustrated with a bioinformatics workflow characteristic of those currently available with string-based inputs. Future workflows with more complex inputs are to be expected, which our architecture will also readily support.

Concerning the integration of our approach with workflow tools, we defined two possible architectures. The first, using AutoMed as a mediation service, can be used from within a workflow tool by invoking AutoMed as a service and does not require XMLDSS or XQuery support. On the other hand, in the shim generation architecture AutoMed is used to statically generate shims, which can then be incorporated into any workflow tool that supports XQuery.

Our current implementation has supported testing of the transformation pathways underpinning the service reconciliation examples presented within the AutoMed toolkit. Ongoing work is aimed at integrating our approach with the Taverna workflow tool. The resulting implementation will be evaluated within the proteomics grid infrastructure being developed in the ISPIDER project [24].

In future work, we will investigate the necessary conditions under which a set of correspondences, transformed by a 'semantic bridge' defined between multiple ontologies, adheres to the required format of our schema conformance algorithm. Other extensions to our work include investigating the effect on our approach of constraints on XMLDSS schemas and/or the ontologies, and also considering the effect of the evolution of the inputs and outputs of services.

**Acknowledgements.** The work presented in this paper is part of the BBSRC-funded ISPIDER project. The authors would also like to thank the ISPIDER members and especially Khalid Belhajjame, Suzanne Embury and Norman Paton for the fruitful discussions that helped shape the work presented in this paper.

# References

1. Amann, B., Beeri, C., Fundulaki, I., et al.: Ontology-based integration of XML web resources. In: Proc. of Int. Semantic Web Conference, pp. 117–131 (2002)
2. Benatallah, B., et al.: Declarative composition and peer-to-peer provisioning of dynamic web services. In: Proc. of ICDE'02, pp. 297–308 (2002)
3. Bowers, S., Ludäscher, B.: An ontology-driven framework for data transformation in scientific workflows. In: Proc. of Data Integration in the Life Sciences (DILS'04), pp. 1–16 (2004)
4. Cruz, I.F., Xiao, H., Hsu, F.: An ontology-based framework for XML semantic integration. In: Proc. IDEAS'04, pp. 217–226 (2004)
5. Friedman, M., Levy, A., Millstein, T.: Navigational plans for data integration. In: National Conference on Artificial Intelligence, pp. 67–73. AAAI Press, Stanford, California, USA (1999)
6. Goldman, R., Widom, J.: DataGuides: Enabling Query Formulation and Optimization in Semistructured Databases. In: Proc. VLDB'97, pp. 436–445 (1997)
7. Hull, D., et al.: Treating shimantic web syndrome with ontologies. In: Proc. of Advanced Knowledge Technologies workshop on Semantic Web Services (2004)
8. Hull, D., Stevens, R., Lord, P.: Describing web services for user-oriented retrieval. In: Proc. of W3C Workshop on Frameworks for Semantics in Web Services (2005)

9. Jasper, E., Poulovassilis, A., Zamboulis, L.: Processing IQL queries and migrating data in the AutoMed toolkit. AutoMed Technical Report 20 (July 2003)
10. Jasper, E., Tong, N., McBrien, P.J., Poulovassilis, A.: View generation and optimisation in the AutoMed data integration framework. In: Proc. of 6th Baltic Conference on Databases and Information Systems (2004)
11. Lenzerini, M.: Data integration: A theoretical perspective. In: Proc. PODS'02, pp. 233–246 (2002)
12. Lord, P., Alper, P., Wroe, C., Goble, C.: Feta: A light-weight architecture for user oriented semantic service discovery. In: Gómez-Pérez, A., Euzenat, J. (eds.) ESWC 2005. LNCS, vol. 3532, pp. 17–31. Springer, Heidelberg (2005)
13. Lord, P., Bechhofer, S., Wilkinson, M., Schiltz, G., Gessler, D., Hull, D., Goble, C., Stein, L.: Applying Semantic Web services to bioinformatics: experiences gained, lessons learnt. In: Proc. of Int. Semantic Web Conference, pp. 350–364 (2004)
14. Madhavan, J., Halevy, A.Y.: Composing mappings among data sources. In: Aberer, K., Koubarakis, M., Kalogeraki, V. (eds.) Databases, Information Systems, and Peer-to-Peer Computing. LNCS, vol. 2944, pp. 572–583. Springer, Heidelberg (2004)
15. McBrien, P., Poulovassilis, A.: Data integration by bi-directional schema transformation rules. In: Proc. ICDE'03, pp. 227–238 (March 2003)
16. McBrien, P., Poulovassilis, A.: Defining peer-to-peer data integration using both as view rules. In: Aberer, K., Koubarakis, M., Kalogeraki, V. (eds.) Proc. Workshop on Databases, Information Systems and Peer-to-Peer Computing (at VLDB'03). LNCS, vol. 2944, Springer, Heidelberg (2004)
17. McBrien, P.J., Poulovassilis, A.: P2P query reformulation over Both-as-View data transformation rules. In: Proc. of Databases, Information Systems and Peer-to-Peer Computing (at VLDB'06), page TBC. Springer (2006)
18. Medjahed, B., Bouguettaya, A., Elmagarmid, A.K.: Composing web services on the Semantic Web. VLDB Journal 12(4), 333–351 (2003)
19. Oinn, T., Addis, M., Ferris, J., et al.: Taverna: a tool for the composition and enactment of bioinformatics workflows. Bioinformatics 20(17), 3045–3054 (2004)
20. Srivastava, B., Koehler, J.: Web Service composition - current solutions and open problems. In: Proc. of Workshop on Planning for Web Services (ICAPS'03), pp. 28–35 (2003)
21. Stein, L.: Creating a bioinformatics nation. Nature 417, 119–120 (2002)
22. Thakkar, S., Ambite, J.L., Knoblock, C.A.: Composing, optimizing, and executing plans for bioinformatics web services. VLDB Journal 14(3), 330–353 (2005)
23. Zamboulis, L.: XML data integration by graph restructuring. In: Williams, H., MacKinnon, L.M. (eds.) Proc. British National Conferences on Databases(BNCOD' 04), LNCS, vol. 3112, pp. 57–71. Springer, Heidelberg (2004)
24. Zamboulis, L., Fan, H., Belhajjame, K., Siepen, J., Jones, A., Martin, N.J., Poulovassilis, A., Hubbard, S.J., Embury, S., Paton, N.W.: Data access and integration in the ISPIDER proteomics grid. In: Leser, U., Naumann, F., Eckman, B. (eds.) DILS 2006. LNCS (LNBI), vol. 4075, pp. 3–18. Springer, Heidelberg (2006)
25. Zamboulis, L., Martin, N., Poulovassilis, A.: Bioinformatics service reconciliation by heterogeneous schema transformation. Birkbeck TR BBKCS-07-03 (March 2007)
26. Zamboulis, L., Poulovassilis, A.: Using AutoMed for XML Data Transformation and Integration. In: Proc. International Workshop on Data Integration over the Web (at CAiSE'04), pp. 58–69 (2004)
27. Zamboulis, L., Poulovassilis, A.: Information sharing for the Semantic Web - a schema transformation approach. In: Proc. International Workshop Data Integration and the Semantic Web (at CAiSE'06), pp. 275–289 (2006)

# A Formal Model of Dataflow Repositories

Jan Hidders[1], Natalia Kwasnikowska[2], Jacek Sroka[3], Jerzy Tyszkiewicz[3],
and Jan Van den Bussche[2]

[1] University of Antwerp, Belgium
[2] Hasselt University and Transnational University of Limburg, Belgium
[3] Warsaw University, Poland

**Abstract.** Dataflow repositories are databases containing dataflows and
their different runs. We propose a formal conceptual data model for such
repositories. Our model includes careful formalisations of such features
as complex data manipulation, external service calls, subdataflows, and
the provenance of output values.

## 1 Introduction

Modern scientific research is characterized by extensive computerized data
processing of lab results and other scientific data. Such processes are often complex, consisting of several data manipulating steps. We refer to such processes as
*dataflows*, to distinguish them from the more general *workflows*. General workflows also emphasize timing, concurrency, and synchronization aspects of a complex process, whereas in this paper we are less interested in such aspects, and
our focus is mainly on data manipulation and data management aspects.

Important data management aspects of scientific dataflows include:

- Support of complex data structures, such as records containing different
  attributes of a data object, and sets (collections) of data objects. When
  combining, merging, and aggregating data, complex compositions of records
  and sets can arise.
- It must be possible to iterate operations over all members of a set.
- It must be possible to call external resources and services, like GenBank.
- Subdataflows must be supported, i.e., one dataflow can be used as a service
  in another dataflow.
- Dataflows must be specified in a clean, high-level, special purpose programming formalism.
- A dataflow can be run several times, often a large number of times, on
  different inputs.
- The data of these different runs must be kept, including input parameters,
  output data, intermediate results (e.g., from external services), and metadata (e.g., dates).

The last item above is of particular importance and leads to the notion of
a *dataflow repository*: a database system that stores different dataflows together with their different runs. Dataflow repositories can serve many important
purposes:

S. Cohen-Boulakia and V. Tannen (Eds.): DILS 2007, LNBI 4544, pp. 105–121, 2007.

- Effective management of all experimental and workflow data that float around in a large laboratory or enterprise setting.
- Verification of results, either within the laboratory, by peer reviewers, or by other scientists who try to reproduce the results.
- Tracking the provenance (origin) of data values occurring in the result of a dataflow run, which is especially important when external service calls are involved.
- Making all data and stored dataflows available for complex decision support or management queries. The range of such possible queries is enormous; just to give two examples, we could ask "did an earlier run of this dataflow, using an older version of GenBank, also have this gene as a result?", or "did we ever run a dataflow in which this AA sequence was also used for a BLAST search?"

The idea of dataflow repository is certainly not new. It has been repeatedly emphasized in the database and bioinformatics literature, and practical dataflow systems such as Taverna [1] or Kepler [2] do accommodate many of the features listed above. What is lacking so far, however, is a formal, conceptual data model of dataflow repositories. This paper contributes towards this goal.

A conceptual data model for dataflow repositories should offer a precise specification of the types of data (including the dataflows themselves) stored in the repository, and of the relationships among them. Such a data model is important because it provides a formal framework that allows:

- Analyzing, in a rigorous manner, the possibilities and limitations of dataflow repositories.
- Comparing, again in a rigorous manner, the functionalities of different practical systems.
- Highlighting differences in meaning of common notions as used by different authors or in different systems, such as "workflow", "provenance", or "collection".

For the dataflow programming language, our model uses the nested relational calculus (NRC), enhanced with subtyping and external functions. NRC [3] is a well-studied language with exactly the right set of operations that are needed for the manipulation of the types of complex data that occur in a dataflow [4]. The suitability of NRC (in the form of a variant language called CPL) for scientific data manipulation and integration purposes has already been amply demonstrated by the Kleisli system [5,6]. We have confirmed this further by doing some case studies ourselves (e.g., of a proteomics dataflow [7]). A detailed report on several case studies of bioinformatics dataflows modeled using our formalism will be presented in a companion paper.

In this paper we provide formalisations of a number of fundamental notions related to dataflow repositories, such as:

- the notion of run of a dataflow;
- the provenance tracking of dataflow results;

- the binding of service names to external functions or to subdataflows; and
- the relationship between a run of a dataflow and the runs of its subdataflows.

## 2   Example

In this section we provide a simple example to illustrate different aspects involved in modelling of both dataflows and dataflow repository.

We begin by showing two dataflows, expressed in the nested relational calculus, for the following protocol: "Given two organisms A and B, extract all messenger RNA sequences from GenBank belonging to A. Then for each found sequence, search for similar sequences belonging to B."

dataflow *findSimilar*($A$ : **Organism**, $B$ : **Organism**) : MatchedSeqs is
  $\bigcup$ for $s$ in *entrez*($A$, genbank) return
    if $s.moltype = $ mRNA
      then $\{\langle a\colon s, b\colon \mathit{filter}(\mathit{blast}(s, 1e - 4), 300, B)\rangle\}$
      else $\varnothing$

dataflow *filterBlastRep*($rep$ : **BlastRep**, $min$ : **Int**, $org$ : **Organism**) : Seqs is
  $\bigcup$ for $a$ in *accDb*($rep$, $min$) return
    let $seq := \mathit{getSeq}(a.accessionnr, a.database)$ in
      if $seq.organism = org$
        then $\{seq\}$
        else $\varnothing$

These dataflows use the following complex types:

MatchedSeqs = $\{\langle a\colon$ Seq, $b\colon$ Seqs$\rangle\}$,
Seqs = $\{$ Seq $\}$,
Seq = $\langle organism\colon$ **Organism**, $moltype\colon$ **MolType**, $content\colon$ **NCBIXML**$\rangle$,
AccNrDB = $\langle accesionnr\colon$ **AccessionNr**, $database\colon$ **Database**$\rangle$.

The dataflows also contain various service calls, with the following signatures:

    *entrez*($org\colon$ **Organism**, $db\colon$ **Database**): Seqs,
    *filter*($rep\colon$ **BlastRep**, $score\colon$ **Int**, $org\colon$ **Organism**): Seqs,
    *blast*($seq\colon$ Seq, $evalue\colon$ **String**): **BlastRep**,
    *accDb*($rep\colon$ **BlastRep**, $score\colon$ **Int**): $\{$AccNrDB$\}$,
    *getSeq*($acc\colon$ AccNrDB): Seq.

Before we can execute the dataflows, we must bind the service names used to express service calls to actual services. We bind *entrez* and *blast* to external services provided by NCBI. We bind *filter* to dataflow *filterBlastRep*, which thus becomes a subdataflow of *findSimilar*. Now we have to bind all service names appearing in *filterBlastRep*, i.e., *accDb* and *getSeq*. We choose to bind both of them to some external service. The binding process stops here, as *filterBlastRep* does not have any subdataflows.

Suppose now that we have executed *findSimilar* with value cat for parameter $A$, and mouse for $B$. Suppose the following value has been returned:

$\{\langle a\colon \langle organism\colon \mathtt{cat}, moltype\colon \mathtt{mRNA}, ncbiXML\colon \mathtt{AY800278}\rangle,$
  $b\colon \{\langle organism\colon \mathtt{mouse}, moltype\colon \mathtt{mRNA}, ncbiXML\colon \mathtt{XM\_908677}\rangle,$

      $\ldots,$

      $\langle organism\colon \mathtt{mouse}, moltype\colon \mathtt{DNA}, ncbiXML\colon \mathtt{NW\_042634}\rangle\}\rangle,$

  $\ldots,$

  $\langle a\colon \langle organism\colon \mathtt{cat}, moltype\colon \mathtt{mRNA}, ncbiXML\colon \mathtt{NM\_001079655}\rangle,$
    $b\colon \{\langle organism\colon \mathtt{mouse}, moltype\colon \mathtt{mRNA}, ncbiXML\colon \mathtt{NM\_053015}\rangle,$

        $\ldots,$

        $\langle organism\colon \mathtt{mouse}, moltype\colon \mathtt{DNA}, ncbiXML\colon \mathtt{NT\_078297}\rangle\}\rangle\}\},$

where values like AY800278 are used as place holders for the corresponding XML documents. We denote this complex value by *finalresult*.

In our dataflow repository model, however, not just the final result value will be kept, but also information about the service calls that have happened during the run. Such information would look as follows:

$(entrez, [(A, \mathtt{cat}), (B, \mathtt{mouse}), (org, \mathtt{cat}), (db, \mathtt{genbank})], catseqs),$
$(blast, [(A, \mathtt{cat}), (B, \mathtt{mouse}), (s, cat_1), (seq, cat_1), (evalue, \mathtt{1e}-4)], rep_1),$
$(filter, [(A, \mathtt{cat}), (B, \mathtt{mouse}), (s, cat_1),$
        $(rep, rep_1), (score, 300), (org, \mathtt{mouse})], foundcat_1),$
$(blast, [(A, \mathtt{cat}), (B, \mathtt{mouse}), (s, cat_2), (seq, cat_2), (evalue, \mathtt{1e}-4)], rep_2),$
$(filter, [(A, \mathtt{cat}), (B, \mathtt{mouse}), (s, cat_2),$
        $(rep, rep_2), (score, 300), (org, \mathtt{mouse})], foundcat_2),$

Here, *catseqs* is a set containing the following tuples (among many others):

$$cat_1 = \langle organism\colon \mathtt{cat}, moltype\colon \mathtt{mRNA}, ncbiXML\colon \mathtt{AY800278}\rangle,$$
$$cat_2 = \langle organism\colon \mathtt{cat}, moltype\colon \mathtt{mRNA}, ncbiXML\colon \mathtt{NM\_001079655}\rangle.$$

Also, $rep_i$ would be documents of type **BlastRep**, and, for instance, $foundcat_2$ would be a set containing the following tuples (among several others):

$$m_1 = \langle organism\colon \mathtt{mouse}, moltype\colon \mathtt{mRNA}, ncbiXML\colon \mathtt{NM\_053015}\rangle,$$
$$m_2 = \langle organism\colon \mathtt{mouse}, moltype\colon \mathtt{DNA}, ncbiXML\colon \mathtt{NT\_078297}\rangle.$$

Since all information needed to reconstruct the entire run is available, we can trace the provenance (origin) of a particular subvalue appearing in *finalresult*, say $m_1$. We produce a back-trace of the entire run by using subexpression occurrences and their respective input values, as follows:

$\{(\bigcup, [(A, \mathtt{cat}), (B, \mathtt{mouse})], m_1),$
  $(\mathtt{for}, [(A, \mathtt{cat}), (B, \mathtt{mouse}), (s, cat_2)], m_1),$
  $(\mathtt{if}, [(A, \mathtt{cat}), (B, \mathtt{mouse}), (s, cat_2)], m_1),$
  $(\{\,\}, [(A, \mathtt{cat}), (B, \mathtt{mouse}), (s, cat_2)], m_1),$
  $(\langle\,\rangle, [(A, \mathtt{cat}), (B, \mathtt{mouse}), (s, cat_2)], m_1),$
  $(filter, [(A, \mathtt{cat}), (B, \mathtt{mouse}), (s, cat_2), (rep, rep_2), (min, 300), (org, \mathtt{mouse})], m_1)\}.$

Note that *filter* is bound to a subdataflow, so if desired, we can further track the provenance of $m_1$ in the corresponding run of *filterBlastRep*.

The main idea of a dataflow repository is that dataflows, as well as their runs, type declarations, input values, intermediate results, and subdataflow links, are all stored together in a database system. Here is an illustration of this idea:

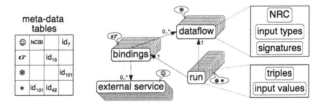

The database should also contain meta-data stored in various additional tables. By meta-data we understand annotating information such as author of a dataflow, date of a run, or version of an external service. Through dataflow identifiers, run identifiers, and referential integrity, these meta-data tables are linked to the repository tables. In the above illustration, we use call-outs to represent these links.

## 3    Dataflows, Runs, and Provenance

In this section, we present the formal dataflow model. Due to space limitations, we will only be able to give a sample of the formal definitions, and all proofs of mathematical properties will be omitted.

*Complex Values.* We model the complex data structures occurring in a workflow using *complex values*. Complex values are constructed, using record and set constructions, from *base values*. Base values can be numbers or strings, but can also be XML files; it is essentially up to the application to decide which kinds of values are considered to be "atomic", and of which kinds of values we want to explicitly model the internal structure within the dataflow.

For example, consider the report returned by a BLAST search. One can consider the entire report as a base value, e.g., an XML file, and use an XQuery operation to extract information from it. This implies that the dataflow will model the XQuery operation as a single step: we will model such single steps by *service calls.* On the other hand, one can consider the structure of the report, modeled as a long record with various attributes, including a set of search results, and model this explicitly as a combination of record and set structures. Since our dataflow model includes the operations of the nested relational calculus, the process of extracting information from the report can be fully modeled in the dataflow. It always depends on the designer of the workflow application which data manipulation aspects of the dataflow need to be explicitly modeled, and which can be modeled as a single step: a good formal model should not enforce this choice in a particular direction.

Formally, we assume a given countably infinite set $\mathcal{A}$ of *base values*. To label attributes in *tuples* (records), we also need a countably infinite set $\mathcal{L}$ of *labels.* Then the set $\mathcal{V}$ of *complex values* is the smallest set satisfying the following: $\mathcal{A} \subseteq \mathcal{V}$; if $v_1, \ldots, v_n \in \mathcal{V}$, then the finite set $\{v_1, \ldots, v_n\}$ is a complex value; if $v_1, \ldots, v_n \in \mathcal{V}$, and $l_1, \ldots, l_n \in \mathcal{L}$ are distinct labels, then the tuple $\langle l_1 : v_1, \ldots, l_n : v_n \rangle$ is also a complex value. The positioning of elements within a named tuple is arbitrary.

Note that we work with sets as the basic collection type, because other kinds of collections can be modeled as sets of records. For an ordered list, for example, one could use a numerical attribute that indicates the order in the list.

*Complex Types.* Types are a basic mechanism in computer programming to avoid the application of operations to inputs on which the operation is not defined. Thus, all data occurring in our dataflow model is strongly typed. We use a type system for complex objects with tuples and sets, known from database theory, that includes a form of subtyping.

Our type system starts from a finite set $\mathcal{B}$ of *base types*. Then the set $\mathcal{T}$ of *complex types* is the smallest set satisfying the following: $\bot \in \mathcal{T}$; $\mathcal{B} \subseteq \mathcal{T}$; if $\tau \in \mathcal{T}$, then the expression $\{\tau\}$ is also a complex type, called a *set type*; if $\tau_1, \ldots, \tau_n \in \mathcal{T}$, and $l_1, \ldots, l_n \in \mathcal{L}$ are distinct labels, then the expression $\langle l_1 : \tau_1, \ldots, l_n : \tau_n \rangle$ is also a complex type, called a *tuple type*. The positioning of elements within a tuple type is arbitrary.

The purpose of base types is obviously to organize the base values in classes. The purpose of $\bot$ is to have a generic type for the empty set; that type is the set type $\{\bot\}$. More generally, the semantics of types is that for each type $\tau$ we have a set $[\![\tau]\!]$ of *values of type* $\tau$, defined in the obvious manner (omitted).

Reasons of flexibility require that the type system is equipped with a form of subtyping [8]. Base types provide an organization of the different types of base values into different classes, and it is standard to allow for classes and subclasses. For example, base types "Protein" and "Peptide" could be subclasses of a base type "AminoAcidSeq", which in turn could be a subtype of "BioSeq". Moreover, subtyping allows a flexible typing of if-then-else statements in dataflows. Thus, the type system of our dataflow model, while guaranteeing safe execution of operations, does not impede flexible specification of dataflows. Due to space limitations, however, we omit all details concerning subtyping.

*Abstract Services.* A common and general view of dataflows is that of a complex composition of atomic actions. In our model, the composition is structured using the programming constructs of the nested relational calculus (NRC). Moreover, the basic data manipulation operators of the NRC are already built in as atomic actions. Any further atomic actions are modeled in our formalism as service calls. Service calls can be really calls to external services, such as NCBI BLAST, but can also be calls to library functions provided by the underlying system, such as addition for numbers or concatenation for strings, or the application of an XQuery to an XML file. Moreover, one dataflow can appear as a service call in another dataflow, thus becoming its subdataflow.

In our model, dataflows use abstract *service names* to denote services. The type system requires *signatures* to be attached to these names. Only at the time a dataflow needs to executed we provide meaning to the service names by assigning them *service functions*. In this section, service functions are merely abstract non-deterministic functions, as this is already sufficient to formally define a run of a dataflow. In Section 4, we will need to be more specific and distinguish between external services (or library functions) on the one hand, and subdataflows on the other hand.

Formally, a *signature* is an expression of the form $\tau_1, \ldots, \tau_n \rightarrow \tau_{\text{out}}$, where the $\tau$'s are types. Likewise, a *service function* is an (infinite) relation $L$ from $[\![\tau_1]\!] \times \cdots \times [\![\tau_n]\!]$ to $[\![\tau_{\text{out}}]\!]$, that is total in the sense that for any given values $v_1, \ldots, v_n$ of types $\tau_1, \ldots, \tau_n$ respectively, there must exist at least one value $v_{\text{out}}$ of type $\tau_{\text{out}}$ such that $(v_1, \ldots, v_n, v_{\text{out}}) \in L$. We denote the universe of all possible signatures by $\mathcal{S}$, and that of all possible service functions by $\mathbb{F}$.

Service functions thus model the input-output behavior of services. Note that service functions can be non-deterministic, in that there may be more than one output related to a given input. This is especially important for modeling external services over which we have no control. The internal database of an on-line service (e.g., BLAST) may be updated, or the service may from time to time fail and produce an error value instead of the actual output value.

Note that we assume service functions to be total. For external services over which we have no control, or to model system failures, totality can always be guaranteed using wrappers. We also assume that wrappers take care of all compatibility issues between used services, as data integration aspects are beyond the scope of this paper.

*The Nested Relational Calculus.* NRC is a simple functional programming language [3], built around the basic operations on records and sets, with for-loops and if-then-else (and let-expressions) as the only programming constructs. We naturally augment NRC with service calls.

Formally, we assume countably infinite sets $\mathcal{X}$ of variables and $\mathcal{N}$ of service names. Then the *NRC expressions* are defined by the following BNF grammar:

$Expr \rightarrow BaseExpr \mid CompositeExpr$
$BaseExpr \rightarrow Constant \mid Variable \mid \text{``}\varnothing\text{''}$
$CompositeExpr \rightarrow \text{``}\{\text{''} \; Expr \; \text{``}\}\text{''} \mid Expr \; \text{``}\cup\text{''} \; Expr \mid \text{``}\bigcup\text{''} \; Expr \mid$
$\qquad\qquad\qquad \text{``}\langle\text{''} \; Element \; (\text{``},\text{''} \; Element)^* \; \text{``}\rangle\text{''} \mid Expr\text{``}.\text{''} Label \mid$
$\qquad\qquad\qquad \text{``for''} \; Variable \; \text{``in''} \; Expr \; \text{``return''} \; Expr \mid$
$\qquad\qquad\qquad Expr \; \text{``}=\text{''} \; Expr \mid Expr \; \text{``}= \varnothing\text{''} \mid$
$\qquad\qquad\qquad \text{``if''} \; Expr \; \text{``then''} \; Expr \; \text{``else''} \; Expr \mid$
$\qquad\qquad\qquad \text{``let''} \; Variable \; \text{``}:=\text{''} \; Expr \; \text{``in''} \; Expr \mid$
$\qquad\qquad\qquad ServiceName \; \text{``(''} \; Expr \; (\text{``},\text{''} \; Expr)^* \; \text{``)''}$
$Element \rightarrow Label \; \text{``:''} \; Expr$
$Constant \rightarrow \mathbf{a} \in \mathcal{A}$
$Variable \rightarrow x \in \mathcal{X}$
$Label \rightarrow l \in \mathcal{L}$
$ServiceName \rightarrow f \in \mathcal{N}$

The variables in an expression that are introduced by a for- or a let-construct are said to occur *bound*; all other occurrences of variables in an expression constitute the *free variables* of an expression. For simplicity of exposition, we disallow that different for- or let-subexpressions bind the same variable. We also disallow that a free variable occurs bound at the same time. We denote the set of free variables of an expression $e$ by $FV(e)$. Naturally, the free variables are the input parameters of the dataflow expressed by the expression. We also use the notation $SN(e)$ for the set of service names used in expression $e$.

When we want to run an expression, we need to assign input values to the free variables, and we need to assign service functions to the service names used in the expression. Formally, a *value assignment* is a mapping $\sigma$ from $FV(e)$ to $\mathcal{V}$, and a *function assignment* is a mapping $\zeta$ from $SN(e)$ to $\mathbb{F}$. The evaluation of expressions is then defined by a system of rules (one rule for each construct of NRC) by which one can infer judgments of the form $\sigma, \zeta \models e \Rightarrow v$, meaning "value $v$ is a possible final result of evaluating $e$ on $\sigma$ and $\zeta$". Recall that there can be more than one possible final result value, if non-deterministic service functions are involved in the evaluation.

Since these inference rules are known from the literature [3], we just present a sample of them, using big union, if-then-else, and for-loops as examples. We also show the rule for service calls:

$$\frac{\sigma, \zeta \models e \Rightarrow \{v_1, \ldots, v_n\}}{\sigma, \zeta \models \bigcup e \Rightarrow v_1 \cup \cdots \cup v_n}$$

$$\frac{\sigma, \zeta \models e_1 \Rightarrow \{w_1, \ldots, w_n\} \qquad \forall i \in \{1, \ldots, n\}: add(\sigma, x, w_i), \zeta \models e_2 \Rightarrow v_i}{\sigma, \zeta \models \text{for } x \text{ in } e_1 \text{ return } e_2 \Rightarrow \{v_1, \ldots, v_n\}}$$

$$\frac{\sigma, \zeta \models e_1 \Rightarrow \text{true} \qquad \sigma, \zeta \models e_2 \Rightarrow v}{\sigma, \zeta \models \text{if } e_1 \text{ then } e_2 \text{ else } e_3 \Rightarrow v} \qquad \frac{\sigma, \zeta \models e_1 \Rightarrow \text{false} \qquad \sigma, \zeta \models e_3 \Rightarrow v}{\sigma, \zeta \models \text{if } e_1 \text{ then } e_2 \text{ else } e_3 \Rightarrow v}$$

$$\frac{\forall i \in \{1, \ldots, n\} : \sigma, \zeta \models e_i \Rightarrow v_i \qquad (v_1, \ldots, v_n, w) \in \zeta(f)}{\sigma, \zeta \models f(e_1, \ldots, e_n) \Rightarrow w}$$

In the rule for for-loops, by $add(\sigma, x, w_i)$ we mean the value assignment obtained from $\sigma$ by updating the value of $x$ to $w_i$.

In order to guarantee that expression evaluation will not fail, we must typecheck the expression. The typechecker requires that we declare types for the free variables, and that we declare *signatures* for the service names. Formally, a *type assignment* for $e$ is a mapping $\Gamma$ from $FV(e)$ to $\mathcal{T}$, and a *signature assignment* is a mapping $\Theta$ from $SN(e)$ to $\mathcal{S}$. Typechecking is then defined by a system of rules (omitted due to space limitations) by which one can infer judgments of the form $\Gamma, \Theta \vdash e : \tau$, meaning "$e$ is well typed given $\Gamma$ and $\Theta$, with result type $\tau$". The rules are such that there can be at most one possible result type for $e$ given $\Gamma$ and $\Theta$.

The following property now states that the type system assures safe execution of expressions.

*Property 1.* If $\Gamma, \Theta \vdash e : \tau$ can be inferred, and $\sigma$ and $\zeta$ are value and function assignments consistent with $\Gamma$ and $\Theta$, then there always exists a value $v$ of type $\tau$ such that $\sigma, \zeta \models e \Rightarrow v$ can be inferred.

*Runs.* In a dataflow repository, we want to keep the information about the different runs we have performed of each dataflow. For this, it is not sufficient to just keep the input values. Indeed, if external services are called in the dataflow, merely rerunning the dataflow on the same inputs may not produce the same result as before, because the behavior of the external service may have changed in the meantime, or because it may even fail this time. It is also not sufficient to keep only the final result value of every run in addition to the input values. Indeed, the repository should support provenance tracking of output values, by which the system can show how certain output values were produced during the dataflow execution. Again, as before, merely rerunning the dataflow will not do here.

We conclude that it is necessary to keep, for each run of an expression $e$, the information about the service calls that have happened during the run. We can naturally represent this information as a number of triples of the form $(e', \sigma', v')$, where: $e'$ is a service call subexpression of $e$; $\sigma'$ is the value assignment constituting the input values of the service call; and $v'$ is the output produced by the service call. Note that there can be many such triples, even if $e$ contains only one service call subexpression, because that service call may occur inside a for-loop.

From that information, the *entire* run can then be reconstructed. We can represent the entire run equally well as a set of such triples, where now $e'$ is not restricted to just service calls, but where we consider all subexpressions instead.[1] Specifically, we have defined a new system of inference rules (one rule for each construct of NRC) that allow to infer judgments of the form $\sigma, \zeta \approx e \Rightarrow R$, meaning that $R$ is a possible run of $e$ on $\sigma$ and $\zeta$. Recall that service functions may be non-deterministic, so that for the same value and function assignments, there may be several different runs. The rules also define the final result value of the run. Moreover, because we will need this for provenance tracking, our rules define the set of *subruns* of a run $R$ — these are runs of subexpressions of $e$ that happened as part of $R$. Formally, each subrun is represented by a triple of the form $(e', \sigma', R')$, where $e'$ is a subexpression of $e$ and $\sigma', \zeta \approx e' \Rightarrow R'$ holds.

Like before, we only show a sample of the rules:

$$\frac{e = \bigcup e' \qquad \sigma, \zeta \approx e' \Rightarrow R' \qquad v = \bigcup result(R') \qquad R = R' \cup \{(e, \sigma, v)\}}{\sigma, \zeta \approx e \Rightarrow R \qquad result(R) \stackrel{def}{=} v \qquad Subruns(R) \stackrel{def}{=} Subruns(R') \cup \{(e, \sigma, R)\}}$$

---

[1] For simplicity of exposition, in the present version of this paper, we will ignore the complication that a subexpression may have several different occurrences in an expression. We know how to incorporate this in the formalism.

$$\frac{\begin{array}{c} e = \text{for } x \text{ in } e_1 \text{ return } e_2 \qquad \sigma, \zeta \approx e_1 \Rightarrow R' \\ result(R') = \{w_1, \ldots, w_n\} \qquad \forall i \in \{1, \ldots, n\} \colon add(\sigma, x, w_i), \zeta \approx e_2 \Rightarrow R_i \\ v = \{result(R_1), \ldots, result(R_n)\} \qquad R = R' \cup R_1 \cup \cdots \cup R_n \cup \{(e, \sigma, v)\} \end{array}}{\sigma, \zeta \approx e \Rightarrow R \qquad result(R) \stackrel{def}{=} v}$$

$$Subruns(R) \stackrel{def}{=} Subruns(R') \cup Subruns(R_1) \cup \cdots \cup Subruns(R_n) \cup \{(e, \sigma, R)\}$$

$$\frac{\begin{array}{c} e = f(e_1, \ldots, e_n) \qquad \forall i \in \{1, \ldots, n\} \colon \sigma, \zeta \approx e_i \Rightarrow R_i \\ (result(R_1), \ldots, result(R_n)), v) \in \zeta(f) \qquad R = R_1 \cup \cdots \cup R_n \cup \{(e, \sigma, v)\} \end{array}}{\sigma, \zeta \approx e \Rightarrow R \qquad result(R) \stackrel{def}{=} v}$$

$$Subruns(R) \stackrel{def}{=} Subruns(R_1) \cup \cdots \cup Subruns(R_n) \cup \{(e, \sigma, R)\}$$

Let us explain the rule for the flatten expression $e = \bigcup e'$. We see that, in order to be able to derive a possible run $R$ of $e$ on given $\sigma$ and $\zeta$, we must first derive a possible run $R'$ for $e'$ on $\sigma$ and $\zeta$. From this particular $R'$, we construct a final result value $v$ for $e$, and a run $R$ of which $v$ is the final result value. This $R$ is one of the possible runs of $e$ on $\sigma$ and $\zeta$, in particular the one that has $R'$ as its subrun. Therefore all subruns of $R'$ are also subruns of $R$.

The run inference rules have the following property.

*Property 2.* Given a run $R$, for each subexpression $e'$ and each $\sigma'$ there is at most one $R'$ such that $(e', \sigma', R') \in Subruns(R)$. We denote this run $R'$ by $Subrun(e', \sigma', R)$.

*Provenance.* We are now ready to consider provenance tracking. We define provenance tracking *for any occurrence of a subvalue* of the final result value of a run. The following simple example will illustrate what we mean by subvalue occurrences. Consider the simple expression $e = \langle a : x, b : f(5) \rangle$, where we declare $x$ to be of type int, and assign the signature int $\rightarrow$ int to service name $f$. Suppose now that we run $e$ on the value assignment where $x = 3$, and on a function assignment $\zeta$ by which $(5, 3) \in \zeta(f)$. Then the tuple $\langle a : 3, b : 3 \rangle$ is a final result value of $e$. Note that 3 occurs twice as a subvalue in this result, but both occurrences have a quite different provenance: the first occurrence is simply a copy of the input value $x = 3$, whereas the second occurrence was produced by the service call $f(5)$.

Formally, we define a *subvalue path* of some complex value $v$ as a path from the root in $v$, viewing $v$ as a tree structure in the obvious manner. Space limitations prevent us from giving the detailed definition. We will use the notation $\varphi \leftarrowbullet v$ to denote that $\varphi$ is a subvalue path of $v$. Note that if $v$ is a set value, and $\varphi$ is not just $v$ itself, i.e., $\varphi$ leads to a proper subvalue, then $\varphi$ is of the form $v; \varphi'$, with $\varphi' \leftarrowbullet u$ for some $u \in v$. We will use that observation in the inference rules below.

Indeed, we have designed a new system of inference rules that defines, for any run $R$, the provenance $Prov(\varphi, R)$ for any subvalue path $\varphi$ in $result(R)$. Intuitively, the provenance is the restriction of $R$ to all subexpressions and subvalues of intermediate results that have contributed to the production of $\varphi$ in

$R$. Formally, considering that $R$ is a set of triples of the form $(e', \sigma', v')$, we will define $Prov(\varphi, R)$ as a set of triples of the form $(e', \sigma', \varphi')$, where $\varphi'$ is a sub-value path of $v'$. Intuitively, such a triple represents the information that the intermediate result $v'$ (resulting from an evaluation of the subexpression $e'$) has partly contributed to $\varphi$ in the output—$\varphi'$ then indicates which part.

We give a sample of the provenance inference rules next.

$$\frac{\begin{array}{cccc} & e = e'.l & \sigma, \zeta \approx e \Rightarrow R & \\ v = result(R) & \varphi \longmapsto\!\bullet\, v & S = Subrun(e', \sigma, R) & v' = result(S) \end{array}}{Prov(\varphi, R) \stackrel{def}{=} Prov(v'; l; \varphi, S) \cup \{(e, \sigma, \varphi)\}}$$

$$\frac{\begin{array}{cccc} e = \langle l_1 \colon e_1, \ldots, l_n \colon e_n \rangle & \sigma, \zeta \approx e \Rightarrow R & v = result(R) \\ i \in \{1, \ldots, n\} & S = Subrun(e_i, \sigma, R) & \varphi = v; l_i; \varphi' & \varphi' \longmapsto\!\bullet\, result(S) \end{array}}{Prov(\varphi, R) \stackrel{def}{=} Prov(\varphi', S) \cup \{(e, \sigma, \varphi)\}}$$

$$\frac{\begin{array}{ccc} e = \text{if } e_1 \text{ then } e_2 \text{ else } e_3 & \sigma, \zeta \approx e \Rightarrow R \\ v = result(R) & \varphi \longmapsto\!\bullet\, v & result(Subrun(e_1, \sigma, R)) = \text{true} \end{array}}{Prov(\varphi, R) \stackrel{def}{=} Prov(\varphi, Subrun(e_2, \sigma, R)) \cup \{(e, \sigma, \varphi)\}}$$

$$\frac{\begin{array}{ccc} e = \text{for } x \text{ in } e_1 \text{ return } e_2 & \sigma, \zeta \approx e \Rightarrow R & v = result(R) \\ w = result(Subrun(e_1, \sigma, R)) & \forall w' \in w \colon S_{w'} = Subrun(e_2, add(\sigma, x, w'), R) \\ \varphi = v; \varphi' & \varphi' \longmapsto\!\bullet\, u & u \in v \end{array}}{Prov(\varphi, R) \stackrel{def}{=} \bigcup_{\{w' \in w \,|\, result(S_{w'}) = u\}} Prov(\varphi', S_{w'}) \cup \{(e, \sigma, \varphi)\}}$$

$$\frac{\begin{array}{cccc} e = \bigcup e' & \sigma, \zeta \approx e \Rightarrow R & v = result(R) \\ S = Subrun(e', \sigma, R) & w = result(S) & \varphi = v; \varphi' & \varphi' \longmapsto\!\bullet\, u & u \in v \end{array}}{Prov(\varphi, R) \stackrel{def}{=} \bigcup_{\{w' \in w \,|\, u \in w'\}} Prov(w; w'; \varphi', S) \cup \{(e, \sigma, \varphi)\}}$$

The rule for tuple field selection delegates the provenance to the immediate subexpression. Note that the rule for tuple construction includes only information from the subrun of the subexpression corresponding to the tuple field in which the subvalue path $\varphi$ occurs. The rule for if-then-else (only given for the then-case) is similar in this respect; only the then-branch is tracked. The rule for for-loops shows how provenance is tracked in all subruns that contributed a value in which $\varphi$ occurs. The rule for big union is again similar in this respect.

## 4   Binding Trees

In a dataflow repository, different dataflows are stored together with their runs. An important feature is that the same dataflow may have been run several times, on distinct inputs (value assignments), but also with different function assignments. Recall that a function assignment binds the service names occurring in the dataflow expression to concrete service functions. While some of the service names will be bound to external functions, in a dataflow repository, it should also be possible to use an existing dataflow as the functionality of some service name. In other words, one dataflow can be used as a "subdataflow" of another. A complication now is that subdataflows may in turn contain service names, so those must be bound as well. In order to avoid non-terminating executions, we must pay attention not to create cycles in this binding process. This is taken care by the notion of *binding tree*, which we formally introduce in the present section.

Formally, consider a set $D$ of *dataflow identifiers*. Each dataflow id has an associated NRC-expression that serves as the dataflow expression. Formally, this corresponds to a given mapping $expr : D \to$ NRC. Moreover, consider a set $Ext$ of *external service identifiers*. We now define:

**Definition 1.** *A* binding tree *is a finite tree, where the nodes are labeled with dataflow identifiers or external service identifiers, and the edges are labeled with service names, with the following properties:*

- *The root is labeled with a dataflow identifier.*
- *Only leaves can be labeled with external service identifiers.*
- *Suppose a node $x$ is labeled with a dataflow identifier $d$. Let $f_1$, ..., $f_n$ be the different service names used in $expr(d)$. Then $x$ has precisely $n$ children, with edges labeled by $f_1$, ..., $f_n$, respectively.*

Intuitively, a binding tree specifies, for the dataflow mentioned in the root, which service names in the dataflow expression are bound to external services, and which to subdataflows. For these subdataflows, the binding tree again specifies a binding for their own service names, and so on. Indeed, note that in a binding tree, a subtree rooted in a node labeled with a dataflow id is itself a binding tree. Note also that the same dataflow id can appear several times in a binding tree (and with different binding subtrees), and that also the same external service name can appear several times.

In order to define this formally, to begin with, we need an assignment of service functions to external service identifiers, i.e., a mapping $func : Ext \to \mathbb{F}$. We can then define the function assignment specified by a binding tree by induction on the height of the tree:

**Definition 2.** *Let $\beta$ be a binding tree. Let the root of $\beta$ be labeled with $d$, and let $expr(d) = e$. We define a function assignment $\zeta_\beta$ for $e$ as follows. Let $h$ be the height of $\beta$.*

- *If $h = 0$, then $\zeta_\beta$ is empty.*
- *If $h > 0$, then $\zeta_\beta(f)$, for any service name $f$ used in $e$, is defined as follows. Let $x$ be the $f$-child of the root of $\beta$.*
  - *If $x$ is labeled with an external service id $z$, then we define $\zeta_\beta(f) := func(z)$.*
  - *If $x$ is labeled with a dataflow id $d'$, then let $e' = expr(d')$, and consider the subtree $\beta'$ of $\beta$ rooted at $x$. By induction, we already have a function assignment $\zeta'_{\beta'}$ for $e'$. Then we define $\zeta_\beta(f)$ to be the relation that associates input value assignments for $e'$ to final result values, given $\zeta'_{\beta'}$.*

In the above, due to space limitations, we have ignored the signatures of service names and of external service identifiers. Incorporating these signatures requires that we enrich a binding tree with mappings that associate parameter positions of service calls to free variables of subdataflow expressions.

## 5    Repository Data Model

We are now in a position to give a formal definition of a dataflow repository. A conceptual schema illustrating the different entities that play a role in a repository, and their relationships, is given in Fig. 1. We use the following notation: $\mathbb{G}$ for all possible type assignments; $\mathbb{S}$ for all possible value assignments; *Runs* for all possible runs; $\mathbb{T}$ for all possible signature assignments; $\mathbb{B}$ for the set of all possible binding trees; and *Triples* for all possible triples in runs. Also, as in the previous section, we assume a given set *Ext* of external service identifiers and a mapping *func*: $Ext \to \mathbb{F}$. (External service identifiers also have signatures, but we ignore these due to space limitations.)

**Definition 3.** *A dataflow repository consists of two finite, pairwise disjoint sets D and R, whose elements are called* dataflow identifiers *and* run identifiers, *respectively, together with eight mappings of the following signatures:*

$$expr\colon D \to \mathcal{E} \qquad\qquad inputtypes\colon D \to \mathbb{G}$$
$$servicesigs\colon D \to \mathbb{T} \qquad\qquad dataflow\colon R \to D$$
$$inputvals\colon R \to \mathbb{S} \qquad\qquad binding\colon R \to \mathbb{B}$$
$$run\colon R \to Runs \qquad\qquad internalcall\colon R \times Triples \to R$$

*The first seven mappings are standard, total, many-to-one mappings; the last mapping, however, is partial but must be one-to-one.*

*Moreover, the mappings must satisfy the following integrity constraints, for any $d \in D$ and any $r \in R$:*

- *inputtypes(d) is defined on $FV(expr(d))$.*
- *servicesigs(d) is defined on $SN(expr(d))$.*
- *expr(d) is well-typed under inputtypes(d) and servicesigs(d).*
- *inputvals(r) is defined on $FV(expr(dataflow(r)))$, and is compatible with inputtypes(dataflow(r)).*
- *The root of binding(r) is labeled with dataflow(r).*
- *run(r) is a run of expr(dataflow(r)) on inputvals(r), given $\zeta_{binding(r)}$.*
- *The repository is* closed *by the mapping internalcall.*

We still have to explain the last item in the above definition (closure). Closure is an important integrity constraint that corresponds to the following intuition: if the repository contains a run of some dataflow, then it also contains all corresponding runs of its subdataflows. (Note that if a subdataflow is inside a for-loop, the subdataflow may be run several times.) This is precisely the function of the

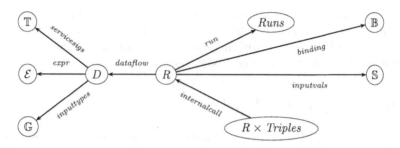

**Fig. 1.** Conceptual dataflow repository schema

mapping *internalcall*, which given a run and a call in that run to a service name bound to a subdataflow, will indicate the run identifier of the corresponding subdataflow run. Formally, we define:

**Definition 4.** *A repository is closed by internalcall if for any* $r \in R$ *and any* $t = (\Phi, \sigma, v) \in Triples$, *the following holds:*

- *internalcall*$(r, t)$ *is defined if and only if* $t \in run(r)$ *and* $\Phi$ *is (an occurrence of) a service call to some service named* $f$ *such that the* $f$-*child of the root of binding*$(r)$ *is labeled with a dataflow identifier* $d'$.
- *If internalcall*$(r, t) = r'$ *is indeed defined, then*
  - *dataflow*$(r') = d'$;
  - *binding*$(r')$ *equals the subtree of binding*$(r)$ *rooted in the* $f$-*child of the root of binding*$(r)$;
  - *inputvals*$(r') = \sigma$; *and*
  - *the final result value of run*$(r')$ *equals* $v$.

Note that we do not explicitly model meta-data in the repository data-model. However, it is possible to extend the conceptual data model with meta-data, for instance by adding mappings from various entities in the repository to annotation identifiers, which represent diverse meta-data entities. The actual content of meta-data is beyond the scope of this paper.

## 6    Related Work

Several researchers advocate integration of workflows and DBMSs [9, 10, 11], as they provide mechanisms for planning, scheduling, and logging. We believe that to properly design a dataflow repository, you need a formal model for dataflows and runs. Although there are several dataflow specification languages [9, 12, 13, 1, 2], to our knowledge, none of them presents a formal model of repository storing dataflows and runs. With increasing importance of provenance [14,15,16,17], often with different interpretations for this term, it is essential that our model includes a formal definition of the kind of provenance that our work targets. For instance, our notion of provenance largely covers the queries of the Provenance Challenge (http://twiki.ipaw.info/bin/view/challenge/). Pioneers in integration of workflows and DBMSs are ZOO [9] and OPM [10]. The ZOO system, implemented on top of an OODBMS, uses database schemas to model dataflows: object classes model data, and relations between them model operations (associated rules specify their execution). An instance contains run information. OPM uses schemas for workflow design with separate object and protocol classes. Protocol classes employ attributes for input, output and connections, and constrains are used to enforce various rules. OPM is implemented in RDBMS, and runs are stored as instances of relational schemas. More recently, Shankar et al. [11] have proposed dataflow specification integrated with SQL and relational DBMS. Dataflows are modeled as active relational tables, and invoked through SQL queries. The Taverna [1] workflow system focuses on practical workflow design and integration of bioinformatics tools and databases. They store runs and associated meta-data in a provenance store implemented as a Web Service  et al. [16]. There are also systems with dataflow design repositories, e.g., WOODSS [18], mainly focusing on workflow reuse. Tröger et al. [12] present a language for workflow design, similar to *in vitro* experiments. Although the compiler produces a persistent repository of workflow specifications and meta-data, it does not include (intermediate) results. Another well-known workflow system is Kepler [2]. Workflow design is actor-oriented and supports collections through an abstract data model for actor design [19]. Intermediate results are recorded through automatic report generation.

## 7    Towards a Dataflow Repository System

A dataflow repository system, following the conceptual model presented in this paper, could be implemented in various ways. An approach that seems promising to us, and which is the object of our current work, is to build the system on top of a modern relational DBMS using SQL/PSM and SQL/XML. A similar approach was also advocated by Shankar et al. [11]. Base values are implemented using SQL datatypes; more complicated base types such as NCBIXML can be implemented using the XML column type, or as large objects (LOBs). Complex values can be decomposed into tables using standard techniques. NRC expressions can be compiled into SQL procedures that, when run, will insert not only the final result value in the repository, but also the intermediate results of external service calls. Service calls can be implemented using SQL user-defined functions. The conceptual data model of the dataflow repository is readily mapped to the relational data model. All semi-structured data belonging to the repository, such as NRC expressions, type assignments, signatures, or binding trees, can be stored using XML columns.

Last but not least, the database may include various additional tables, which contain meta-data such as author of a dataflow, date of a run, version of an external databases, etc. Through dataflow identifiers, run identifiers, and referential integrity, these tables are linked to the repository tables.

# 8    Conclusions

In this paper we have presented an attempt to lay the formal groundwork of dataflow repository systems. Now that we have a precise specification of the various data stored in such repositories, we can start envisaging ways of querying all this data. Note that computing provenance information can already be considered as a kind of query computed over a single run stored in the repository. But clearly much more is possible, given that many different dataflows, with many different runs, are in the database. Two examples of potential decision support queries were already given in the introduction. It remains to be investigated whether special-purpose query language mechanisms must be designed, or whether SQL/XML, where XQuery and SQL can be freely combined, provides enough flexibility and expressive power.

# References

1. Oinn, T., et al.: Taverna: A tool for the composition and enactment of bioinformatics workflows. Bioinformatics 20(17), 3045–3054 (2004)
2. Ludäscher, B., et al.: Scientific workflow management and the Kepler system. Concurrency and Computation: Practice And Experience 18(10), 1039–1065 (2006)
3. Buneman, P., Naqvi, S., Tannen, V., Wong, L.: Principles of programming with complex objects and collection types. Theor. Computer Science 149, 3–48 (1995)
4. Stevens, R., Goble, C., Baker, P., Brass, A.: A classification of tasks in bioinformatics. Bioinformatics 17(1), 180–188 (2001)
5. Chen, J., Chung, S.-Y., Wong, L.: The Kleisli query system as a backbone for bioinformatics data integration and analysis. In: Bioinformatics: Managing Scientific Data, pp. 147–187. Morgan Kaufmann, San Francisco (2003)
6. Davidson, S., et al.: The Kleisli approach to data transformation and integration. In: The Functional Approach to Data Management, pp. 135–165. Springer, Heidelberg (2004)
7. Gambin, A., Hidders, J., Kwasnikowska, N., et al.: NRC as a formal model for expressing bioinformatics workflows. Poster at ISMB, Detroit, MI, USA (2005)
8. Pierce, B.: Types and Programming Languages. MIT Press, Cambridge (2002)
9. Ailamaki, A., Ioannidis, Y., Livny, M.: Scientific workflow management by database management. In: Proceedings of SSDBM, pp. 190–199. IEEE Computer Society, Los Alamitos (1998)
10. Chen, I., Markowitz, V.: An overview of the object protocol model (OPM) and the OPM data management tools. Information Systems 20(5), 393–418 (1995)
11. Shankar, S., Kini, A., DeWitt, D., Naughton, J.: Integrating databases and workflow systems. SIGMOD Record 34(3), 5–11 (2005)
12. Tröger, A., et al.: A language for comprehensively supporting the In Vitro experimental process. In: Silico Proceedings of BIBE, pp. 47–56. IEEE Computer Society, Los Alamitos (2004)
13. Zhao, Y., et al.: A notation and system for expressing and executing cleanly typed workflows on messy scientific data. SIGMOD Record 34(3), 37–43 (2005)
14. Cohen, S., Cohen Boulakia, S., Davidson, S.: Towards a model of provenance and user views in scientific workflows. In: Leser, U., Naumann, F., Eckman, B. (eds.) DILS 2006. LNCS (LNBI), vol. 4075, pp. 264–279. Springer, Heidelberg (2006)

15. Bose, R., Frew, J.: Lineage retrieval for scientific data processing: A survey. ACM Computing Surveys 37(1), 1–28 (2005)
16. Wong, S., Miles, S., Fang, W., et al.: Provenance-based validation of e-science experiments. In: Proceedings of ISWC, LNCS 3729, pp. 801–515 (2005)
17. Mutsuzaki, M., et al.: Trio-One: Layering uncertainty and lineage on a conventional DBMS. Proceeding of CIDR Januari, Asilomar, California (2007)
18. Medeiros, C., et al.: WOODSS and the Web: annotating and reusing scientific workflows. SIGMOD Record 34(3), 18–23 (2005)
19. McPhillips, T., et al.: Collection-oriented scientific workflows for integrating and analyzing biological data. In: Proceedings of DILS, LNCS, vol. 4075, pp. 248–263 (2006)

# Project Histories: Managing Data Provenance Across Collection-Oriented Scientific Workflow Runs[*]

Shawn Bowers[1], Timothy McPhillips[1], Martin Wu[1], and Bertram Ludäscher[1,2]

[1] UC Davis Genome Center, University of California, Davis
[2] Department of Computer Science, University of California, Davis
{sbowers, tmcphillips, martinwu, ludaesch}@ucdavis.edu

**Abstract.** While a number of scientific workflow systems support data provenance, they primarily focus on collecting and querying provenance for single workflow runs. Scientific research projects, however, typically involve (1) many interrelated workflows (where data from one or more workflow runs are selected and used as input to subsequent runs) and (2) tasks between workflow runs that cannot be fully automated. This paper addresses the need for recording data dependencies across multiple workflow runs and accommodating data management activities performed between runs. We define a new conceptual model for representing project-level provenance based on the notion of project histories and folders, and describe mechanisms to support this model in the collection-oriented modeling and design framework of KEPLER. Our approach allows users to conveniently organize their projects and data using the familiar folder-hierarchy metaphor, while at the same time integrating this information with detailed provenance of data products generated via automated scientific workflows.

## 1 Introduction

Scientific workflows promise to *automate* complex and repetitive operations, *model* (*i.e.*, clarify for the scientist) the tasks being automated, and *record* how results of workflow runs were computed from input data. However, few results of great significance are likely to emerge from a single run of one scientific workflow. Novel research involves project organization, data exploration, decision making, and trial-and-error activities that cannot be automated in advance. Researchers employing scientific workflow automation generally wish to run a number of distinct workflows in the context of a single project, apply workflows multiple times on different data or with different parameter settings, modify workflows, and compose completely new workflows as needed. As a result, data provenance support in scientific workflow systems is likely to be of limited usefulness unless the flow of data can be tracked rigorously across multiple workflow runs. Because researchers require the freedom to organize their data and projects as they see fit, comprehensive workflow provenance support also necessitates the recording of data management activities performed manually by researchers between workflow runs. This paper describes how the Collection-Oriented Modeling and Design (COMAD) paradigm

---

[*] This work supported in part by NSF grants DBI-053368, EAR-0225673, IIS-0630033, IIS-0612326, and EF-0228651; and DOE grant DE-FC02-01ER25486.

S. Cohen-Boulakia and V. Tannen (Eds.): DILS 2007, LNBI 4544, pp. 122–138, 2007.

[11] may be extended to provide project-scale data provenance support while enabling users to freely organize their projects and data using the familiar nested-folder metaphor.

## 1.1 Nested Collections for Scientific Data Management

Nested collections are a ubiquitous organizational scheme for scientific data. The large number of standards based on XML testify to the naturalness of representing data as nested collections. Digital libraries and data archives also commonly exploit this metaphor to provide a primary organization for discovering data and exploring collections. While technological inertia may explain some use of the nested collection metaphor, there are more fundamental reasons for its predominance in the realm of scientific data management.

- **Hierarchical structure of natural systems.** Much scientific data can be represented readily as nested collections due to the hierarchical structure of the natural world. For example, the structure of a protein molecule can be described as successively nested collections of polypeptide chains, amino acid residues, and atoms each with their own identifiers and attributes.
- **Intuitive nested file folder metaphor.** Many individuals (including scientists) find the nested collection metaphor an intuitive way to organize information [9]. Folders are often nested and named in ways that reflect significant associations between files. Further, the meaning attached to a particular folder generally cascades to subfolders, where the resulting hierarchy represents a nested collection of metadata that annotates contained files.
- **Projects, tasks, and subtasks.** Projects are generally structured hierarchically. Project tasks can be broken down into (sometimes ordered) collections of subtasks. For this reason, files created during a project often are organized according to a nested folder scheme reflecting the task-hierarchy of the project [9]. The contents of the folders represent data used or created while carrying out the project and together reflect the state of the project at a particular point in time. For scientists in particular, storing information in nested collections is a natural way of persisting data between research tasks.
- **Operations that generate lists of lists.** Many experiments and calculations generate lists of results. For example, a BLAST search can take one molecular sequence and return a list of genes containing similar sequences. A search within the promoter for a gene can reveal a list of over-represented sequences or motifs. When tasks are performed in series the results often are more deeply nested hierarchies, *e.g.*, collections of genes with sub-collections of sequence motifs.

For the above reasons it is important that systems for automating scientific workflows respect, preserve, and ideally exploit the hierarchical structure of scientific data and project information. Workflow systems that lack such support require users to repeatedly map back and forth between a project data organization intuitive to them and the data models employed by the automation system. In contrast, COMAD was envisioned precisely for the purpose of managing data organized within nested collections during workflow execution and facilitating this relatively unrecognized yet common component of doing science.

## 1.2    Leveraging Nested Collections in Scientific Workflow Automation

The COMAD framework in KEPLER [10] extends the conventional actor-oriented framework provided by PTOLEMY II[1] to support collection-oriented workflows. Unlike workflows built from conventional actors, workflows employing collection-aware actors, or *co-actors*, transparently manage data organized as nested collections, maintain the associations implied by the collections, and exploit generic approaches for operating on streams of collections. Abstracting the generic data management tasks associated with collections from the workflow definition in this way often drastically reduces the complexity of the workflow graph (*i.e.*, fewer actors and actor connections are required) with the result that collection-oriented workflows generally are simpler to compose, understand, and maintain than their conventional counterparts. The approach also largely decouples the structure of the workflow from the structure of the data flowing through it, producing workflows that are more reusable, *i.e.*, workflow definitions need not change when the structure of input data changes. In KEPLER, COMAD leverages the nested structure of scientific data to: (1) enable concurrency-safe pipelined execution across actors connected in series or parallel; (2) associate annotations with particular collections and data; (3) dynamically deliver customized parameter values to actors operating on specific collections; and (4) efficiently capture the detailed provenance of all data and collections created during the course of workflow execution. These advantages derive directly from the COMAD approach of exploiting the hierarchical properties of scientific data and projects, much as scientists have been doing manually for many years.

The COMAD implementation in KEPLER delivers the above capabilities by streaming nested collections of data through co-actors as "flat" token sequences where collections are delimited using paired (opening and closing) control tokens. COMAD provides services to actors for managing collections, *e.g.*, for constructing internal representations of collections from input token sequences, inserting and deleting collection elements, and (re-)serializing collections to output token sequences. Co-actors can declare the types of collections and data they process via *read scope* expressions. The CO-MAD framework iteratively invokes actors over portions of the input stream matching these expressions. Data and collections that fall outside of an actor's read scope are automatically forwarded by the framework to succeeding actors, enabling "assembly-line" style data processing. Annotations (*e.g.*, represented as name-value pairs) are modeled explicitly in COMAD, and may be used to represent data and collection metadata that actors create and access during workflow execution. Annotations may also be used to override actor parameters, *e.g.*, allowing co-actor behavior to be changed at runtime within the context of particular collections. Like data and collections, annotations are represented as tokens and are automatically streamed through co-actors by the COMAD framework.

## 1.3    Data Provenance Within Single Workflow Runs

Accurately recording the provenance of workflow products is a critical step towards enabling scientists to incorporate workflow systems into their day-to-day research processes. The COMAD framework records the events (*i.e.*, actor invocations) involved

---

[1] http://ptolemy.eecs.berkeley.edu/ptolemyII/

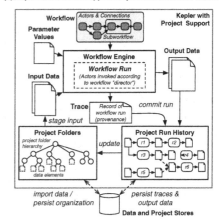

**Fig. 1.** Proposed extensions to Kepler for supporting project-scale provenance: (a) data management between workflow runs occurs outside of the current implementation of Kepler; and (b) envisioned support for managing data and provenance between workflow runs

in computing each workflow product along with the dependencies of these products on input and intermediate data. However, the current COMAD implementation supports provenance for events that occur within a single run of a workflow. In general, the scope of most scientific workflow systems, *e.g.*, [5,14,15,6], is limited to single workflow runs. An exception is [4], which tracks changes to workflow definitions and caches output data to optimize subsequent workflow runs. In COMAD, users may specify the collections, data, and metadata to be applied as input to a workflow run using an XML input file. Similarly, the results of a run, including references to any new data, collections, metadata and provenance records may be persisted as an XML trace file. However, these XML files, and any referenced external data, must be managed directly by the user. Thus, like most other scientific workflow systems, KEPLER and the CO-MAD implementation are largely ignorant of data management tasks carried out between workflow runs. Figure 1a depicts the current state of affairs and emphasizes that workflow definitions, input and output data, and records of workflow runs are generally maintained outside the workflow system. The remainder of this paper describes extensions to the COMAD framework to capture and manage provenance information throughout research projects employing scientific workflows.

## 2   Project Histories and Folders

Figure 1b illustrates our vision for a project-aware version of KEPLER. The KEPLER system boundary is expanded to include workflow definitions and project data as part of the system state. Data may be imported or exported from the project folders, whereupon this data may be supplied to workflow runs. Data and traces produced by a workflow run are retained within the system and are added to the project history, together with the workflow definition and input data sets, when the run is committed. This section introduces the primary use cases of the envisioned system through a realistic usage scenario.

A researcher $\mathcal{R}$ wishes to build an improved phylogenetic tree (*i.e.*, evolutionary history) of the bacterial kingdom using molecular sequences common to representatives of major taxa in the bacterial tree of life. Figure 2 shows the state of the project history and project folders at three points in time.

**Creating the organizational structure for the project.** $\mathcal{R}$ creates a new project and within it a folder named 'Genomes' for holding the annotated genomes of each of seventy bacterial species. $\mathcal{R}$ downloads files representing these genomes from NCBI[2] and The Institute for Genomic Research[3] (TIGR), imports these files into the project, and stores them in sub-folders of Genomes. $\mathcal{R}$ now creates another folder to hold information related to the thirty phylogenetic markers he intends to use to infer the evolutionary relationships between the bacterial taxa. Each marker is a protein sequence corresponding to a highly conserved gene expected to be present and identifiable in every bacterial genome. $\mathcal{R}$ creates a folder under Markers for each marker and data associated with it. Into each he imports a single reference protein sequence that will be used as a search pattern for identifying homologous sequences in each bacterial genome. The system is now aware of the raw data to be used throughout the rest of the project. Subsequent manual and automated operations will fill out the Markers sub-folders and create new top-level folders corresponding to attempts to build an updated bacterial tree of life based on available genome sequences.

**Identifying markers in each genome.** $\mathcal{R}$ is now ready to employ the first scientific workflow, *wf1*, which takes each reference protein sequence stored in the Marker folders, and then locates (via BLAST searches) and refines (using the HMMER [7] program suite) likely candidates for these genes in each of the bacterial genomes. The workflow accepts a stream of collections, each corresponding to a single marker and containing a reference sequence and sub-collection for each genome to search. The products of the workflow are candidate protein sequences for each marker-genome combination.

Once $\mathcal{R}$ has selected this workflow for execution, he stages the input data and workflow parameters. He drags a visual representation of the Markers collection onto a data staging widget, then requests that the contents of the Genomes collection be copied into each Marker sub-collection in the staging area. These interactive operations have no effect on the project collections themselves. After specifying values for workflow parameters, $\mathcal{R}$ starts the workflow. After the workflow run completes, $\mathcal{R}$ browses the output of the run via a workflow-product evaluation area. $\mathcal{R}$ inspects the results of the run, *e.g.*, skimming log files generated by BLAST and HMMER. Noting that the results appear reasonable, he commits the results of the workflow run to the project history and then updates the project folders with the workflow outputs. Once the run is committed, it appears in the project history panel (*wf1:r1* on left side of Figure 2a). The input and output collections of the run can also be accessed at any point in the future via the project history panel, regardless of whether the project folders are updated with the run's output data. Because $\mathcal{R}$ also updates the project folders, the protein sequences

---

[2] http://www.ncbi.nlm.nih.gov/

[3] http://www.tigr.org/

(a) Project history and folders after identifying markers in each genome.

(b) Project history and folders after performing sequence alignments.

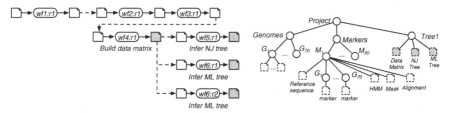

(c) Project history and folders after inferring maximum likelihood trees.

**Fig. 2.** Evolution of project history (left) and project folders (right) during the course of the hypothetical project described in the text

discovered during the workflow run are pasted into the corresponding project collections (right side of Figure 2a). Note that the copies of the genomes placed into the staging area are not duplicated in the project tree. Only new data is propagated to the project collections during an update.

$\mathcal{R}$ must now confirm that the markers have been unambiguously identified in each of the genomes. He browses the project folders and notes how many candidate marker sequences were discovered for each genome-marker pair. If more than one candidate was identified for a particular pair, he compares these sequences to each other and the reference sequence, and inspects statistics produced by the workflow run in an attempt to determine which sequence is orthologous to the reference sequence. $\mathcal{R}$ discards the non-orthologous sequences where this distinction can be made, and annotates any folders that still contain multiple candidate markers with metadata that will prevent these sequences from being used in downstream computations.

**Creating the data matrix and maximum likelihood tree.** $\mathcal{R}$ now selects a second workflow that will build a hidden Markov model (HMM) for each marker using the sequences identified in the previous run. He drags the Markers collection from the project folders panel into the staging area, enters parameter values, and starts the workflow. He evaluates the results of the run, commits the run to the project history, and updates the project collections. Each marker folder now contains a file representing the HMM.

$\mathcal{R}$ reviews the HMMs and then creates, edits, and imports a *mask* for each, indicating unreliable segments of the models. He then selects a third workflow for aligning each marker sequence to the corresponding HMM and again specifies the Markers collection as the workflow input. After committing and updating, each Markers sub-collection contains a multiple sequence alignment (see project history and folders in Figure 2b).

$\mathcal{R}$ runs a fourth workflow that concatenates the alignments into a single data matrix for further analysis. After updating the project folders, $\mathcal{R}$ visualizes the data matrix interactively to check for problems. Satisfied with its quality, $\mathcal{R}$ moves the data matrix from the Markers collection to a new, top-level collection named 'Tree1'. He then selects a fifth workflow for rapidly calculating a neighbor-joining (NJ) tree from the data matrix, updates the tree to the Tree1 collection, and visualizes the tree in an interactive application. Noting that no unusual groupings are evident in the tree, $\mathcal{R}$ stages the input data to a sixth workflow that calculates a maximum-likelihood (ML) tree from the data matrix, where $\mathcal{R}$ again visualizes the tree interactively. To check the tree topologies under different evolutionary models, $\mathcal{R}$ re-runs workflow six specifying a slightly different model of evolution via the workflow parameters. The project history and folders now appear as shown in Figure 2c.

**Re-running the analysis with additional bacterial genomes.** $\mathcal{R}$ periodically re-runs the above analysis to take advantage of new bacterial genome sequences that have become available. Usually he does not find it necessary to recompute the hidden Markov models or alignments from scratch. Instead, $\mathcal{R}$ simply imports the new genomes and re-runs the alignment workflow specifying (via a parameter value) that the alignments should be updated with the new sequences, and then computes the data matrix, the NJ tree, and ML tree as before, storing the results in a new top-level collection.

**Querying provenance for maximum likelihood tree.** One year after starting the project $\mathcal{R}$ has computed eight maximum likelihood trees, each based on more bacterial genomes than the previous analyses. During this time, $\mathcal{R}$ has updated the hidden Markov models three times. At this point a collaborator asks for the hidden Markov model for the third marker used in computing the maximum likelihood tree produced by the fourth iteration of the analysis. Because $\mathcal{R}$ has been replacing the hidden Markov models stored in the project folders each time a new HMM is computed, he cannot simply browse the folders to answer this question. Instead, he right-clicks on the maximum likelihood tree in question and selects *Show dependencies* from the context menu, whereupon a data dependency graph appears (*e.g.*, see Figure 5). $\mathcal{R}$ selects the requested HMM from the visual display and exports it for sharing with the collaborator.


## 3   Collection-Oriented Workflows and Provenance

Here we describe the basic COMAD data model and an approach for recording and querying provenance in *single workflow runs*. We extend the provenance approach to support multiple workflow runs (via project histories and folders) in the next section. Figure 3 shows a collection-oriented workflow in KEPLER for inferring phylogenetic trees. This workflow is similar in intent to the project described in the previous section, *i.e.*, it combines simpler approaches analogous to those used in the workflows described in Section 2 into a single automated workflow (for the purpose of describing COMAD for single workflow runs). The Collection Reader actor inputs a nested data collection containing DNA sequences for homologous genes from a number of taxa. The Align Sequences actor performs an initial alignment of the sequences, and the Refine Alignment actor refines this initial alignment. The Infer Trees actor infers a set of phylogenetic trees from the aligned sequences, and the Compute Consensus actor computes the consensus of these trees (see Figure 4). Finally, the Collection Writer actor stores the output of workflow runs, including the provenance information described below.

### 3.1  Basic COMAD Data Model

A COMAD nested data collection forms a node-labeled, ordered tree. Within workflow runs, these trees are flattened into sequences of data, metadata, and collection delimiter tokens (similar to SAX-based parsing[4] of XML documents). Co-actors work concurrently on nested data collections via these token sequences, *inserting* new data items and collections into the token stream and *deleting* (i.e., not forwarding) existing data items and collections from the token stream. Insertions and deletions always occur within the context of a co-actor's read scope.

The following relations can be used to represent the types of nodes that occur within nested data collections.

- Data($label, id_{node}, id_{obj}$)
- Collection($label, id_{node}, [node]$)
- Metadata($label, id_{node}, id_{obj}$)
- Parameter($label, id_{node}, id_{actor}, id_{obj}$)

Every node in a nested data collection has a label and is assumed to have a unique token identifier (denoted $id_{node}$ above). *Data nodes* are used to encapsulate values. The label of a data node is used to represent the type of value encapsulated. We distinguish between primitive data values (*e.g.*, strings and integers) and complex values (*e.g.*, Java objects and external resources such as files). We generally refer to both primitive and complex values as objects. All object values are assumed to have unique identifiers (denoted $id_{obj}$ above), *e.g.*, represented using URIs. A value can either be stored directly within a data-node token (*i.e.*, inlined), or stored externally and resolved on demand via the object identifier. Inlining complex data values can be used to cache values within a workflow run (*e.g.*, reducing the number of derefences made during workflow execution). *Collection nodes* contain (possibly empty) ordered sequences of "child" nodes, shown as a node list above. *Metadata nodes* can be used to assign annotations to data nodes and collections. Metadata nodes always precede the collections or data items being annotated. When metadata is assigned to a collection, it applies ("cascades") to all collection descendents (*i.e.*, collection data and sub-collections). *Parameter nodes* are used to configure actors during workflow execution. The actor identifier and label arguments of a parameter node specify a target actor and corresponding parameter name. The value of a parameter node is used to set the target actor's parameter value prior to invocation. Parameters are embedded within collections and can be overridden by parameters within sub-collections. In this way, parameter nodes provide a convenient mechanism to dynamically change the default configuration of a workflow.

---

[4] http://www.saxproject.org/

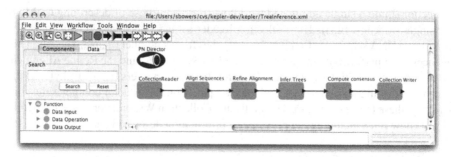

**Fig. 3.** A collection-oriented workflow for computing phylogenetic trees from DNA sequences

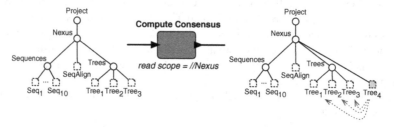

**Fig. 4.** An example invocation of Compute Consensus from Figure 3

### 3.2 Representing and Recording Provenance in Workflow Runs

The main goals of the current COMAD provenance implementation are to (1) enable scientists to ask "scientific" questions about a workflow run by providing convenient queries against the run's execution trace, and (2) have the system track the true data and actor invocation dependencies within a run so that answers to such scientific questions may be as accurate as possible. For example, in most collection-oriented workflows only a portion of an actor's input data is used to produce output data. In addition, workflow input collections are often organized into distinct data sets that enable *independent* "sub-runs" of the workflow. Each sub-run corresponds to one workflow execution over an input data set, and the set of sub-runs are executed concurrently via pipelining. In each of these cases, assuming that all output data from either an actor invocation or a workflow depends on all input data would result in false dependencies [2,3].

The approach described here for single-workflow runs extends our previous work on recording and querying provenance in conventional KEPLER scientific workflows [3]. The approach was successfully applied to the Provenance Challenge [13], as briefly reported in [2]. In the following we describe the basic COMAD model of provenance, the provenance-related annotations recorded during workflow execution, and examples of using this provenance information for querying COMAD workflow traces.

**Model of provenance.** We adopt a simple model of provenance for capturing data dependencies and corresponding source events (such as actor invocations) related to creating and modifying nodes of nested data collections. Instances of the provenance model can be represented using the following relation

– $Dependency(node_w, \{node_r\}, \{event\})$

This relation records the set of nodes ($\{node_r\}$) and events that were used to produce a particular node ($node_w$). Note that $node_w$ is a primary key for the relation, *i.e.*, a given node has at most one associated set of dependency nodes and events. Here we only consider events that correspond to actor invocations, where each $node_r$ was read (*i.e.*, was in the actor's read scope) and each $node_w$ was written (*i.e.*, inserted) by the invocation. Actor invocations are represented via their actor identifiers and invocation number, *e.g.*, where the first invocation of actor $a_1$ is written $a_1:1$. Figure 4 shows an invocation of the Compute Consensus actor. The invocation creates a new tree, inserting it under the input Nexus collection. This tree was derived from three of the input trees. For data, metadata, and parameter nodes, a dependency always represents a *one-step derivation* (*i.e.*, via one actor invocation) with respect to a workflow run. Thus, for non-collection nodes, dependencies may be due to at most one actor invocation. For collections, multiple actor invocations may contribute to distinct portions of intermediate "versions" of the collection. Thus, dependency relations for collections record their changes within a workflow run. We can view a set of dependency relations for a workflow run as a directed acyclic graph, where vertices correspond to nested data-collection nodes and edges correspond to dependencies labeled by their invocation events. Figure 5 shows a portion of the dependency graph for a consensus tree output from a run of the workflow in Figure 3. Each vertex in the tree is labeled with its corresponding node type and identifier.

**Recording provenance annotations.** We infer data dependencies for a workflow run from lower-level *provenance annotations* that are directly embedded into the token stream by co-actors. Three different types of annotations are recorded:

- Insertion($id_{ins}, \{id_r\}, id_{invoc}$)
- Deletion($id_{del}, id_r$)
- InvocationDependency($id_{invoc_1}, id_{invoc_2}$)

Each of these relations store only node *identifiers* as opposed to entire nodes, as in dependency relations. An *insertion annotation* records that a particular node ($id_{ins}$) was inserted by an actor invocation based on the presence of a given set of nodes ($\{id_r\}$). A *deletion annotation* records that a particular node was deleted (*i.e.*, input to, but not forwarded) by an actor invocation. Nodes can be inserted and deleted at most once within a workflow run. An *invocation-dependency annotation* records that an actor invocation ($id_{invoc_1}$) modified a collection (*i.e.*, inserted into or deleted from the collection), and the modified collection was used by another actor invocation ($id_{invoc_2}$). The set of invocation dependencies induces a partial ordering of actor invocations in a workflow run.

Our approach of including provenance annotations within the data stream contrasts with existing workflow systems (*e.g.*, [16,1,12]) that maintain separate provenance stores. While our approach does not prevent the use of a separate provenance database, it simplifies the overall implementation by not requiring additional communication protocols between the workflow engine and the provenance store (possibly requiring synchronization overhead), and allows provenance and run results to be easily viewed and archived outside of any given system implementation. The result of a workflow run is serialized into a single, self-contained XML *trace* file containing all run output and provenance annotations. Figure 6 shows a portion of a trace file generated from a run of the workflow in Figure 3.

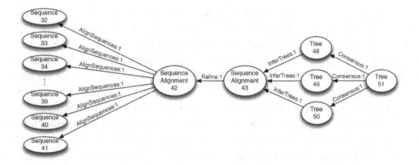

**Fig. 5.** A portion of a simple dependency graph starting from an output consensus tree

The current implementation of COMAD requires actors to declare dependencies when new items are inserted into collections during workflow execution. The COMAD framework validates declared dependencies (*e.g.*, checking that each of the items referred to are within the actor's current read scope), and inserts appropriate provenance annotations into the output token stream of the actor. COMAD can automatically infer dependencies in certain cases. For example, *composite co-actors* [11] are composed from sub-workflows comprising conventional KEPLER actors, enabling data dependencies to be automatically inferred based on the read scope of the co-actor and the data actually accessed by the contained sub-workflow. The COMAD framework inserts appropriate deletion annotations for input items not forwarded by an actor invocation. The framework ensures that items having deletion annotations are inaccessible to subsequent downstream actor invocations. Retaining deleted items is essential to inferring complete data dependencies when input or intermediate items are deleted. Invocation dependencies are automatically inferred by the framework based on insertion and deletion dependencies. For example, when a new item is inserted into a collection, an invocation dependency is inferred between the current invocation and each invocation used to create the item's immediate insertion dependencies. These invocation dependencies are then inserted into the token stream.

**Inferring dependencies from provenance annotations.** Following a workflow run, data dependencies can be *inferred* from a workflow's trace file, *i.e.*, from the provenance annotations generated by the COMAD system together with the nested data collections output by the workflow run. In general, we aim at minimizing the number and size of provenance annotations that must be recorded to compute data dependencies. For example, when an actor inserts a new collection node, a single insertion annotation is created for the collection that cascades to all descendents of the collection.[5] Similarly, if an inserted node is derived from an entire collection (*i.e.*, the collection structure including all subnodes), an insertion annotation is created that refers just to the collection identifier (and not the various subnodes). Shorthands are also used to specify dependen-

---

[5] If items are inserted within this collection by subsequent actors, the insertion annotations for these items override the collection insertion annotation.

```
<Trace id="1">
    <Collection type="Project" id="4">
        <Collection type="Nexus" id="19">
            <Collection type="Sequences" id="20">
                <Data type="Sequence" id="32" objectId="1"/>
                    ...
                <Data type="Sequence" id="41" objectId="10"/>
            </Collection>
            <Deletion item="42" invocation="RefineAlignment:1"/>
            <Insertion item="42" dep="32 33 34 35 36 37 38 39 40 41" actor="AlignSequence:1"/>
            <Data type="SequenceAlignemnt" id="42" objectId="22"/>
            <Insertion item="43" dep="42" actor="RefineAlignment:1"/>
            <Data type="SequenceAlignment" id="43" objectId="23"/>
            <InvocationDependency from="AlignSequence:1" to="RefineAlignment:1"/>
        </Collection>
            ...
    </Collection>
</Trace>
```

**Fig. 6.** An example portion of an XML trace file for Figure 3

cies on subsets of collections, *e.g.*, by referring to the collection node identifier and one or more of its subnode identifiers.

Informally, the dependencies of a data node are computed by (1) identifying the insertion annotation for the node; (2) obtaining the set of node identifiers given as dependencies by the insertion annotation; (3) for those dependency nodes that are collections, pruning away nodes inserted after the actor invocation and nodes deleted prior to the actor invocation; and (4) pruning away non-selected subnodes of collection dependencies (if any such subnodes are specified). Collection node dependencies are computed in a similar way, except that collection dependencies may span multiple actor invocations, since invocation dependencies are in general partially ordered.

Finally, the following relations are computed to store the input and output collection structure of a trace.

- input(*Trace*, [*Node*])
- output(*Trace*, [*Node*])

These relations are computed directly using insertion and deletion annotations (as opposed to first computing the dependency graph for the run). In this case, the input structure is computed by removing all nodes inserted by the run, and the output structure is computed by removing all nodes deleted by the run. We use these operations in the following section for managing multiple-run provenance via project histories.

### 3.3 Querying Provenance

The current COMAD implementation includes a prototype subsystem for querying traces. The prototype is implemented as an SWI-Prolog[6] program, and operates over XML trace files output by workflow runs. The system provides basic operations (*i.e.*, Prolog rules) for accessing trace nodes, constructing dependency relations, and querying corresponding dependency graphs. The operations are defined as views over the underlying COMAD XML schema (as shown in Figure 6). Dependency graphs are constructed by applying various inference rules. Methods also are provided to reconstruct

---

[6] http://www.swi-prolog.org/

parameter settings and metadata annotations attributed to data and collection nodes. The main operation in our current implementation computes dependency edges and is defined as follows.

dependencyEdges $(Trace, [Node], [Edge])$

This operation takes an XML trace file and a list of nodes, and returns a list of dependency edges denoting paths that start from each of the given nodes. For example, the following Prolog query selects the set of sequence alignments used to compute the output consensus tree of Figure 5.

$q(S) :-$ $traceId(1,T),$ nodeForId$(T,51,N),$ dependencyEdges$(T,[N],E),$ edge$(E,N_1,S,I),$
$\quad$ nodeType$(S,$ 'SequenceAlignment'$)$.

In this query, we (1) select the trace with identifier 1 (traceId); (2) get the Tree node having identifier 51 (nodeForId); (3) get the set of dependency edges that start from the Tree node; (4) select an edge; and (5) return the edge node $S$ if it is of type Sequence Alignment (nodeType). As another example, the following query gives the set of actor invocations involved in creating the final consensus tree.

$q(I) :-$ traceId$(1,T),$ nodeForId$(T,51,N),$ dependencyEdges$(T,[N],E),$ edge$(E,N_1,N_2,I)$.

A number of additional operations for querying traces, along with more complex query examples are given in [2]. As ongoing work, we are developing a specialized query language built upon a minimal set of low-level graph-based operations, as well as KE-PLER-based tools for displaying and navigating COMAD workflow run results.

## 4   Extensions for Supporting Project Histories and Folders

Here we present extensions to COMAD for enabling provenance support across scientific workflow runs via project histories and folders. The extensions are designed to address the following challenges in supporting the project-history approach.

- *Staging input data from project collections.* Staging involves the selection of relevant data and sub-collections from the project-folders view, and organizing the selected items to conform to the desired collection schema of the target workflow. Selected items must be tracked and appropriately associated as input to the run, given that items may be organized in new ways, *e.g.*, data may be copied into different collections, new collections may be introduced, collection nesting may be inverted, and so on.
- *Updating project folders from workflow run results.* Once a user has committed a run to the project history, they have the option of updating the project-folders view using the run result. Updating requires the identification of new items generated by the run and determining where these items should be placed within the project folders (given the restructuring described above). The update process should be semi-automatic, *e.g.*, users should be asked whether to apply deletions and where to put items unrelated to the existing structure of the project folders.
- *Tracking dependencies between workflow runs.* In general, workflow runs (like actor invocations) are partially ordered, where all or a subset of data output by one run can be used as input to another run. The system must track these "run dependencies", *i.e.*, the order of runs and their data dependencies, for display within the project run history.

- *Querying data dependencies spanning multiple workflow runs.* Latent data dependencies currently exist between workflow runs, where an output of one run may have depended on input that was generated from a previous run, and so on. To expose these dependencies, provenance queries must be extended to leverage run dependencies (*e.g.*, to obtain the set of markers used to generate a phylogenetic tree in the scenario of Section 2).

These tasks are supported by the following additions to the CoMAD provenance framework.

**Collection identifiers.** As described in the previous section, we require every node in a nested data collection to have a unique identifier. However, as a result of staging data, nodes of project folders may be copied into multiple input sub-collections. For instance, in the marker identification example of Section 2, the contents of each genome collection are copied into each marker sub-collection. Each marker sub-collection will contain nodes that have the same identifiers as nodes in each of the other marker sub-collections. To address these problems, we create new node identifiers for all nodes in a staged nested data collection. Note that with new node identifiers, collections can no longer be tracked back to the project folder. Thus, similar to object identifiers for data values, we add collection identifiers as a mechanism to distinguish, track, and merge collection nodes. These new collection nodes can be represented using the following relation

- Collection($label, id_{node}, id_{col}, [node]$)

where the argument $id_{col}$ is the collection identifier for the node. Unlike object identifiers, collection identifiers are used only to uniquely identify a particular occurrence of a collection, and do not prescribe a collection structure. Thus, two collection nodes that refer to the same collection identifier may have different content.

**Run dependencies.** To construct the project-history graph (see Figure 2), we explicitly track the order of runs for those runs with data dependencies, *i.e.*, where the output of one run is used as input to another run. The following relation is used to represent run order

- RunDependency($id_{run_1}, id_{run_2}, type$)

Each run is assigned a unique identifier (denoted $id_{run}$). Run identifiers can be associated with additional metadata, *e.g.*, with the version of the workflow used in the run, the date and time of the run, and so on. As shown in Figure 2, we distinguish between partial and full run dependencies (where partial dependencies are shown using dashed arrows). A *partial run dependency* between a run A and B means that (1) run B occurred after run A; and (2) some of the collection and object identifiers *produced* (*i.e.*, inserted) by A were included in the input to B. In a *full run dependency*, the input to B is identical to the output of A, *i.e.*, there is a one-to-one correspondence between node identifiers and nesting is preserved. The *type* argument above denotes whether the run dependency is full or partial, which can be easily computed directly from the associated output and input collections (via the input and output operations described in Section 3). There are a number of ways we envision users staging workflows to produce data dependencies between runs. In particular, a user can select data from project folders, can select some or all of the output data from the project history, or can select from both.

**Integrated trace storage.** In the current COMAD implementation, each trace generated from a workflow run is stored in a separate XML document. We extend this approach by allowing each trace result to be stored in a single, integrated "project store" (see Figure 1), providing central database storage and access to project traces. The project store supports both the project-history and project-folders views, as well as queries against one or multiple traces. Individual trace XML documents can be generated from the project store if needed, *e.g.*, to exchange run results between workflow systems or for archival purposes. We add the following relation to represent traces in the project store

– Trace($id_{run}$, [$node$])

This relation maps run identifiers to the nodes of the trace. When a user decides to *commit* a run (*i.e.*, save the run in their project history), a new trace record is constructed in the project store, the run's XML trace file is used to update the project store, and associated run dependencies are created.

Although related to the problem of schema matching [8], the introduction of unique collection identifiers significantly simplifies the task of updating project folders from a committed run result. Figure 7 shows an example update operation, based on applying the following update rules for modifying project folders from run outputs.

- Only new data and collection nodes inserted by the run are added to project folders. The new items in Figure 7c include $D_7$, $D_8$, and $D_9$. Collections $C_6$ and $C_7$ are not considered new since they were input to the run (*i.e.*, introduced during staging).
- New data and collection nodes are added to the collection in the project-folder corresponding to their nearest ancestor collection in the run output. For example, $D_7$ is added to collection $C_4$ in Figure 7d.
- Data and collection nodes without a corresponding ancestor collection in the project-folder are added directly under the root collection of the project folders. For example, data items $D_8$ and $D_9$ are added directly under $C_1$ in Figure 7d.
- Data and collection nodes marked as deleted in the run output are removed from project folders. Project-folder nodes that are not marked as deleted, but with parents that are marked as deleted, are nested under their nearest non-deleted ancestor collection in the project folders. For example, $D_4$ is placed within $C_3$ in Figure 7d, since $C_5$ was deleted by the run.

We intend to allow users to *incrementally* accept or reject modifications to project folders resulting from an update operation. In addition, users can freely modify and rearrange project folders and their contents as needed.

Finally, the addition of run dependencies can facilitate provenance queries across multiple runs. One of our requirements is to allow users to specify the workflow runs to use to answer a particular provenance query. That is, without modifying the query expression, the user should be able to easily specify the run or set of runs to query over. We extend the rules used to infer dependency graphs (as described in Section 3.2) to include run dependency information. In this way, an input node of a workflow run may

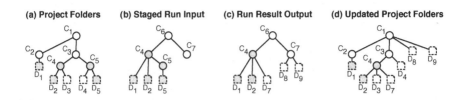

**Fig. 7.** An example update operation: (a) the initial project folders; (b) the staged run input, where collection and object identifiers taken from project folders are shaded; (c) the result of the run on the staged input; and (d) the resulting updated project folders

depend on an intermediate node created within a different run. For example, assume the set of data dependencies have been computed for a workflow run $B$, and a run dependency exists between $A$ and $B$ (*i.e.*, the output of $A$ was used as input to $B$). For each input node to $B$, we additionally compute the data dependencies for corresponding nodes (*i.e.*, nodes with the same collection or data identifier) in workflow run $A$.

## 5   Conclusion

This paper introduces the notion of project histories and folders as a natural model for managing provenance information across scientific-workflow runs. The model leverages the file-folder metaphor for organizing project data, provides a simple and intuitive project-history view of workflow runs that emphasizes run dependencies, and leverages our previous work on collection-oriented workflows and provenance. In addition, we propose extending the single-run provenance support in COMAD with new constructs to support project histories, *e.g.*, for tracking collections and data across runs, and updating project folders with run results. These extensions allow multiple workflow traces to be stored in a single, integrated repository, *e.g.*, to better support provenance queries across runs.

## References

1. Barga, R.S., Digiampietri, L.S.: Automatic generation of workflow provenance. In: Moreau, L., Foster, I. (eds.) IPAW 2006. LNCS, vol. 4145, Springer, Heidelberg (2006)
2. Bowers, S., McPhillips, T.M., Ludäscher, B.: Provenance in collection-oriented scientific workflows. Concurrency and Computation: Practice and Experience (To appear 2007)
3. Bowers, S., McPhillips, T.M., Ludäscher, B., Cohen, S., Davidson, S.B.: A model for user-oriented data provenance in pipelined scientific workflows. In: Moreau, L., Foster, I. (eds.) IPAW 2006. LNCS, vol. 4145, Springer, Heidelberg (2006)
4. Callahan, S.P., Freire, J., Santos, E., eidegger, C.E.S., Silva, C.T., Vo, H.T.: Managing the evolution of dataflows with VisTrails. In: IEEE Workshop on Workflow and Data-Flow for Scientific Applications (SciFlow) (2006)
5. Churches, D., Gombas, G., Harrison, A., Maassen, J., Robinson, C., Shields, M., Taylor, I., Wang, I.: Programming scientific and distributed workflow with Triana services. Concurrency and Computation: Practice and Experience, Special Issue on Scientific Workflows (2005)

6. Deelman, E., Blythe, J., Gil, Y., Kesselman, C., Mehta, G., Patil, S., Su, M.-H., Vahi, K., Livny, M.: Pegasus: Mapping scientific workflows onto the grid. In: European Across Grids Conference (2004)
7. Eddy, S.R.: Profile hidden markov models. Bioinformatics 14(9), 755–763 (1998)
8. Fuxman, A., Hernández, M.A., Ho, C.T.H., Miller, R.J., Papotti, P., Popa, L.: Nested mappings: Schema mapping reloaded. In: VLDB, pp. 67–78 (2006)
9. Jones, W., Phuwanartnurak, A.J., Gill, R., Bruce, H.: Don't take my folders away!: Organizing personal information to get things done. In: CHI Extended Abstracts (2005)
10. Ludäscher, B., Altintas, I., Berkley, C., Higgins, D., Jaeger-Frank, E., Jones, M., Lee, E., Tao, J., Zhao, Y.: Scientific workflow management and the Kepler system. Concurrency and Computation: Practice & Experience, Special Issue on Scientific Workflows (2005)
11. McPhillips, T.M., Bowers, S., Ludäscher, B.: Collection-oriented scientific workflows for integrating and analyzing biological data. In: Leser, U., Naumann, F., Eckman, B. (eds.) DILS 2006. LNCS (LNBI), vol. 4075, pp. 248–263. Springer, Heidelberg (2006)
12. Miles, S., Groth, P., Branco, M., Moreau, L.: The requirements of recording and using provenance in e-science experiments. Journal of Grid Computing (To appear 2006)
13. Moreau, L., Ludäscher, B., et al.: The first provenance challenge (editorial). Concurrency and Computation: Practice and Experience (To appear 2007)
14. Oinn, T., Addis, M., Ferris, J., Marvin, D., Senger, M., Greenwood, M., Carver, T., Glover, K., Pocock, M., Wipat, A., Li, P.: Taverna: A tool for the composition and enactment of bioinformatics workflows. Bioinformatics Journal, 20(17) (2004)
15. Thain, D., Tannenbaum, T., Livny, M.: Distributed computing in practice: the condor experience. Concurrency and Computation: Practice and Experience 17(2-4), 323–356 (2005)
16. Zhao, J., Wroe, C., Goble, C., Stevens, R., Quan, D., Greenwood, M.: Using semantic web technologies for representing e-Science provenance. In: McIlraith, S.A., Plexousakis, D., van Harmelen, F. (eds.) ISWC 2004. LNCS, vol. 3298, Springer, Heidelberg (2004)

# Fast Approximate Duplicate Detection for 2D-NMR Spectra

Björn Egert[1], Steffen Neumann[1], and Alexander Hinneburg[2]

[1] Leibniz Institute of Plant Biochemistry, Department of Stress and Developmental Biology, Germany
{begert,sneumann}@ipb-halle.de
[2] Institute of Computer Science, Martin-Luther-University of Halle-Wittenberg, Germany
hinneburg@informatik.uni-halle.de

**Abstract.** 2D-Nuclear magnetic resonance (NMR) spectroscopy is a powerful analytical method to elucidate the chemical structure of molecules. In contrast to 1D-NMR spectra, 2D-NMR spectra correlate the chemical shifts of $^1$H and $^{13}$C simultaneously. To curate or merge large spectra libraries a robust (and fast) duplicate detection is needed. We propose a definition of duplicates with the desired robustness properties mandatory for 2D-NMR experiments. A major gain in runtime performance wrt. previously proposed heuristics is achieved by mapping the spectra to simple discrete objects. We propose several appropriate data transformations for this task. In order to compensate for slight variations of the mapped spectra, we use appropriate hashing functions according to the locality sensitive hashing scheme, and identify duplicates by hash-collisions.

## 1   Motivation

Nuclear magnetic resonance (NMR) spectra are important to analyze unknown natural products. In contrast to standard one-dimensional NMR spectroscopy, advanced two-dimensional NMR spectroscopy is able to capture the influences of two different atom types at the same time, e.g. $^1$H (hydrogen) and $^{13}$C (carbon).

The result of a 2D-NMR measurement can be seen as an intensity function measured over two independent variables[1]. Regions of the plane with high intensity are called peaks, which contain the real information about the underlying molecular structure. The usual visualizations of 2D-NMR spectra are contour plots as shown in figure 1 ($^1$H,$^{13}$C-HSQC NMR spectrum). [2] Contour lines in low intensity regions are clipped away, because they are produced by irreproducable fluctuations. An ideal peak would register as small dot. In the biochemical literature, peaks are noted by their two-dimensional positions.

However, due to the limited resolution available (depending on the strength of the magnetic field) multiple peaks may appear as a single merged object with

---

[1] The measurements are in parts per million (ppm).
[2] HSQC: Heteronuclear Single Quantum Coherence.

S. Cohen-Boulakia and V. Tannen (Eds.): DILS 2007, LNBI 4544, pp. 139–155, 2007.

non-convex shape, and after thresholding two different peaks, which are close together, may be merged and so both are represented by a single point. This is usually accepted. The pattern of peaks is very characteristic and specific for a particular substance.

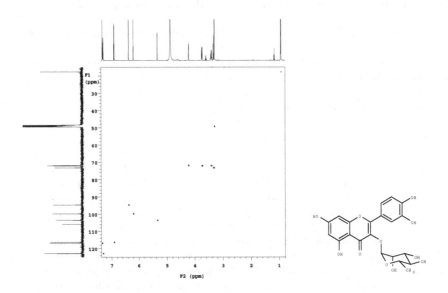

**Fig. 1.** 2D-NMR (HSQC) spectrum of Quercetrin, the one-dimensional plots at the axes are projections of the original two-dimensional intensity function including the respective signal intensities. Each peak captures characteristic $^{13}C,^{1}H$- atomic resonance interactions present in the specific molecule.

As modern NMR devices allow the automatic analysis of many samples per day, the number of a spectra in a database can be up to several thousands per laboratory. Yet, manual work is needed to deduce the chemical structure of a complex organic substance from the spectrum. Thus, most of the NMR data is unpublished but contains a lot of experimental knowledge. Duplicate detection is needed for a use case where two or more libraries are merged, and the experimental knowledge for a pair of duplicates needs to be manually merged and curated. The matching has to be robust against merged peaks and measurements deviations between the two laboratories.

The problem is, given an automatically measured spectrum find all matching spectra on the basis of their peaks with annotations. We cast the specific problem in a more general setting: given a set of spectra find all pairs which are near-duplicates.

Our approach is based on a similarity measure with the desired robustness properties. In [15], we describe heuristics which guarantee no false negatives and reduce the average run time. However, the runtime complexities of those heuristics are still quadratic and the run times for very large data sets are still unacceptable.

In this paper, we propose to map the spectra to simple discrete objects like fixed length integer vectors or discrete sets, for which duplicates can be found much easier. The mapping may cause false negatives, as duplicate spectra may be mapped to discrete objects with slight variations. The effect is compensated by searching similar discrete objects instead of identical ones. We use 1) manhattan distance and 2) the Jaccard coefficient for this task. For both similarity measures exist instances of the locality sensitive hashing scheme (LSH) [16], which uses a proper set of hashing functions to identify duplicate spectra by hash-collisions. The effectiveness of the proposed transformations are evaluated on real data with respect to quality and run time.

The remainder of the paper is organized as follows: after a discussion of related work in the next section, we introduce a simple definition of similarity and define fuzzy duplicates in section 3. Based on the exact method we discuss the transformation of spectra into discrete space in section 4, followed by the application of LSH to the problem. Our experiments are based on real data, their setup and results are shown in section 6. With the summary in section 7 we conclude the paper.

## 2   Related Work

Duplicate detection can be seen as a special case of content-based similarity search, where pairs of spectra are considered duplicates if their similarity exceeds a certain cutoff value. While content-based similarity search is already in use for 1D-NMR spectra [19, 1, 22, 18, 2], to the best of our knowledge, no effective similarity search method is known for 2D-NMR-spectra. Besides technical details (like how to choose the particular cutoff values for similarity) the problem of an approach purely based on similarity is, that the similarities between all pairs of spectra have to be computed. This leads to quadratic run time in the number of spectra, which is prohibitive for large spectra databases. In case of duplicate detection, more efficient algorithms exist.

Various aspects of detecting duplicates have received a lot of attention in database and information retrieval research. The closest type of approaches is near-duplicate detection of documents. The efficient detection of near-duplicate documents has been studied by several authors [5, 24]. In particular, near-duplicate detection of web documents is a quite active research area [13, 8, 12]. The difference between near-duplicate documents and fuzzy duplicates of 2D-NMR spectra is that documents are composed of discrete entities, namely words or index terms, but 2D-NMR spectra consists of continuous 2D points. The crucial difference is that the matching operation is transitive for words but not for 2D points. An extension of near-duplicate documents are duplicates in XML documents [23], where the set of terms is organized as tree.

Duplicates are often found by using a similarity measure. Such measures can be manually defined, but in case of strings suitable similarity measures can be learned automatically using a support vector machine [3], which improves the detection accuracy. Another example of very difficult duplicates are those found in the WHO drug safety database [21]. In this case, a classification problem was solved in order to find a measure for comparison of the records. As those duplicates themselves are very difficult to detect, it seems unlikely to find subquadratic algorithms for this problem class. Fortunately, fuzzy duplicates of 2D-NMR spectra have a more simple definition, which does not require advanced learning techniques.

The detection of duplicate records in data streams [9] or click streams [20] are new variants of the problem. Here, duplicates have simple definitions and the records have fixed length. NMR spectra have not that simple nature, e.g. the number of peaks may differ between spectra (due to the experimental setup even for chemical duplicates). Also the streaming scenario does not appear naturally for 2D-NMR spectra. However, the used technique, namely Bloom filters, are very promising and we will investigate in future research, whether Bloom filters can be applied in our scenario as well.

The detection of duplicates in images [17] is slightly related to our research, as 2D-NMR spectra could be thought as images as well. However, the used techniques in [17] ensure invariance wrt. scaling, shifting and rotation, which is not meaningful in case of 2D-NMR spectra.

The detection of duplicates is slightly related to collision detection in computer graphics [7]. The problem in this concern is to find 2D or 3D objects with overlapping boundaries in real time. The algorithms make the assumption, that only a few bounding boxes of the objects are overlapping. However, in our setting almost all bounding boxes of the spectra overlap. So, collision detection is not applicable to our problem.

Record linkage and especially the sorted neighborhood method [14] is also related to our approach. Sorted neighborhood determines for every object, in our case a 2D NMR spectrum, a key by which the objected are ordered. A sliding window is moved over the sorted sequence and objects within a window are checked for duplicates. The assumption behind the method is, that duplicates have the keys, which are close in the sorted object sequence. Key selection is crucial for the method. The sorted neighborhood method has been successfully used for identifying duplicates in customer databases with data objects consisting mainly of discrete attributes. Since those attributes ensure transitivity of duplicates, the key generation consists of selecting subsets of the discrete attributes. As 2D-NMR spectra do not have discrete attributes, the construction of a key is much more difficult. So far no promising technique is known for numeric attributes.

## 3  Definition of Similarity and Fuzzy Duplicates

A 2D-NMR spectrum of an organic compound captures characteristics of the chemical structure like rings and chains. As the shape of the measured peaks varies between experiments (even with the same substance!), we use centroid peak positions for the representation of the spectra. So, we define a spectrum as a set of two-dimensional points:

**Definition 1.** *A 2D-NMR spectrum A is defined as a set of points $\{x_1, \ldots, x_n\} \subset \mathbb{R}^2$. The $|\cdot|$ function denotes the size of the spectrum $|A| = n$.*

The number of peaks per spectrum is typically between 4 and 60. Our definition of duplicates is based on the idea that peaks can be matched. As spectra are measured experimentally, peak positions can differ even between technical replicates[3]. For that reason, peaks cannot be matched by their exact positions, but rather some slight deviations have to be allowed. A simple but effective approach is to match peaks only within a small spatial neighborhood, The neighborhood is defined by the ranges $\alpha$ and $\beta$:

**Definition 2.** *A peak x from spectrum A **matches** a peak y from spectrum B, iff $|x.c - y.c| < \alpha$ and $|x.h - y.h| < \beta$, where .c and .h denote the NMR measurements for carbon and hydrogen respectively.*

Based on the notion of matching peaks, we are ready to define a set-oriented similarity measure, from which in turn we derive the definition of duplicates as a special case. Note, that a single peak of a spectrum can match several peaks from another spectrum. Given two spectra $A$ and $B$, the subset of peaks from $A$ which find matching partners in $B$ is denoted as $matches(A, B) = \{x : x \in A, \exists y \in B : x \text{ matches } y\}$. The function $matches$ is not symmetric, but helps to define a symmetric similarity measure

**Definition 3.** *Let be A and B two spectra and $A' = matches(A, B)$ and $B' = matches(B, A)$, so **similarity** is defined as*

$$sim(A, B) = \frac{|A'| + |B'|}{|A| + |B|}$$

The measure is close to one if most peaks of both spectra are matching peaks. Otherwise, the similarity drops towards zero.

An important special case of similarity search is the detection of duplicates to increase the data quality of a collection of 2D-NMR-spectra. In addition to the measurement inaccuracies, in case a substance is measured twice with a high and low resolution, it may happen that neighboring peaks are merged to a single one. A restriction to one-to-one relationships between matching peaks can not handle such cases. This means that a single peak from spectrum $A$ can be matching partner for two close peaks from spectrum $B$.

We propose a definition of fuzzy duplicates based on the similarity measure which can deal with the problems mentioned, namely deviances in peak measurements as well as splitted/merged peaks.

**Definition 4.** *A pair of 2D-NMR-spectra A and B are **fuzzy duplicates**, iff $sim(A, B) = 1$.*

---

[3] A technical replicate is the same substance/molecule under the same experimental conditions subjected to the measurement device at least twice.

By that definition it is only required that every peak of a spectrum finds at least one matching peak in the other spectrum. The parameters $\alpha$ and $\beta$ can be set with the application knowledge of typical variances of single peak measurements. For our application, we chose $\alpha = 3$ ppm ($^{13}$C coordinate) and $\beta = 0.3$ ppm ($^1$H coordinate) if not stated otherwise.

### 3.1   Why Is the Problem Difficult?

The duplicate definition is not transitive, that means if $A$ is duplicate of $B$ and $B$ is duplicate of $C$ that not necessarily $A$ is duplicate of $C$. An example for this fact is sketched in figure 2. The reason is the nature of continuous measurements of the peak coordinates. The lack of transitivity has the consequence that a set

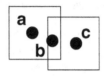

**Fig. 2.** The peak $a$ from spectrum $A$ matches peak $b$ from spectrum $B$ and $b$ matches $c$ from spectrum $C$. However $a$ and $c$ are not matching.

of duplicate spectra (where all spectra are pairwise duplicates) cannot be represented by a single spectrum. Such a representative would ease the detection of duplicates, since all duplicates of the representative are also pairwise duplicates. Because fuzzy duplicates of 2D-NMR spectra do not have this property, all pairs of the set have to be checked in order to calculate a set of duplicates. Thus, the complexity of an algorithm which finds all duplicates in a set of spectra has a quadratic worst case runtime $O(n)^2$ in the number of spectra $n$. Therefore, we have to resort to heuristics which reduce the experimental runtime on typical data sets.

## 4   Spectra Transformation

The exact methods [15], which are guaranteed to have no false negatives, do not scale to very large data sets, even when using peak selecting heuristics. Therefore, we investigate methods which have significantly lower run time. The price for the lower runtime is the possibility of false negatives, that means some duplicate pairs could be missed. We will discuss later how to avoid false negatives.

The problem of finding fuzzy duplicates of 2D-NMR spectra is, that the duplicate relation lacks transitivity. The reason is the continuous nature of the peak measurements. So, the idea is to map the peaks to some discrete objects. Among the many possibilities to do that, we will explore two principal alternatives of those mappings. First, the peak coordinates are discretized and then those integers are concatenated to a fixed length vector. Second, the peaks of a spectrum are mapped to discrete objects so that a spectrum is represented by a set of those objects.

The task of finding duplicate spectra is then reduced to finding duplicates of integer vectors and duplicate sets of discrete objects respectively. Both of the latter duplicate relations are transitive, so that a set of duplicates can be specified by a single representative vector or set. In order to check whether a new mapped spectrum belongs to a set of duplicates, it suffices to test the duplicate relation with the representative of the set.

False negatives occur in this approach, when duplicate spectra are mapped to different discrete objects. We propose mappings which map duplicate spectra to discrete objects which are – if not identical – at least very similar.

### 4.1   Mapping to Integer Vectors

The first proposed mapping of 2D-NMR spectra maps transformed peaks to coordinates of the discrete integer vectors. Such a mapping involves three issues, namely (1) how to handle possible splits/merges of peaks, (2) how to order the transformed peaks to a vector, and (3) how to chose the overall dimensionality of the vectors.

**Robustification:** In order to handle the problem of peak splitting, some peak $x$ of a spectrum is selected and those peaks $y$ are deleted from the same spectrum which are in the neighborhood of $x$. The neighborhood is given by $N(x) = \{y: y \neq x, |x.c - y.c| \leq \alpha$ and $|x.c - y.c| \leq \beta\}$. The peaks are selected in decreasing order of $|N(x)|$, so that the peak with the largest number of neighbors is selected first. The iteration stops when each peak in the spectrum is a singleton, i.e. the neighborhoods of the remaining peaks are empty. The remaining peaks are called the *representative peak set* of a spectrum. After this step, a one to one relation between between peaks of duplicate spectra can be assumed.

**Peak Ordering:** The coordinates of the representative peaks of a spectrum are discretized by binning. The question remains how to order the discretized peak coordinates to form a vector, so that the order is not affected by small measurement errors. The most robust order is to sort $^{13}$C- and $^{1}$H-coordinates independently and discretize afterwards. The vector consists of a block of $^{13}$C-coordinates followed by a block of $^{1}$H-coordinates. However, this procedure would entirely ignore the joint distribution of $^{13}$C- and $^{1}$H-measurements but resorting to the marginal distributions only. So, quite different spectra could be mapped to the same integer vector.

The other extreme is to sort the peaks by one coordinate – say $^{13}$C – only, and form a vector of alternating discretized $^{13}$C- and $^{1}$H-coordinates. The information of the joint distribution of $^{13}$C- and $^{1}$H- coordinates is retained in this mapping. In case of two peaks with close $^{13}$C-coordinates but different $^{1}$H-coordinates, measurement errors in the $^{13}$C-coordinate of a duplicate spectrum could result in swapped order of the two peaks, which in effect also swaps the

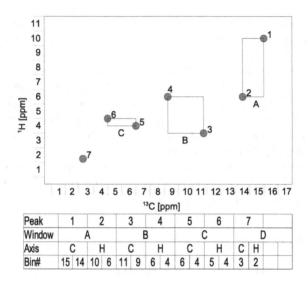

| Peak | 1 | | 2 | | 3 | | 4 | | 5 | | 6 | | 7 | | |
|------|---|---|---|---|---|---|---|---|---|---|---|---|---|---|---|---|
| Window | A | | | | B | | | | C | | | | D | | | |
| Axis | C | H | C | H | C | H | C | H | C | H | C | H | C | H | | |
| Bin# | 15 | 14 | 10 | 6 | 11 | 9 | 6 | 4 | 6 | 4 | 5 | 4 | 3 | 2 | | |

**Fig. 3.** Mapping of peaks from a spectrum to integer vectors for $w = 2$. The blocks of the peaks are indicated by rectangles. The resulting integer vector of the discretized spectrum is shown in the table underneath (last row). The windows and C and H blocks within a window are shown in the second and third row respectively.

positions of the $^1$H-coordinates. In case of two spectra being duplicates their integer vectors could be quite dissimilar, because of the difference in the swapped $^1$H-coordinates.

We propose an intermediate approach, which combines the robustness of the first with the discrimination power of the second. The representative peaks of a spectrum are sorted by one coordinate, say $^{13}$C. Starting with the peak of the largest $^{13}$C-coordinate, we use a jumping window of $w$ consecutive peaks. We sort the $^{13}$C- and $^1$H- coordinates independently for the $w$ peaks inside a window, and arrange them in blocks as in the first approach. The last window might contain less than $w$ peaks if $\#peaks$ mod $w \neq 0$. The important aspect of this technique is, that peaks in the close neighborhood from another spectrum map to the same sorted blocks, regardless of their order in the $^{13}C$- axis. The problem of the second extreme approach can only occur at the jump positions. So, by choosing $w$ we can search for a tradeoff between robustness and retained information. The process is illustrated in figure 3.

Although some peaks of duplicate spectra might map to different integer vectors due to the binning process, i.e. close peaks coordinates are mapped to different bins, the difference is at most one bin per coordinate.

**Overall dimensionality:** The overall dimensionality $D$ of the set of resulting spectra vectors $S$ is determined by the spectrum having the largest set of representative peaks $D = max(\#peaks(S_i))$. Since the spectra have different numbers of representative peaks, we need to pad their integer vectors up to the fixed dimensionality D. Padding the vectors with zeroes increases their overall similarity, whereas padding by random values would decrease their overall similarity. Therefore we pad a vector by repeating the vector itself until the the length of the maximal vector is reached, thereby retaining the similarity of the original vectors.

## 4.2   Mapping to Discrete Sets

We introduce a simple grid-based mapping to map a spectrum to a set of discrete objects, on which we will build a more sophisticated method.

**Simple Grids.** A simple grid-based method is to partition each of the both axis of the two-dimensional peak space into intervals of same size. Thus, an equidistant grid is induced in the two-dimensional peak space and a peak is mapped to exactly one grid cell it belongs to. When a grid cell is identified by a discrete integer vector consisting of the cells coordinates the mapping of a peak $x \in \mathbb{R}^2$ is formalized as

$$g(x) = (g_c(x.c), g_h(x.h)) \text{ with } g_c(x.c) = \left\lfloor \frac{x.c}{\alpha} \right\rfloor, \ g_h(x.h) = \left\lfloor \frac{x.h}{\beta} \right\rfloor$$

The quantities $\alpha$ and $\beta$ are the extensions of a cell in the respective dimensions. The grid is centered at the origin of the peak space.

**Shifted Grids.** A problem of the simple grid-based method is that peaks which are very close in the peak space may be mapped to different grid cells, because a cell border is between them. So proximity of peaks does not guaranty that they are mapped to the same discrete cell.

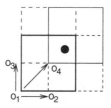

**Fig. 4.** The four grids are marked as follows: base grid is bold, $(1,0)$, $(0,1)$ are dashed and $(1,1)$ is normal

Instead of mapping a peak to a single grid cell, we propose to map it to a set of overlapping grid cells. This is achieved by several shifted grids of the same granularity. In addition to the base grid some grids are shifted into the three directions $(1,0)(0,1)(1,1)$. An illustration of the idea is sketched in figure 4. In figure 4, one grid is shifted in each of the directions by half of the extent of a cell. In general, there may be $s-1$ grids shifted by fractions of $1/s, 2/s, \ldots, s-1/s$ of the extent of a cell in each direction respectively. For the mapping of the peaks to words which consist of cells from the different grids, two additional dimensions are needed to distinguish (a) the $s-1$ grids in each direction and (b) the directions themselves. The third coordinate represents the fraction by which a cell is shifted and the fourth one represents the directions by the following coding: value 0 is (0,0), 1 is (1,0), 2 is (0,1) and 3 is (1,1). So each peak is mapped to a finite set of four-dimensional integer vectors. A nice property of the mapping is that there exists at least one grid cell for every pair of matching peaks both peaks are mapped to.

## 5   Approximate Methods as Filter

The proposed mappings of the 2D-NMR data to discrete objects cannot guarantee, that duplicate spectra are mapped exactly to the same discrete objects. However, the mappings are designed in a way, that the mapped duplicate spectra are at least very similar discrete objects. In this section we focus on methods, which approximate similarity measures for those discrete objects (i.e. integer vectors and discrete sets).

### 5.1   Locality Sensitive Hashing

A general approximation scheme is locality sensitive hashing (LSH) [16], which is a distribution on a family of hash functions $F$ on a collection of objects, such that for two objects $x$, $y$

$$Pr_{h \in F}[h(x) = h(y)] = sim(x, y)$$

The idea is to construct $k$ hash functions $h$ on the set of objects according to the family $F$. The percentage of collisions among the $k$ pairs of hash values for two objects estimates the probability of a collision and gives an approximative similarity score. In general, the outcome of a hash function can be thought of as an integer. So, the LSH-scheme maps each object to a $k$-dimensional integer vector.

In case, two objects $x, y$ are very similar, their integer vectors agree on all $k$ coordinates with high probability. Let be $s = sim(x, y)$, $s \in [0, 1]$ the similarity between $x, y$, then the probability is $s^k$ that $h_i(x) = h_i(y)$ agree for all $1 \leq i \leq k$. To amplify that probability, the sampling process is repeated $L$ times [10]. So, after $L$ repetitions the probability that their integer vectors agree on all $k$ coordinates at least once is

$$Pr[1 \leq i \leq k \colon h_i(x) = h_i(y) \text{ at least once}] = 1 - (1 - s^k)^L$$

Thus, the duplicate detection consist of finding $L$ times the duplicates among integer vectors and union the results. Finding groups of equal integer vectors can be done by sorting, which has lower run time complexity than the naive algorithm.

There are locality sensitive hashing schemes known for the following similarity functions, Manhattan distance between fixed length integer vectors [11], and

Jaccard coefficient for set similarity [4,6]. We briefly review the hashing schemes for the similarity measures.

## 5.2   Manhattan Distance

Given a set of $d$-dimensional integer vectors with coordinates in the set $\{1, \ldots, C\}$, the Manhattan distance between two vectors is $x, y \in X$, $d_1(x,y) = \sum_{i=1}^{d} |x_i - y_i|$. Let be $x = (x_1, \ldots, x_d)$ a vector from $X$ and $u(x) = \mathrm{Unary}_C(x_1) \ldots \mathrm{Unary}_C(x_d)$ a transformation of $x$ into a bit string, where $\mathrm{Unary}_C(a)$ is the unary representation of $a$ with $C$ bits, i.e. a sequence of $a$ ones followed by $C - a$ zeros. For any two vectors $x, y \in X$ there is $d_a(x,y) = d_H(u(x), u(y))$ with $d_H$ is the Hamming distance, which gives the number of different bits between bit strings. An appropriate family of hash functions with the LSH property consists of $h_i(b)$, $1 \le i \le \mathrm{length}(b)$, where $h_i(b)$ returns the $i$th bit from $b$.

Sampling uniformly from those hash functions and testing for collisions reduces to probabilistically counting the number of equal bits:

$$d_1(x,y) = d_H(u(x), u(y)) = dC(1 - Pr[h_i(u(x)) = h_i(u(y))])$$

with random $h_i$, $1 \le i \le dC$.

For the implementation of this LSH scheme, $k$ random indices $i_1, \ldots i_k$ are picked. The transformation into the Hamming space, which can be quite large, is in practice not necessary. In order to find the value of $h_i(u(x))$ we have to look to which coordinate of the integer vector the index $i$ belongs and if $(i - 1 \mod C) + 1$ is larger than the integer value of that coordinate. So the hash function for index $i$ is

$$h_i(u(x)) = \begin{cases} 1 & \text{if } (i - 1 \mod C) + 1 \le x_{\lfloor \frac{i}{C} \rfloor + 1} \\ 0 & \text{else} \end{cases}$$

## 5.3   Jaccard Coefficient

Given two subsets $A, B \subset U$ of a universe $U$ the Jaccard coefficient is

$$sim_J(A, B) = \frac{|A \cap B|}{|A \cup B|}$$

The hash functions for the LSH scheme are constructed by random orderings of the universe $U$. Such a random ordering can by viewed as a random permutation $\pi$ of the elements of $U$, where $\pi(\cdot)$ delivers the position of an element according to $\pi$. The hash function $h_\pi(A) = \min\{\pi(x) \colon x \in A\}$ returns the smallest position of an element of $A$ with respect to the ordering $\pi$. Then for two sets $A, B$ :

$$Pr[h_\pi(A) = h_\pi(B)] = sim_J(A, B)$$

The probability is estimated by sampling from the set of possible permutations.

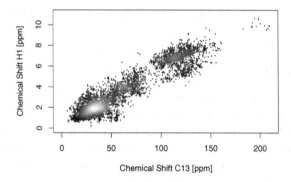

**Fig. 5.** Density of the peaks of all spectra. Light gray means higher density. Note that when plotting a spectrum with $^{13}$C as x-axis (0-220)ppm and $^1$H as y-axis (0-12)ppm, aromatic structures are located in the upper right region and aliphatic structures are located in lower left region.

## 6    Results

In this section we evaluate the proposed definition of duplicates and conduct experiments to investigate the tradeoff between costs for candidate filtering of the approximative methods and candidate checking of the exact methods.

### 6.1    2D-NMR Database

The substances included in the database are mostly secondary metabolites of plants and fungi. They cover a representative area of naturally occurring compounds and originate either from experiments or from simulations[4] based on the known structure of the compound. The database includes 1524 spectra with 2 to 60 peaks each, for a total of about 20,000 peaks. The density in the peak space for all peaks in the database is shown in figure 5.

### 6.2    Performance Results of the Approximate Methods

We implemented the approximate methods as single SQL statements[5] using the SQL 1999 standard. The used data are the 1524 original spectra, which contain 118 fuzzy duplicates. The run times of the approximate methods are below 20 seconds for all methods. That is a large speedup with respect to the exact methods as well as the heuristics proposed in [15], since those methods run several minutes on that data. The actual speedup depends on the size of the used data set, since the methods of the two classes have different runtime complexities ($n^2$ versus $n \log n$).

For the approximate methods, we investigate the number of false positives and false negatives for different numbers $k$ of sampled hash functions. First, the

---

[4] ACD/2D NMR predictor, version 7.08, http://www.acdlabs.com/

[5] The code is available at http://users.informatik.uni-halle.de/~hinnebur

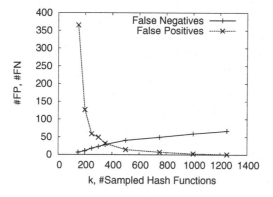

**Fig. 6.** Number of false positives and false negatives FP,FN for Manhattan with LSH ($L = 5$) and diferent k for four repeated experiments

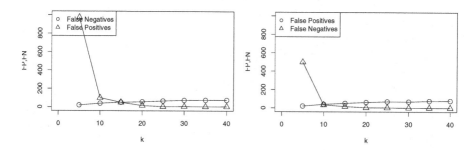

**Fig. 7.** Number of false positives and false negatives for Jaccard coefficient with Minhashing ($L = 5$), simple grids (left) and shifted grids (right)

parameter $L = 5$ is fixed. For small $k$ more spectra are likely to be reported as similar. The larger $k$, the more the reported integer vectors as well as the discrete sets have to be identical. Since our mapping to discrete integer vectors and discrete sets respectively may cause false negatives, we want to allow a some variability of the detected spectra.

A relevant performance measure is the number of false positives for very small false negatives. At this point, the reported similar spectra can be subsequently checked with the naive exact method to exclude the false positives. In that respect, the approximate method acts as a strong filter while only few true duplicates are missed. The results for Manhattan distance with LSH are shown in figure 6. Here the number of false positives is about 390 without any false negative. For Jaccard coefficient with Minhashing we tested the mapping to simple grids and shifted grids. The number of false positives are about 900 and 500 respectively, as shown in figure 7.

As Jaccard coefficient with Minhashing gives more false negatives than the Manhattan distance, additionally, we experimented with different values for $L$. The results are shown in table 1. The table shows (especially in the two blocks

**Table 1.** Number of false positives and false negatives for Jaccard coefficient with Minhashing for different setting for $L$ and $k$

| k | L | Minhashing | | Minhashing+Shift | |
|---|---|---|---|---|---|
| | | FN | FP | FN | FP |
| 2 | 1 | 42 | 9352 | 46 | 2918 |
| 3 | 1 | 59 | 252 | 55 | 558 |
| 4 | 1 | 67 | 170 | 57 | 168 |
| 5 | 1 | 69 | 57 | 66 | 47 |
| 2 | 5 | 19 | 15167 | 11 | 13828 |
| 3 | 5 | 32 | 2626 | 31 | 1540 |
| 4 | 5 | 39 | 514 | 36 | 547 |
| 5 | 5 | 46 | 199 | 47 | 183 |
| 5 | 10 | 35 | 444 | 31 | 285 |
| 5 | 15 | 26 | 654 | 17 | 481 |
| 5 | 20 | 25 | 836 | 16 | 584 |
| 5 | 50 | 20 | 1445 | 12 | 1119 |

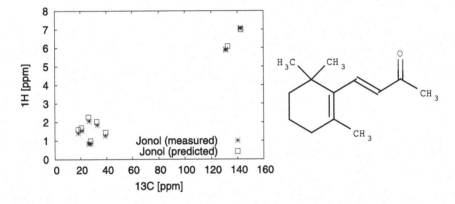

**Fig. 8.** Two spectra as an example for a detected duplicate in our database: Peaks as simple points from an experimental and predicted spectrum of $\beta$ –Jonol. Note, that each peak in A has matching peak in B according to $\alpha = 3.0ppm$ and $\beta = 0.3ppm$.

at the bottom) that increasing $L$ produces more false positives while the number of false negatives is reduced at the same time.

All reported measurements are averages of five runs. The main point is that merely several hundreds of spectra must be explicitly checked as putative duplicates compared to two millions ($1524 \cdot (1524 - 1)/2$) for the naive method. For comparison, the best exact heuristic reported in [15] still needs to check about 30,000 duplicate pairs with the naive method. So, approximate methods are a huge performance gain.

In conclusion, the mapping to integer vector in combination with Manhattan distance and LSH turned out to be the best method, delivering the least number of false positives and no false negatives. The mapping to shifted grids is better than the mapping to simple grids, but the number of false positives is higher. However, the minhashing method has a slight runtime advantage, since less hash functions need to be sampled. This might be useful in case of very large data sets.

### 6.3   Detected Duplicates

There were no duplicates intentionally included in the database. With a setting of $\alpha = 3$ppm and $\beta = 0.3$ppm, which are reasonable tolerances, 118 of 2,322,576 (naive method) possible pairs are reported as fuzzy duplicates.

The found duplicate pairs revealed the following types of classes of duplicates occurring in practice: (i) accidental entry of the same spectra/substance with different names, (ii) spectra prediction software ignoring stereochemical quaternary carbon configurations, (iii) some pairs consist of an experimental and a simulated spectrum (see figure 8) of the same substance (which speaks for both our duplicate definition and the simulation software), (iv) same chemical compound in different measurement conditions (measurement frequency, solvent).

Due to the deletion of peaks in the preprocessing step, different substitutional patterns are also candidates for near duplicates because a discrimination between a peak splitting event or an additional substituent peak is not possible.

## 7   Conclusion

We proposed a simple and robust definition for fuzzy duplicates of 2D-NMR spectra on the basis of co-matching peaks. Considering peak splitting as well as inherent measurement errors are crucial to respect for in NMR- Data. We described ideas and heuristics to embed 2D- spectra data into vector spaces and discrete objects, to suitably interface NMR- data to data mining algorithms. A scale up to large data volumes is achieved by applying approximate and fast algorithms as preliminary filters prior to the computation of the exact duplicates, avoiding the quadratic nature of searching for duplicates in sets of spectra.

We found that our mapping to integer vectors in combination with LSH and Manhattan distance is more suitable for the task than mappings to discrete set in combination with Jaccard coefficient and minhashing. A conservative choice of the parameters guarantees no false negatives. The developed methods are the foundation to start and manage a large collection of NMR spectra, which is part of an ongoing metabolomics project at the IPB in Halle (Saale).

## Acknowledgements

Thanks to Andrea Porzel for valuable discussions and access to the NMR data collection. Steffen Neumann is supported under BMBF grant 0312706G.

# References

1. Tsipouras, A., Ondeyka, J., Dufresne, C., et al.: Using similarity searches over databases of estimated c-13 nmr spectra for structure identification of natural products. Analytica Chimica Acta. 316, 161–171 (1995)
2. Barros, A.S., Rutledge, D.N.: Segmented principal component transform-principal component analysis. Chemometrics & Intelligent Laboratory Systems 78, 125–137 (2005)
3. Bilenko, M., Mooney, R.J.: Adaptive duplicate detection using learnable string similarity measures. In: KDD '03: Proceedings of the ninth ACM SIGKDD international conference on Knowledge discovery and data mining, pp. 39–48. ACM Press, New York, NY, USA (2003)
4. Broder, A.Z., Glassman, S.C., Manasse, M.S., Zweig, G.: Syntactic clustering of the web. In: Selected papers from the sixth international conference on World Wide Web, pp. 1157–1166. Elsevier Science Publishers, Essex, UK (1997)
5. Chowdhury, A., Frieder, O., Grossman, D., McCabe, M.C.: Collection statistics for fast duplicate document detection. ACM Trans. Inf. Syst. 20(2), 171–191 (2002)
6. Cohen, E.: Size-estimation framework with applications to transitive closure and reachability. J. Comput. Syst. Sci. 55(3), 441–453 (1997)
7. Cohen, J.D., Lin, M.C., Manocha, D., Ponamgi, M.K.: I-COLLIDE: An interactive and exact collision detection system for large-scale environments. Symposium on Interactive 3D Graphics 218, 189–196 (1995)
8. Conrad, J.G., Guo, X.S., Schriber, C.P.: Online duplicate document detection: signature reliability in a dynamic retrieval environment. In: CIKM '03: Proceedings of the twelfth international conference on Information and knowledge management, pp. 443–452. ACM Press, New York, NY, USA (2003)
9. Deng, F., Rafiei, D.: Approximately detecting duplicates for streaming data using stable bloom filters. In: SIGMOD '06: Proceedings of the 2006 ACM SIGMOD international conference on Management of data, pp. 25–36. ACM Press, New York, NY, USA (2006)
10. Gionis, A., Gunopulos, D., Koudas, N.: Efficient and tunable similar set retrieval. In: SIGMOD '01: Proceedings of the 2001 ACM SIGMOD international conference on Management of data, pp. 247–258. ACM Press, New York, NY, USA (2001)
11. Gionis, A., Indyk, P., Motwani, R.: Similarity search in high dimensions via hashing. In: VLDB'99: Proceedings of the 25th International Conference on Very Large Data Bases, pp. 518–529. Morgan Kaufmann Publishers, CA USA (1999)
12. Gomes, D., Santos, A.L., Silva, M.J.: Managing duplicates in a web archive. In: SAC '06: Proceedings of the 2006 ACM symposium on Applied computing, pp. 818–825. ACM Press, New York, NY, USA (2006)
13. Henzinger, M.: Finding near-duplicate web pages: a large-scale evaluation of algorithms. In: SIGIR '06: Proceedings of the 29th annual international ACM SIGIR conference on Research and development in information retrieval, pp. 284–291. ACM Press, New York, NY, USA (2006)
14. Hernandez, M.A., Stolfo, S.J.: Real-world data is dirty: Data cleansing and the merge/purge problem. Data. Mining and Knowledge Discovery 2(1), 9–37 (1998)
15. Hinneburg, A., Egert, B., Porzel, A.: Duplicate detection of 2d-nmr spectra. Journal of Integrative Bioinformatics 4(1), 53 (2007)
16. Indyk, P., Motwani, R.: Approximate nearest neighbor - towards removing the curse of dimensionality. In: Proceedings of the 30th Symposium on Theory of Computing, pp. 604–613 (1998)

17. Ke, Y., Sukthankar, R., Huston, L.: An efficient parts-based near-duplicate and sub-image retrieval system. In: MULTIMEDIA '04: Proceedings of the 12th annual ACM international conference on Multimedia, pp. 869–876. ACM Press, New York, NY, USA (2004)
18. Krishnan, P., Kruger, N.J., Ratcliffe, R.G.: Metabolite fingerprinting and profiling in plants using nmr. Journal of Experimental Botany 56, 255–265 (2005)
19. Farkas, M., Bendl, J., Welti, D.H., et al.: Similarity search for a h-1 nmr spectroscopic data base. Analytica Chimica Acta. 206, 173–187 (1988)
20. Metwally, A., Agrawal, D., Abbadi, A.E.: Duplicate detection in click streams. In: WWW '05: Proceedings of the 14th international conference on World Wide Web, pp. 12–21. ACM Press, New York, NY, USA (2005)
21. Noren, G.N., Orre, R., Bate, A.: A hit-miss model for duplicate detection in the who drug safety database. In: KDD '05: Proceeding of the eleventh ACM SIGKDD international conference on Knowledge discovery in data mining, pp. 459–468. ACM Press, New York, NY, USA (2005)
22. Steinbeck, C., Krause, S., Kuhn, S.: Nmrshiftdb-constructing a free chemical information system with open-source components. J. chem. inf. & comp. sci. 43, 1733–1739 (2003)
23. Weis, M., Naumann, F.: Detecting duplicate objects in xml documents. In: IQIS '04: Proceedings of the 2004 international workshop on Information quality in information systems, pp. 10–19. ACM Press, New York, NY, USA (2004)
24. Yang, H., Callan, J.: Near-duplicate detection by instance-level constrained clustering. In: SIGIR '06: Proceedings of the 29th annual international ACM SIGIR conference on Research and development in information retrieval, pp. 421–428. ACM Press, New York, NY, USA (2006)

# Ontology – Supported Machine Learning and Decision Support in Biomedicine

Alexey Tsymbal[1], Sonja Zillner[2], and Martin Huber[1]

[1] Corporate Technology Div., Siemens AG, Erlangen, Germany
{Alexey.Tsymbal, Martin.Huber}@siemens.com
[2] Corporate Technology Div., Siemens AG, Munich, Germany
Sonja.Zillner@siemens.com

**Abstract.** Nowadays, ontologies and machine learning constitute two major technologies for domain-specific knowledge extraction which are actively used in knowledge-based systems of different kind including expert systems, decision support systems, knowledge discovery systems, etc. While the aim of these two technologies is the same – the extraction of useful knowledge – little is known about how the two sources of knowledge can be successfully integrated. Today the two technologies are used mainly separate; even though the knowledge extracted by the two is complementary and significant benefits can be obtained if the technologies were integrated. This problem is especially important for biomedicine where relevant data are often naturally complex having large dimensionality and including heterogeneous features, and where a large body of knowledge is available in the form of ontologies. In this paper we propose one approach for improving the performance of machine learning algorithms by integrating the knowledge provided by ontologies. The basic idea is to redefine the concept of similarity for complex heterogeneous data by incorporating available ontological knowledge, creating a bridge between the two technologies. Potential benefits and difficulties of this integration are discussed, two techniques for empirical evaluation and fine-tuning of feature ontologies are described, and an example from the field of paediatric cardiology is given

## 1 Introduction

Ontologies and machine learning constitute two major technologies for domain-specific knowledge extraction actively used in knowledge-based systems of different kind including expert systems, decision support systems, knowledge discovery systems, etc. By establishing an explicit formal specification of the concepts in a particular domain and relations among them, *ontologies* provide the basis for reusing and integrating valuable domain knowledge within applications [13]. *Machine learning* algorithms are also actively applied in order to extract useful knowledge in different problem domains by searching for interesting patterns (dependencies) in large volumes of data [21].

The principal difference between the two technologies is that the first is usually expert-driven (ontologies are a result of the knowledge elicitation process from a

S. Cohen-Boulakia and V. Tannen (Eds.): DILS 2007, LNBI 4544, pp. 156–171, 2007.

domain expert by knowledge engineers, and data is not necessarily involved in this process); while the latter is data-driven (the search for patterns is usually automatic and does not involve substantial interaction with the expert). While the aim of these two technologies is the same – the extraction of useful knowledge – little is known about how the two sources of knowledge can be successfully integrated.

Traditional machine learning algorithms are not able to incorporate background domain knowledge, but instead work with a sequence of instances, where each instance is represented by a single feature (attribute) vector describing the instance [21]. This limitation of traditional machine learning techniques is widely acknowledged today. The issue of learning from more complex data, and in particular similarity for complex heterogeneous data with rich background knowledge was, for example, in focus at the recent International Workshop on Mining Complex Data, MCD 2006 [31].

The principle of instance similarity is fundamental to the vast majority of machine learning algorithms. The main assumption in supervised, unsupervised and semi-supervised machine learning algorithms is that the instances of the same class (cluster) are more similar to each other than the instances of different classes (clusters). In this paper, we propose an approach to improving the performance of machine learning by redefining the concept of similarity with incorporating constraints provided by ontological domain knowledge, that is instead of simply providing the machine learning algorithm with unrelated features in the form of a single vector or a vector set, we will semantically enhance them by integrating the graph structures of relevant domain ontologies.

We see two main benefits that can be obtained from this procedure: first and the most important, is that the performance of machine learning algorithms will be improved by incorporating knowledge provided by domain ontologies. For example, the predictive accuracy of $k$-nearest neighbour classification can be improved. Second, a more practical and application-oriented advantage, is that an ontology, describing the interrelations between the features in a machine learning problem, can be presented to the user of a knowledge-based system via a Graphical User Interface, and provide an effective means of feature control and manipulation for decision support. Thus, the ontology will not be fixed, but will rather be integrated as a flexible wrapper for more efficient machine learning and knowledge discovery. Changes in the feature ontology, initiated by the user and leading to an increase in machine learning performance, may serve as an important source of novel knowledge in the domain.

This paper is organised as follows. In Section 2 we briefly analyse major existing medical ontologies. In Section 3 we give an overview of related work in mining complex data with taking into account feature semantics. In Section 4 we introduce the concept of feature ontology, consider how instance similarity can be redefined with it and discuss potential benefits of its use, and in Section 5 we consider one example application – the problem of predicting Atrial Septal Defect development. In Section 6 we present two techniques for the empirical evaluation of distance functions that can be used for the validation and fine-tuning of feature ontologies, and in Section 7 we conclude with a summary and directions for future research.

## 2   Ontologies in the Biomedical Domain

Clinical and biomedical applications often have to deal with large volumes of complex information originating from different sources, with different structures and different semantics. There is a long tradition of structuring clinical and biomedical information producing a vast number of standards and conceptual vocabularies that are reused in various medical applications. The efficient reuse of medical information requires the automatic processing, semantic integration, and semantic enhancement of medical knowledge resources enabled by an efficient and adequate knowledge organisation mechanism. There exists a variety of knowledge organisation systems that can be used for capturing semantic knowledge, including taxonomies, thesauri, and ontologies. All of these knowledge organisation systems express, either implicitly or explicitly, a more or less detailed semantic model of the world [14].

A taxonomy establishes a classification hierarchy of terms [25] by subsuming similar objects under distinct classes and subclasses. In contrast to taxonomies, thesauri provide additional means for refining the established classification hierarchies by constituting a fixed set of predefined relations between the concepts, enabling, for instance, the specification of similar or synonymic concepts [22]. Thus, by specifying a terminology of a particular domain, thesauri allow for the sophisticated and detailed annotation of objects of interest.

In computer science, an ontology is defined as "an explicit, formal specification of a shared conceptualisation" [13]. Through the specification of rules, ontologies enable the formulation of constraints, negations, logical functions, and mathematical operations. As taxonomies and thesauri are less expressive than ontologies, their captured content can easily be represented with ontological structures.

As already mentioned, in the domain of healthcare and biomedical informatics, a number of different knowledge repositories have been developed. Figure 1 provides an overview of relevant medical knowledge bases ordered by their size, i.e. the number of concepts. As one can see, the knowledge bases vary in the size (from 900,000 in UMLS to 40 in BioPax), in the way of knowledge organisation (ontology, meta-thesauri, thesauri, and taxonomy), in the covered subject domain, and in the format.

The Unified Medical Language System (UMLS) [8] originated in 1986 at the US National Library of Medicine (NLM) as a terminology integration project. It is a controlled compendium of medical vocabularies enhanced by mappings between them, with over 900 thousand concepts and 12 million relations between them. UMLS has three major components:

- the UMLS Meta-thesaurus being a repository of interrelated biomedical concepts integrating more than 60 families of biomedical vocabularies;
- the UMLS Semantic Network providing high-level categories for classifying every concept from Meta-thesaurus;
- the SPECIALIST lexicon yielding lexical resources and programmes for generating lexical variants of biomedical terms that enable the identification of lexically similar concepts.

| Name | Domain | Size | Format | Type | Licensing | Institution |
|------|--------|------|--------|------|-----------|-------------|
| UMLS | Biomedical and Health domain | 900 000 concepts | Relational files, OO-model, web access | Meta-Thesarus | UMLS Licensing | NLM/NIH |
| SNOMED CT | Healthcare Terms | 400 000 concepts | proprietary format, web access | Thesaurus | Commerical & in UMLS | SNOMED International |
| ICD | Diseases and Injuries | 60 000 concepts | book format, proprietary format, web access | Taxonomy | Commerical & in UMLS | WHO |
| FMA | Human Anatomy | 70 000 concepts | Protege-Frames, web access | Ontology | Free | Univ. of Washington |
| MeSH | Medical Terms | 22500 concepts | XML, ASCII, Tree, MARC, RDF/OWL, web access | Thesaurus | Free | NLM/NIH |
| Gene Ontology | Genetics | 22 000 concepts | OWL, XML, OBO, Text, MySQL, web access | Thesaurus | Free | Collaborative |
| MGED | Microarray Experiments | 230 concepts | OWL | Ontology | Free | MGED Society |
| BioPax | Biological Pathway Data | 40 concepts | OWL | Ontology | Free | Collaborative |

Fig. 1. Overview of major biomedical knowledge bases

UMLS concepts and relations are captured in a proprietary relational format and can either be accessed online via a web browser or are distributed on a CD-ROM or via FTP for offline usage. Although the access to the UMLS knowledge resources is free of charge, UMLS users have to sign a license agreement authorising them to use the UMLS content for research purposes.

The International Classification of Diseases (ICD) [16] is published by the World Health Organization (WHO). By providing means for the classification of known diseases and other health-related problems, the ICD enables the storage, retrieval and statistical analysis of diagnostic information. It is a taxonomy covering approximately 60 thousand concepts organised in 22 chapters of different classes of diseases. Its focus is to subsume similar diseases under classes, and infrequent diseases are sometimes combined without indicating profound similarity. ICD is commercially available on a CD or as a book. It can also be accessed free of charge with a web browser[1] and is a part of the UMLS knowledge repository.

Medical Subject Headings (MeSH) [23] is a thesaurus used for indexing and annotating journal articles and books in the PubMed database of biomedical literature. It establishes a set of poly-hierarchically structured concepts providing the basis for searching annotated medical literature at various levels of specificity. MeSH is created and maintained by the US National Library of Medicine (NLM). The MeSH Thesaurus establishes approximately 22,500 concepts (e.g., Disease, Cardiovascular Disease, Congenital Heart Defect, Atrial Septal Defect) and 83 qualifiers (e.g., Diagnose or Ultrasonography). Both concepts and qualifiers are hierarchically structured ranging from the most general to the most specific ones. The qualifiers provide means for addressing a particular view of a concept, e.g. by attaching the qualifier Ultrasonography to the concept Atrial Septal Defect (ASD) one can emphasise the ultrasonography-related diagnostic aspects of ASD. The MeSH thesaurus can be downloaded from the US National Library of Medicine[2] in the XML, ASCII, MeSH

---

[1] See www.who.int/classifications/apps/icd/icd10online/
[2] See www.nlm.nih.gov/mesh/MBrowser.html

Tree[3], or MeSH MARC[4] formats; it has also been converted to the RDF/OWL format [1][27]. It is also freely accessible through the UMLS knowledge repository.

The Systematised Nomenclature of Medicine Clinical Terms (SNOMED CT) [29] is a thesaurus of healthcare terms, covering clinical data for various diseases, clinical findings, and procedures. SNOMED CT is supported and maintained by SNOMED International, a division of the College of American Pathologists (CAP). It covers approximately 400 thousand concepts with formal logic-based definitions organised in 18 top-level hierarchies. Besides the classical "is-a" relations, it specifies more than 50 other relation types and encompasses more than 900 thousand instantiated relations. Being a very comprehensive standard, SNOMED CT cannot be provided and used in a classical book format, but has to be integrated into some access software. The SNOMED CT content is commercially distributed on CDs with or without additional access software and can be accessed free of charge via the SMOMED CT Browser[5] or through the UMLS knowledge repository.

The Gene-Ontology (GO) [3] project is a collaborative effort to provide a set of structured vocabularies for labelling gene products in different databases. Aiming to establish a controlled vocabulary for describing the functions of genes in a species-independent manner, the GO comprises of three independent vocabularies establishing terms for annotating molecular functions, cellular components and biological processes in gene products. In short, molecular functions detail what a gene product does at the biochemical level, biological processes capture broad biological objectives and cellular components specify the location of a gene product within cellular structures and within macromolecular complexes. Its approximately 22 thousand concepts are organised as a directed acyclic graph, i.e. a hierarchical structure with concepts having one or more parents, and with two relations, "is-a" and "part-of", linking the concepts. However, the GO specifies no associative relations across its three hierarchies. Being free of charge, the GO can be downloaded[6] in many different formats, such as OWL, XML, OBO, free text, and MySQL, as well as can be accessed online via the GO browser AmiGO[7].

The Microarray Gene Expression Data (MGED) ontology [30] provides standard terms for the annotation of microarray experiments. The ontology was created and is maintained by the MGED Society, an international organisation of biologists, computer scientists, and data analysts whose goal is to facilitate the sharing of microarray data generated by functional genomics and proteomics experiments. It encompasses 229 concepts and 110 properties. The concepts are defined and structured by formal-logic-based constraints, such as existential restrictions (specifying the existence of at least one relation of a given property to an individual being a member of a specific concept). Besides, MGED contains 658 instantiated concepts (instances) covering terms that are common to many microarrray experiments. MGED ontology is available for free in the OWL format.

The Biological Pathway Exchange (BioPAX) project [4] is a collaborative community effort aiming at the developing of a common exchange format for biological

---

[3] See www.nlm.nih.gov/mesh/mtr2007abt.htm
[4] See www.loc.gov/marc/specifications/speccharmarc8.html
[5] See snomed.vetmed.vt.edu/ sct/menu.cfm
[6] See www.geneontology.org
[7] See www.godatabase.org/cgi-bin/amigo/go.cgi

pathway data, capturing the key elements of data models from a wide range of popular pathway databases. The established BioPax ontology covers metabolic pathway information, molecular interactions, protein post-translational modifications, and supports the Proteomics Standards Initiative (PSI). To cope with the complexity of pathway data, the BioPAX working group has decided to use a multi-level development approach, i.e. BioPAX Level 1 is focused on the representation of metabolic pathway data, Level 2 expands the scope of Level 1 by including the representation of molecular binding interaction and hierarchical pathways, and further levels are also planned. The BioPAX Level 2 establishes 40 concepts and 33 properties. BioPAX is freely available and is currently implemented in the OWL format, but other implementations, such as XML Schema may be developed in the future.

The Foundational Model of Anatomy (FMA) is the most comprehensive ontology of human "canonical" anatomy [26]. It is developed and maintained by the School of Medicine of the University of Washington and the US National Library of Medicine (NLM). Beside the specification of anatomy taxonomy, i.e. an inheritance hierarchy of anatomical entities, the FMA provides definitions for conceptual attributes, part-whole, location, and other spatial associations of anatomical entities. By additionally allowing for attributing relations (i.e. relations can be described in more detail by attaching additional attributes) FMA is particularly rich with respect to the specification of relations and, thus, can cope with the requirements for the precise and comprehensive capturing of the structure of the body. FMA covers approximately 70 thousand distinct anatomical concepts and more than 1.5 million relations instances from 170 relations types. The FMA is freely available as a Protégé 3.0 project or can be accessed via the web browser Foundational Model Explorer (FME)[8]. Moreover, there exist research approaches focusing on the conversion of the frame-based Protégé version of FMA to the OWL DL format [12].

# 3 Related Work: Mining Complex Data and Data Mining with Ontologies

Medicine is a domain where large complex heterogeneous data sets are commonplace. Today, a single patient record may include, for example, demographic data, familiar history, laboratory test results, images (including echocardiograms, MRI, CT, angiogram etc), signals (e.g. EKG), genomic and proteomic samples, and history of appointments, prescriptions and interventions. And much if not all of this data may be relevant and may contain important information for decision support [19]. A successful integration of heterogeneous data within a patient record thus becomes of paramount importance. Various techniques for mining complex data that try to take into account feature heterogeneity and inter-feature relations were recently suggested.

Perhaps, the most straightforward way to construct a predictive model from heterogeneous data is simply to merge the heterogeneous features into a single heterogeneous feature-vector, neglect possible inter-relation among the features, and to employ some conventional inductive learning technique that is able to work with features of different types. For example, Berrar et al. [7] integrate clinical and

---

[8] See fme.biostr.washington.edu:8089/FME/index.html

transcriptional data in order to get improved classification performance for lung cancer survival prediction. Different learning algorithms are compared; boosted C5.0 decision trees, SVMs, probabilistic neural networks, $k$-nearest neighbour ($k$-NN), and MLP. MLP proved to be the most sensitive and less efficient with large heterogeneous feature vectors, while $k$-NN (somewhat surprisingly) and SVMs were the most robust classifiers resulting in the best predictive performance. Drawbacks of this "naïve" approach include a high risk of overfitting, the need in relatively low dimensionality ("the curse of dimensionality"), and the fact that not every technique supports feature heterogeneity.

A more sophisticated though not always applicable approach is to build an ensemble of models, one for each type of data. Futschik *et al.* [11] claim to be the first to focus on the combination of clinical and microarray-based classifiers. The hypothesis is that clinical information could be enriched with microarray data such that a combined ensemble predictor would perform better that a classifier based on either microarray data alone or clinical data alone. A Bayesian network was built on clinical data and an Evolving Fuzzy Neural Network (EFuNN) on microarray expression data in order to get an improved prediction accuracy for risk group prognosis in patients with lymphoma cancers. This approach has a number of advantages; 1) the heterogeneous data may be physically located at different sites and the computation can be parallelised; 2) there is relatively less risk of overfitting; 3) there is a possibility to apply more suitable techniques to a particular type of data (e.g., gene expression data), with a larger variety of available techniques. The main drawback of this approach is that it is usually applicable only when the different sources of data are representative enough of the problem, so that two or more relatively strong (better than a random guess) models can be constructed for the problem at hand.

Another common approach to take account of feature semantics for complex data consists in aggregating partial (dis)similarities computed on features of the same type possessing certain conceptual commonality. For example, Camps-Valls *et al.* [10] consider the use of composite kernels in order to combine spatial and spectral information for the enhanced classification of hyperspectral images. The main assumption is that the composite representation will allow modelling the dependencies between the extracted features to some extent and this will lead to a more intuitive definition of similarity between instances. It was demonstrated that the use of such composite kernels leads to a significant increase in predictive performance. However, the representation of feature interrelation is limited here to one-level grouping only (a grouping into non-overlapping feature subsets).

Another important related branch of research is focused on the use of taxonomies and ontologies in order to improve data mining results (normally ontologies are used in order to redefine similarity in data mining). Usually, such studies are based on taxonomies which help to structure the instance space in homogeneously represented classification problems (such as texts, annotated images and genomic data). Normally, in the core of such studies there is a concept of taxonomic or semantic distance which depends on the location of two concepts/instances in the taxonomy (ontology). Perhaps, the most well-studied area in this context is text mining where each document is often represented as a set of concepts (so called "bag-of-words" approach). The ontology used in this case can be a predefined graph-based model that reflects semantic relationship between concepts [24] or it can be derived from the texts themselves using some unsupervised learning (one-level or hierarchical clustering) techniques (perhaps Baker and McCallum [5] were the first to apply this to text classification using so-called distributional clustering). Similar studies are done in order to find semantic similarity between annotated images for improved image retrieval based on the ontological representation of relations between the labels (see e.g. [17]).

With the appearance of the extensive Gene Ontology (GO) and the more and more acknowledged role of personalised genomic medicine, there emerged studies that tried to use the GO in order to define semantic similarity between genes in a similar way as it was done before for texts and images. Thus, Azuaje and Bodenreider [2] demonstrate that there is a significant correlation between the semantic similarity between a pair of genes and the probability of finding them in the same complex (cluster) in the analysis of gene expression data. This is claimed to be an assessment confirming to some extent the quality and consistency of the knowledge represented in the GO. In a related study, Bolshakova et al. [9] suggest to use the GO as the domain knowledge in order to validate clustering results and to determine the number of clusters in gene expression analysis.

A similar attempt to enhance inter-case similarity with the domain semantics, for the field of medicine, was performed by Melton et al. [20]. The ontology used to define semantic similarity was SNOMED CT, and each patient was represented by a "bag of findings" (compared to the "bag-of-words" representation of texts), where findings included SNOMED CT concepts extracted from free texts (clinical notes, discharge summaries, etc) and coded procedure and diagnosis data (ICD9-CM codes), from the Columbia University Medical Center (CUMC) data repository in 1989 – 2003. Patient cases included various disorders treated in the Medical Center. The use of taxonomic distance defined in SNOMED CT helped to improve the similarity assessment a little in comparison with the simple "bag of findings" similarity, checked by the correlation with the expert-perceived similarity. Although being an interesting research about inter-patient similarity, this study is still quite far from its practical application, as long as this similarity assessment is quite noisy and still poorly correlated with the expert-perceived similarity, and, on the other hand, most interesting for data mining medical data sets are rather disease-focused, where the "bag of findings" representation would not be suitable.

In summary, most related studies on the use of ontologies in data mining are focused on homogeneously represented cases and concentrate mainly on taxonomic distance and ontologies with "is_a" relations. These techniques are not particularly suitable for mining complex medical data, as long as medical data are usually heterogeneous and disease-focused, where it does not often make much sense to split the instance space into hierarchical concepts. On the other hand, to the best of our knowledge, no study focuses on mining disease-focused medical records with complex inter-feature relations.

The use of domain semantics in order to improve similarity search and decision support is also under active study in the Case-Based Reasoning community [28]. Although, the focus in the so-called knowledge-intensive similarity measures is on creating a customised distance function for each particular feature, instead of the conventional Euclidean and Manhattan (city-block) metrics, and not on the total aggregated distance (similarity). This research in CBR is rather complementary with regards to the study presented in this paper in that the customised feature distances may be used as components in combination with a feature ontology which structures the feature space.

## 4 Feature Ontology: Redefining the Distance with Complex Data

As discussed above, today the most advanced ways of taking into account feature semantics in complex data consist in one-level feature grouping and either building a separate model for each semantic group (ensemble learning) or aggregating partial distances calculated within each group; or in the use of taxonomic distance over the hierarchical clustering of homogeneous features.

The basic idea in our suggested approach is to improve the performance of machine learning by redefining the concept of similarity with incorporating constraints provided by ontological domain knowledge, that is instead of simply providing a machine learning algorithm with features in the form of a single vector or a set of vectors, they will be semantically enhanced by integrating the graph structures of relevant domain ontologies. This can be achieved through the integration of all related ontological knowledge in a single so-called Feature Ontology, systematically structurising the feature space. The task of ontology integration is lately under active study in the area of ontology mapping [18]. Although a number of different solutions were proposed that may help in automating the integration in some cases, the process still remains routine and largely manual. The idea of structuring the feature space with a Feature Ontology is somewhat similar to object-oriented representation in CBR [6].

The main contribution of the feature ontology in terms of machine learning performance is in a more logical distribution of weights in the feature space, reflecting the semantics of the domain. To give a simple example, imaging features should not outweigh clinical features just because their number can be more than a thousand. They should be considered equally important for determining the distance if they are situated at the same level of the feature ontology (unless the expert intentionally specifies that for the current task a particular branch of features is more important).

A schematic distribution of weights in a feature ontology is shown in Figure 2. Feature ontology is a hierarchical structure in the form of a tree graph, where the nodes ( $N_n^l$, where $l$ is the level at which the node is situated, and $n$ is the ordinal number of the node at level $l$) correspond to a group of features with common semantics, starting from the root node $N^0$ combining all the relevant features, and the leaves ( $f_n^l$ ) include features. The tree structurises the feature space into $k+1$ levels. Leaves (features) can be situated at any level of the tree (although in the figure they are shown at level $k+1$ only for the sake of simplicity). The graph is weighted; weights are assigned to its edges (branches of the tree). Weight $w_n^{lm}$ corresponds to the $n$-th child edge originating from the $m$-th node at level $l$.

The weights of all child branches of a node in such a feature ontology should sum to one: $\sum_n w_n^{lm} = 1$. By default, if no prior knowledge is available, the weights of child branches should be equal. The weight of a particular feature $f_n^l$ is defined as

the product of the weights in the tree on the path towards this feature:
$w(f_n^l) = \prod\limits_{i=o}^{l-1} w_*^i(f_n^l)$ , where $w_*^i(f_n^l)$ is the weight of an ancestor branch of level $i$ for

feature $f_n^l$. According to this definition, the deeper a node (or a feature) is in the hierarchy, the less influence it will have in the similarity assessment.

The weights in the feature ontology can be established by an expert (satisfying to the defined constraints) and/or they can be fine-tuned with some machine learning algorithm (e.g. using a form of genetic search). The resulting feature weights can be used in combination with any distance function supporting feature weighting. In the simplest case, the overall distance can be calculated as the weighted average of contributing partial distances corresponding to each relevant feature. Each partial distance may be different and may take into account the type and semantics of a particular feature but should be normalised (i.e., it should be in the range from 0 to 1).

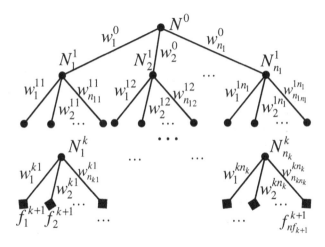

**Fig. 2.** Schematic distribution of weights in a feature ontology

One of the most common machine learning techniques where similarity between instances is explicitly calculated is instance-based learning (e.g., $k$-nearest neighbour classification, $k$-NN) [21]. The distance function that lies in the core of $k$-NN is normally defined for a single set of unrelated features representing the problem. By semantically enhancing the set of relevant features by integrating medical domain knowledge and redefining the distance function, the patient diagnostic (classification) accuracy can be improved. This is the most important expected benefit from the use of the feature ontology. Presumably, feature ontology will also be useful for other learning techniques, implicitly taking instance similarity into account, in order to improve their performance.

Besides the improved predictive performance, the graph-based representation of the feature ontology can be convenient for an expert in order to establish different feature weights by changing the weights of branches corresponding to a certain semantic group of features, instead of assigning importance to each particular feature.

The feature ontology can be presented to the expert as part of the system's GUI and might provide an effective way for feature control and manipulation for decision support. Thus, the ontology will not be fixed, but will rather be integrated as a flexible wrapper for more efficient machine learning and knowledge discovery. Changes in the feature ontology, initiated by the user and leading to an increase in machine learning performance, may also serve as an important source of novel knowledge in the subject domain.

## 5   Example: Prognosis of Atrial Septal Defect Development

The authors of this paper are participants of the EU's 6th Framework Programme's (FP6) Integrated Project "Health-e-Child" (*www.health-e-child.org*), which was started in 2006. The present study is motivated by the main objectives of the project. The focus of the project is on the vertical integration of biomedical data, information and knowledge spanning the entire spectrum from genetic to clinical to epidemiological with the aim of gaining a comprehensive view of a child's health and providing the basis for improving individual disease prevention, screening, early diagnosis, therapy and follow-up of paediatric diseases. Health-e-Child focuses on some carefully selected representative diseases in three different categories: paediatric heart diseases, inflammatory diseases, and brain tumours.

Atrial Septal Defect (ASD) which is characterised by a hole in the atrial septum is a congenital heart defect, is perhaps the most common cause of Right Ventricle Overload (RVO) and is among the most common paediatric heart diseases [15]. Usually, the intervention to treat ASD is performed at a pre-school age (4-6 years of age). However, the size of the hole is constantly changing with time and in some cases the defect may get worse, so that time can be lost to do device closure (trans-catheterisation), and only an open-heart surgery can be performed. On the other hand, in some cases the hole in the septum (even a moderate-sized one, even at the age of 4-6) may close on its own [15]. Up to know the phenomenon of ASD development is rather unclear to physicians and data-driven decision support will be of great help here. Another problem where decision support might be useful is possible complications after trans-catheterisation. E.g., there are cases where tissue erosion and rupture is reported, which might need another trans-catheterisation procedure, or even surgery. Distinguishing potentially high-risk patients in terms of possible complications after ASD treatment is another important task in this context.

Using different examinations and tests, such as echocardiogram, chest X-ray, electrocardiogram, Doppler study, MRI, and cardiac catheterisation, a physician collects all available information for determining the diagnose and the most suitable treatment. As the prognosis of ASD development depends on heterogeneous features of different kind representing clinical data, genetic data, ECG, and imaging data, the resulting feature space becomes quite complex. Therefore, we represent the features in a hierarchical semantically enhanced structure by establishing a feature ontology. By mapping and relating the concepts of the feature ontology to existing medical ontologies (see Section 2), valuable medical background knowledge, such as relations between concepts, constraints, and axioms can be used for refining the feature ontology, providing the basis for improving the predictive performance of decision support.

Integrating machine learning algorithms with feature ontologies is especially important and beneficial in problem domains where the structure of the feature space is complex and significantly unbalanced, where the features are diverse and represent heterogeneous concepts. This is often the case with biomedical problems, and the task of prognosis of cardiological disease (ASD) development is a good representative of such a problem.

Figure 3 demonstrates an excerpt from the feature ontology for the problem of prognosis of ASD development. Some branches are marked with different weights, reflecting the relative importance of corresponding features (these weights are arbitrary and are used for the purpose of illustration only). The weights can be fine-tuned both by the expert and in an automated way using a machine learning technique (e.g., a genetic algorithm). Fine-tuning weights corresponding to different branches in the feature ontology for a particular problem may lead to the discovery of important problem-specific knowledge.

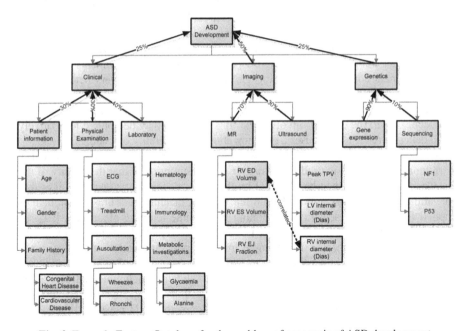

**Fig. 3.** Example Feature Ontology for the problem of prognosis of ASD development

Other information from the existing medical ontologies can be useful as well. For example, normal value ranges for different standard medical features can be extracted from ontologies (a light-blue box at the bottom demonstrates the normal range for Alanine in the figure). The normal value ranges are important for outlier removal and they may influence the distance metric as well. The feature ontology may also represent correlated or redundant features, which may have an influence on determining the inter-patient distance. In the given example, the blood volume of the right ventricle cavity can be determined using both ultrasound and magnetic-resonance images (MRI). The estimate received with MRI is usually more exact, and the ultrasound estimate may be ignored when MRI information is available.

## 6 Two Approaches to Distance Evaluation

There are two basic approaches to distance evaluation that can be applied to the validation and fine-tuning of any distance function in general, and a feature ontology-based distance in particular; evaluation based on expert-perceived similarity, and automatic data-driven wrapper-like evaluation.

The first approach was used in [20]. Its main idea consists in ranking a set of instances by a group of experts in a subject domain, according to the perceived similarity to another control instance (this process can be repeated for a number of control instances). Then the resulting rankings can be compared with the one produced by the distance function under study. For example, Spearman's rank correlation coefficient can be used for the comparison. The quality of the distance function is assumed here to be proportional to the average expert-function rank correlation (the bigger the average correlation between the expert- and distance function-produced ranks the better).

A serious drawback of this approach is the fact that the expert-perceived similarity may be rather subjective and context-dependent. However, experiments show that inter-expert rank correlation is usually significant enough even for very heterogeneous complex domains as in [20], so that such a comparison is appropriate. Inter-expert rank correlation may serve as a measure of expert agreement and partly validity of such an approach. Due to the experts' involvement, this approach may be applied to relatively small data sets only, which raises a question about the generality of findings.

Another approach is to use the distance function under study as an element in a learning algorithm that is used as a wrapper. The assumption is that the quality of the distance function will then be reflected by the performance of the learning algorithm on validation data. This approach is often used in machine learning research; it is applied for parameter selection and tuning in machine learning algorithms [21]. Its advantage is that the distance function is evaluated (or updated) in the context of the task being solved. Thus, for our example from the previous section, a good distance function should result in a better predictive performance (classification accuracy) of ASD classification. Any appropriate data-driven validation technique can be used in combination with an appropriate learning algorithm for wrapper-based distance evaluation. For example, cross validation together with $k$-NN classification can be used in our example.

One drawback of this approach is that enough data is needed in order to avoid potential overfitting and to provide valid evaluation. Thus, in some domains there might be simply not enough data for a separate validation set, and sometimes even for cross validation (for ASD behaviour prognosis, the number of instances is normally of the order of 10, which is significantly exceeded by the number of features, which are of the order of $10^4$ or even $10^5$). When enough data is available, this approach can be applied iteratively, to search for a better distance function in the space of valid distance functions.

# 7 Conclusions

In this paper we identify a problem, give a review of related work and propose one solution for the task of integration of machine learning techniques with existing ontological knowledge, that is especially important for biomedical domains where data is often naturally complex and is represented by a large heterogeneous feature-vector. Our main assumption is that structurising the feature space into a so-called *Feature Ontology* will reflect semantics of the domain and thus may help in improving the performance of machine learning techniques, through the redefined distance (similarity) function. We give an example for the task of prognosis of ASD development in children, and analyse two techniques that can be used for the evaluation and refining of a feature ontology. Beside the benefit in terms of improved predictive performance, the feature ontology may also become an important element of the Graphical User Interface, providing a means to data access and manipulation, in the context of the classification problem under consideration.

We would like to emphasise here that, in the context of this integration, the task of the creation of feature ontology becomes *central*, and this task is, unfortunately, not trivial at all as it may seem (especially taking into account the common complexity of biomedical problem domains). Some techniques were developed, in the area of ontology mapping, that may help to partially automate this process, though this process still remains largely routine and manual, needs a skilful expert and is based on the expert's knowledge and intuition. The feature ontology needs to be carefully developed, and it needs to focus on the classification task under study (the feature space should be structured with the classification task in mind), otherwise it is difficult to expect an improved similarity measure. The usual computer science principle "GIGO" (Garbage In, Garbage Out) works here as well. If enough data is available, data-driven feature ontology refinement may be applied, taking the expert ontology as a starting point in the search.

Our future work includes the creation and evaluation of feature ontologies for the medical problems within the Health-e-Child project. Another interesting direction for further research is the incorporation of various relations available in the existing ontologies in the distance calculation. For example, many ontologies include information about correlation between relevant features.

**Acknowledgements.** This work has been partially funded by the EU project Health-e-Child (IST 2004-027749). The authors wish to acknowledge support provided by all the members of the Health-e-Child consortium in the preparation of this paper.

# References

1. Assem, M., Menken, M., Schreiber, G., Wielemaker, J., Wielinga, B.: A method for converting thesauri to RDF/OWL. In: McIlraith, S.A., Plexousakis, D., van Harmelen, F. (eds.) ISWC 2004. LNCS, vol. 3298, pp. 17–34. Springer, Heidelberg (2004)
2. Azuaje, F., Bodenreider, O.: Incorporating ontology-driven similarity knowledge into functional genomics: an exploratory study. In: Proc. IEEE Symposium on Bioinformatics and Bioengineering, BIBE 2004, pp. 317–324. IEEE Press, Los Alamitos (2004)

3. Ashburner, M., et al.: Creating the gene ontology resource: design and implementation. Genome Research 11(8), 1425–1433 (2001)

4. Bader, G., Cary, M. (eds.): BioPAX – Biological Pathways Exchange Language, Level 2, Version 1.0 Documentation, BioPAX Working Group (2006) available at http://www.biopax.org

5. Baker, L.D., McCallum, A.K.: Distributional clustering of words for text classification. In: Proc. 21st ACM Int. Conf. on Research and Development in Information Retrieval SIGIR'98, pp. 96–103. ACM Press, New York (1998)

6. Bergmann, R., Kolodner, J., Plaza, E.: Representation in case-based reasoning. In: Knowledge Engineering Review, vol. 20, pp. 209–213. Cambridge University Press, Cambridge (2005)

7. Berrar, D., Sturgeon, B., Bradbury, I., Downes, C.S., Dubitzky, W.: Microarray data integration and machine learning methods for lung cancer survival prediction. In: 4th Int. Conf. Critical Assessment of Microarray Data Analysis, CAMDA, pp. 43–54 (2003)

8. Bodenreider, O.: The Unified Medical Language System (UMLS): integrating biomedical terminology. In: Nucleid Acids Research, vol. 31, pp. 267–270. Oxford University Press, Oxford, UK (2004)

9. Bolshakova, N., Azuaje, F., Cunningham, P.: Incorporating biological domain knowledge into cluster validity assessment. In: Rothlauf, F., Branke, J., Cagnoni, S., Costa, E., Cotta, C., Drechsler, R., Lutton, E., Machado, P., Moore, J.H., Romero, J., Smith, G.D., Squillero, G., Takagi, H. (eds.) EvoWorkshops 2006. LNCS, vol. 3907, pp. 13–22. Springer, Heidelberg (2006)

10. Camps-Valls, G., Gomez-Chova, L., Muñoz-Marí, J., Vila-Francés, J., Calpe-Maravilla, J.: Composite kernels for hyperspectral image classification. IEEE Geoscience and Remote Sensing Letters 3(1), 93–97 (2006)

11. Futschik, M.E., Sullivan, M., Reeve, A., Kasabov, N.: Prediction of clinical behaviour and treatment for cancers. Applied Bioinformatics 2(3), 53–58 (2003)

12. Goldbreich, C., Zhang, S., Bodenreider, O.: The foundational model of anatomy in OWL: experiences and perspectives. In: J. of Web Semantics: Science, Services, and Agents on the World Wide Web, vol. 4, pp. 181–195. Elsevier, North-Holland, Amsterdam (2006)

13. Gruber, T.: Towards principles for the design of ontologies used for knowledge sharing, Human and Computer Studies, vol. 43, pp. 907–928. Academic Press, San Diego (1995)

14. Hodge, G.: Systems of Knowledge Organization for Digital Libraries: Beyond Traditional Authority Files, The Digital Library Federation (2000)

15. Hanslik, A., Pospisil, U., Salzer-Muhar, U., Greber-Platzer, S., Male, C.: Predictors of spontaneous closure of isolated secundum atrial septal defect in children: a longitudinal study. Pediatrics 118(4), 1560–1565 (2006)

16. International Statistical Classification of Diseases and Related Health Problems, 10th revision (ICD-10), World Health Organization [classifications/apps/icd/ icd10online/], available at http://www.who.int/

17. Janecek, P., Pu, P.: Searching with semantics: an interactive visualization technique for exploring an annotated image collection. In: Proc. On The Move to Meaningful Internet Systems 2003: OTM 2003 Workshops. LNCS, vol. 2889, pp. 185–196. Springer, Heidelberg (2003)

18. Kalfoglou, Y., Schorlemmer, M.: Ontology mapping: the state of the art. In: Kalfoglou, Y., Schorlemmer, M., Sheth, A., Staab, S., Uschold, M. (eds.) Semantic Interoperability and Integration, Dagstuhl Seminar Proceedings 4391, IBFI (2005) [available at drops.dagstuhl.de/opus/volltexte/2005/40]

19. Louie, B., Mork, P., Martin-Sanchez, F., Halevy, A., Tarczy-Hornoch, P.: Data integration and genomic medicine. Methodological review, Biomedical Informatics 40, 5–16 (2007)
20. Melton, G., Parsons, S., Morrison, F., Rothschild, A., Markatou, M., Hripcsak, G.: Interpatient distance metrics using SNOMED CT defining relationships. Biomedical Informatics 39, 697–705 (2006)
21. Mitchell, T.M.: Machine Learning. McGraw Hill, New York (1997)
22. Moench, E., Ullrich, M., Schnurr, H., Angele, J.: SemanticMiner – ontology-based knowledge retrieval. Universal Computer Science 9(7), 682–696 (2003)
23. Nelson, S., Johnston, D., Humphreys, B.: Relationships in medical subject headings. In: Bean, C., Green, R. (eds.) Relationships in the Organization of Knowledge, pp. 171–184. Kluwer Academic, Boston, MA (2001)
24. Oleshchuk, V., Pedersen, A.: Ontology-based semantic similarity comparison of documents. In: DEXA Workshops 2003, pp. 735–738. IEEE CS Press, Los Alamitos, CA, USA (2003)
25. Panyr, J.: Thesauri, semantic nets, frames, taxonomies, ontologies – conceptual confusion or conceptional diversity? In: Harms, I., Luckhardt, D., Giessen, H. (eds.)Information and Language – Contributions from Computer Science, Computer Linguistics, Librarianship, and Related Disciplines, Saur-Verlag, pp. 139–152 (In German) (2006)
26. Rosse, C., Mejino, J.: A reference ontology for biomedical informatics: the foundational model of anatomy. Biomedical Informatics 36, 478–500 (2003)
27. Soualmia, L.F., Golbreich, C., Darmoni, S.J.: Representing the MeSH in OWL: towards a semi-automatic migration. In: Proc. 1st Int. Workshop on Formal Biomedical Knowledge Representation (KR-MED 2004), Whistler, Canada, pp. 81–87 (2004)
28. Stahl, A.: Learning of Knowledge-Intensive Similarity Measures in Case-Based Reasoning, Ph. D. Thesis, University of Kaiserslautern, Germany (2004)
29. Stearns, M., Price, C., Spackman, K., Wang, A.: SNOMED: clinical terms: overview of the development process and project status. In: Proc. Annual Symposium of American Medical Informatics Association, AMIA 2001, Hanley & Belfus, pp. 662–666 (2001)
30. Whetzel, P., Parkinson, H., Causton, H., Fan, L., Fostel, J., Fragoso, G., Game, L., Heiskanen, M., Morrison, N., Rocca-Serra, P., Sansone, S., Taylor, S., White, J., Stoeckert, C.: The MGED ontology; a resource for semantics-based description of microarray experiments. In: Bioinformatics, vol. 22, pp. 866–873. Oxford University Press, Oxford, UK (2006)
31. Zighed, D.A., Ras, Z.W. (eds.): In: Proc. 2nd IASC Workshop on Mining Complex Data, in conjunction with IEEE Int. Conf. on Data Mining ICDM 2006, Hong Kong (December 2006)

# Instance-Based Matching of Large Life Science Ontologies

Toralf Kirsten[1], Andreas Thor[2], and Erhard Rahm[1,2]

[1] Interdisciplinary Center for Bioinformatics, University of Leipzig, Germany
[2] Dept. of Computer Sciences, University of Leipzig, Germany
tkirsten@izbi.uni-leipzig.de, {thor,rahm}@informatik.uni-leipzig.de

**Abstract.** Ontologies are heavily used in life sciences so that there is increasing value to match different ontologies in order to determine related conceptual categories. We propose a simple yet powerful methodology for instance-based ontology matching which utilizes the associations between molecular-biological objects and ontologies. The approach can build on many existing ontology associations for instance objects like sequences and proteins and thus makes heavy use of available domain knowledge. Furthermore, the approach is flexible and extensible since each instance source with associations to the ontologies of interest can contribute to the ontology mapping. We study several approaches to determine the instance-based similarity of ontology categories. We perform an extensive experimental evaluation to use protein associations for different species to match between subontologies of the Gene Ontology and OMIM. We also provide a comparison with metadata-based ontology matching.

**Keywords:** Ontology matching, instance-based matching, match evaluation.

## 1 Introduction

Ontologies become increasingly important in life sciences application domains. Typically, they are used to semantically describe molecular-biological objects, e.g., to annotate genes and proteins with information on the functions and processes they are involved in. Ontologies also provide controlled vocabularies for a uniform naming of concepts to help reduce variations in terminology. Within an ontology, concepts are usually interrelated with is-a and part-of relationships resulting in specialization/ generalization and aggregation hierarchies (trees) or complex graphs of concepts. A very popular ontology is the Gene Ontology (GO) consisting of three (sub-) ontologies on molecular functions, biological processes and cellular components [7]. Genetic disorders are structured in Online Mendelian in Man (OMIM) [17].

The rapid increase in the number of life science data sources is accompanied by a similar growth in the number of ontologies and mappings between data sources and ontologies. This makes it increasingly valuable to match or align ontologies with each other to determine which of their concepts are semantically related. The resulting ontology mappings can be useful in many ways, in particular for enhanced analysis and annotation of genes, proteins or other objects of interest. For example, such objects may only be assigned to one particular ontology, say GO functions. An ontology

S. Cohen-Boulakia and V. Tannen (Eds.): DILS 2007, LNBI 4544, pp. 172–187, 2007.

mapping between GO functions and GO processes can then be useful to newly assign the objects to the second (process) ontology. Curators could thus use ontology mappings to find missing ontology annotations and get recommendations for possible ontology associations. Conversely, existing ontology associations could be validated against a newly determined ontology mapping in order to locate potential mis-associations reducing data quality. Ontology mappings are also helpful for explorative data analysis, e.g., to find objects with similar ontological properties as interesting targets for a comparative analysis.

Ontology matching is a general problem not limited to life sciences and has become an active research area (see Related Work section). Most previously proposed approaches to determine ontology mappings are metadata-based, i.e., they use the ontology representations themselves to find related concepts, in particular the names of concepts and contextual information like the names of the predecessor and successor concepts within the ontologies. Typically, name similarity is determined using generic (syntactical) string similarity functions on the names. However, in the absence of a globally standardized naming scheme such metadata-based approaches are of potentially little usefulness, especially for life science applications. This is because the same names may refer to completely different concepts while different names may describe the same concept. Furthermore, the concept granularities of different ontologies may widely differ so that comparing names may easily lead to correlations between incomparable concepts.

Figure 1 illustrates some of the problems for sample entries of the GO subontologies on molecular functions (MF) and biological processes (BP). We observe that in both subontologies there are highly similar concept names with partially opposite semantics, e.g., *Ion transporter activity* and *Anion transport* or *Organic anion transporter activity* and *Inorganic anion transport*. A name-based matching between molecular functions with biological processes would probably match these concepts despite potentially opposite semantics, e.g., *Ion* vs. *Anion* and *Organic* vs. *Inorganic*. This fact is also supported by [16] showing that nearly 65% of all concepts found in GO subontologies contain another GO concept as a proper substring. While more sophisticated matchers using helper ontologies like thesauri may somewhat reduce these problems there is no general solution due to the inherent difficulty to agree on common terms and constant creation of new terms.

We therefore advocate for instance-based match approaches which utilize existing associations between ontology concepts and instances, i.e., molecular–biological objects like proteins or genes that are described or annotated by the ontology concepts. This assumes that the real semantics of a concept is often better expressed by such existing object associations rather than metadata like the concept name. The example of Figure 1 shows such associations between species-specific proteins of the Ensembl data source [5] and describing concepts of the GO subontologies MF and BP and genetic disorders (GD) of OMIM. Intuitively, we assume that two concepts of different ontologies are related if their associated instances overlap, i.e., when the same instances are associated to them. The degree of concept similarity should take into account the number of shared associated objects or the relative size of the instance overlap.

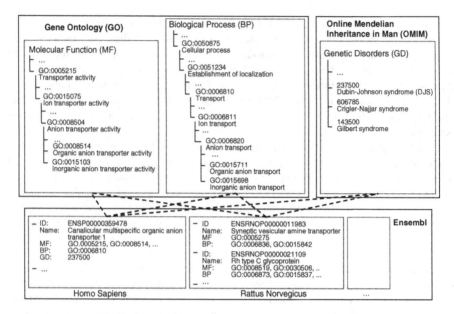

**Fig. 1.** Sample ontology entries and protein associations

We make the following contributions:

- We propose a simple yet powerful methodology for instance-based matching for life science ontologies utilizing existing associations between object data sources and ontologies. We outline several alternatives to determine the instance-based similarity between ontology concepts based on which the ontology mappings are determined. Each data source with associations to the ontologies to match can be used to derive a new ontology mapping. This way the domain-specific knowledge represented by the associations can be utilized to determine semantically meaningful ontology mappings.

- Our approach is flexible and extensible as several mappings between the same ontologies can be combined, e.g., mappings obtained for different data sources, species or similarity metrics. A combination with metadata-based match results is also feasible in order to improve recall and/or precision. Different ways for combining ontology mappings can be employed, e.g., based on intersection or union.

- We provide an extensive experimental evaluation for matching real ontologies, namely the three GO subontologies and OMIM, based on instance data for three species (human, mouse, rat). We consider direct associations between instances and concepts as well as indirect associations which take intra-ontology relationships into account. We also provide a comparison with metadata-based ontology matching. The evaluation utilizes new approximate recall and precision metrics in order to deal with the problem that the perfect ontology mappings are generally unknown.

The rest of the paper is organized as follows. Section 2 introduces the ontologies and instance associations used for our match evaluation. Section 3 presents the similarity metrics we use to derive and evaluate ontology mappings. In Section 4 we discuss the experimental results for instance-based ontology matching while Section 5 provides an experimental comparison with metadata-based ontology matching. Section 6 overviews related work and Section 7 concludes.

## 2 Match Scenario: Ontologies and Instance Associations

For our study, we assume that ontologies form a directed acyclic graph of concept nodes. The directed edges between concept nodes represent either *is-a* or *part-of* relationships. Concepts can have multiple associated instances, i.e., objects that are described or classified by the concept. An instance can be associated with multiple concepts, both leaf-level concepts but also to inner concepts of the ontology graph. Hence, the associations between objects (instances) and ontology concepts are of cardinality n:m.

Our experimental evaluation covers four popular life science ontologies: the three Gene Ontology (GO) subontologies on molecular functions, biological processes and cellular components, and genetic disorders of OMIM[1]. To match these ontologies with each other we use protein associations for three species: Homo Sapiens (human), Mus Musculus (mouse) and Rattus Norvegicus (rat). The protein data and ontology associations are obtained from the Ensembl data source (www.ensembl.org).

Table 1 provides base statistics on the considered ontologies, species-specific instance data sources and protein-concept associations. The number of concepts per ontology is shown on top, the number of proteins per species on the left. For instance,

**Table 1.** Quantity structure of utilized ontologies and instance sources[*]

| #concepts / #proteins / #assoc. | | | Gene Ontology | | | | | | OMIM[**] | |
|---|---|---|---|---|---|---|---|---|---|---|
| | | | Molecular Functions | | Biological Processes | | Cellular Components | | Genetic Disorders | |
| | | | 7,514 | | 12,555 | | 1,848 | | 6,535 | |
| Ensembl (direct assoc.) | Homo Sapiens | 43,605 | 34% | | 24% | | 34% | | 25% | |
| | | | 52% | 58,539 | 45% | 52,536 | 44% | 37,640 | 4% | 2,618 |
| | Mus Musculus | 32,078 | 31% | | 22% | | 32% | | 0% | |
| | | | 61% | 57,997 | 53% | 47,646 | 54% | 36,288 | 0% | 0 |
| | Rattus Norvegicus | 33,745 | 29% | | 22% | | 29% | | 0% | |
| | | | 38% | 29,665 | 33% | 25,703 | 31% | 18,519 | 0% | 0 |
| Ensembl (indirect assoc.) | Homo Sapiens | 43,605 | 39% | | 35% | | 43% | | 25% | |
| | | | 52% | 164,014 | 45% | 209,283 | 44% | 149,548 | 4% | 2,618 |
| | Mus Musculus | 32,078 | 36% | | 33% | | 40% | | 0% | |
| | | | 61% | 145,646 | 53% | 181,583 | 54% | 139,841 | 0% | 0 |
| | Rattus Norvegicus | 33,745 | 34% | | 32% | | 37% | | 0% | |
| | | | 38% | 85,429 | 33% | 107,022 | 31% | 75,919 | 0% | 0 |

[*]  Release states: GO 01/20/2007, OMIM 01/28/2007, Ensembl Release 42 Dec. 2006.
[**]  We focus on phenotype descriptions, i.e., entries marked with #, % and without a mark. Please see http://www.ncbi.nlm.nih.gov/Omim/mimstats.html for more details.

---

[1] OMIM was not originally developed as an ontology but provides a comprehensive set of terms (including term definitions, comments and associated literature) describing genetic disorders which are frequently associated with objects of other data sources. Therefore, OMIM plays an ontology-like role in our evaluation study.

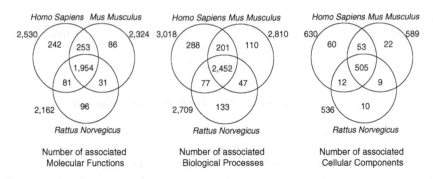

**Fig. 2.** Quantity structure of ontology concepts with at least one associated protein in three selected species

there are 7,514 molecular function concepts in GO and 43,605 human proteins in Ensembl. Furthermore, Table 1 contains the number of associations between proteins and ontology concepts which we separate in direct and indirect associations. Direct associations refer to the original associations recorded in Ensembl and assign objects to the most specific concept of an ontology. For example, there are 58,539 direct associations between human proteins and molecular functions and covering 52% of the human proteins and 34% of the functions. Hence, human protein associations support instance-based matching for up to 34% of the MF concepts. To increase the number of concepts that may be matched we also consider indirect associations which take into account the intra-ontology relationships between concepts. For this we simply assign the direct instances of a concept $c$ also to its parents and grandparents within the ontology graph. In the example this provides human protein instances to 39% of the function concepts, however at the expense of a massive increase in the number of object associations (164,014).

We observe that the available object associations cover significant portions of the ontologies (25-39%) so that instance-based matching promises to provide many correspondences between concepts. While OMIM has associations only for human proteins, the GO ontologies are well connected to all three species. There is a similar number of object associations for human and mouse proteins while the coverage for rat is somewhat reduced. On average, an Ensembl protein is directly assigned to 1.5-3.0 concepts of the GO subontologies. The average number of directly associated proteins per GO concept varies between 9 and 62 per species.

Figure 2 illustrates the species-specific distribution of object associations for the three GO subontologies. For example, we observe that 1,954 molecular functions have protein associations to all three species, whereas merely 86 functions are exclusively associated with mouse proteins. On average over 80% of the matched concepts, i.e., functions, processes, or components, are assigned to all three species. Considering species-specific associations is also helpful to determine species-specific ontology mappings. Furthermore, analysis tasks can benefit from focusing on species-specific concepts, e.g., to analyze an ontology mapping for the 86 mouse-specific GO functions with respect to the 110 mouse-specific processes.

## 3 Similarity and Evaluation Metrics

In order to match two ontologies $O_1$ and $O_2$ we need metrics to determine the similarity between $O_1$ concepts and $O_2$ concepts. All pairs of concepts from $O_1$ and $O_2$ for which the similarity exceeds a certain minimal threshold are called *correspondences* and included in the match result (ontology mapping). The key idea of our instance-based approach to ontology matching is to derive the similarity between concepts from the number of shared instances, i.e., the number of instances associated to both concepts. An important advantage for instance-based ontology matching is that the number of instance associations is typically higher than the number of concepts. This way the match accuracy of the approach can become rather robust against some wrong instance classifications. As discussed, another key advantage is that the instance-based approach is independent from concept names and other metadata.

In the following we first present the used instance-based similarity metrics. We then discuss how to assess the quality of a match result in the absence of a perfect mapping.

### 3.1 Instance-Based Similarity Metrics

In this paper we study four metrics for determining the instance-based similarity between concepts $c_1 \in C_{O1}$ and $c_2 \in C_{O2}$ of different ontologies $O_1$ and $O_2$, namely *baseline*, *minimum*, *dice*, and *kappa* similarity. Most of these metrics are well-know and have already been used in previous match studies (e.g., [8, 21]) however, not yet for an instance-based matching of life science ontologies. To define the similarity of two concepts $c_1$ and $c_2$ we use the number of instances that are (or are not) associated to $c_1$ and $c_2$. Figure 3 illustrates all relevant combinations for the instance cardinalities.

For example, $\cdots\cdots$ is the number of instances which are associated to $c_1$ but not associated to $c_2$. Furthermore, $N_{c_1} \left( N_{\overline{c_1}} \right)$ is the total number of instances that are (not) as $N_{c_1 \overline{c_2}}$ associated to $c_1$. Note that these numbers may be used either for directly associated instances as well as for indirectly associated instances.

The *baseline similarity* metric already matches two concepts $c_1$ and $c_2$ if they share at least one object.

$$Sim_{Base}(c_1, c_2) = \begin{cases} 1 & , \text{ if } N_{c_1 c_2} > 0 \\ 0 & , \text{ if } N_{c_1 c_2} = 0 \end{cases} \in [0...1], \forall c_1 \in C_{O1}, c_2 \in C_{O2}$$

|  | $i \in c_2$ | $i \notin c_2$ | $\Sigma$ |
|---|---|---|---|
| $i \in c_1$ | $N_{c_1 c_2}$ | $N_{c_1 \overline{c_2}}$ | $N_{c_1}$ |
| $i \notin c_1$ | $N_{\overline{c_1} c_2}$ | $N_{\overline{c_1} \overline{c_2}}$ | $N_{\overline{c_1}}$ |
| $\Sigma$ | $N_{c_2}$ | $N_{\overline{c_2}}$ | $N$ |

**Fig. 3.** Matrix of all possible combinations for the number of shared instances $i$ for two concepts $c_1 \in C_{O1}$ and $c_2 \in C_{O2}$

The baseline approach poses minimal requirements to match two concepts so that it can be expected to provide the maximal number of correspondences for instance-based matching. To focus on concept combinations with a higher instance overlap it is necessary to take into account the number of instances per concept.

The *dice similarity metric* [19] considers the concept cardinalities and the number of shared instances:

$$Sim_{Dice}(c_1,c_2) = \frac{2 \cdot N_{c_1 c_2}}{N_{c_1} + N_{c_2}} \in [0...1], \forall c_1 \in C_{O1}, c_2 \in C_{O2}$$

A high dice value indicates a significant instance overlap w.r.t. to both concepts.

A potential limitation of the dice metric is that it can become quite small in case of larger cardinality differences, even if all instances of the smaller concept match to another concept. This aspect is taken care of by the *minimum similarity* metric which determines the relative instance overlap with respect to the smaller-sized concept:

$$Sim_{Min}(c_1,c_2) = \frac{N_{c_1 c_2}}{\min(N_{c_1}, N_{c_2})} \in [0...1], \forall c_1 \in C_{O1}, c_2 \in C_{O2}$$

Our last metric – the *kappa similarity* – is somewhat more complex and adopted from Cohen's kappa coefficient [6]; it has also been adopted in [8] for an e-commerce application. The kappa coefficient measures the agreement of two raters classifying items (e.g., instances) into categories (e.g., concepts). We adopt the kappa coefficient to calculate two probabilities $P$ and $P'$. $P$ is the agreement among both concepts, i.e., the relative number of shared instances combined with the number of instances that do not appear in any of the two concepts. $P'$ is the probability that the agreement that one instance is assigned to both concepts is due to chance. Therefore P and P' are defined as follows:

$$P = \frac{N_{c_1 c_2} + N_{\overline{c_1 c_2}}}{N} \qquad\qquad P' = \frac{N_{c_1} \cdot N_{c_2} + N_{\overline{c_1}} \cdot N_{\overline{c_2}}}{N^2}$$

The kappa similarity for two concepts $c_1$ and $c_2$ is then defined as:

$$Sim_{Kappa}(c_1,c_2) = \frac{P - P'}{1 - P'} \in [0...1], \forall c_1 \in C_{O1}, c_2 \in C_{O2}$$

To test the significance of a match between the two concepts $c_1 \in C_{O1}$ and $c_2 \in C_{O2}$, we can utilize a test distribution Z as proposed in [8]. Z is defined as follows:

$$Z = Sim_{Kappa}(c_1,c_2) \cdot \sqrt{\frac{(N_{c1c2} + N_{c1\overline{c2}} + N_{\overline{c1}c2} + N_{\overline{c1c2}})(1 - P')}{P'}}$$

Z follows a normal distribution so that it can be compared with the standard normal distribution. A significant match correspondence can be assumed if Z exceeds the percentile of the standard distribution for a given significance level.

It can easily been shown that for all correspondences between concepts $c_1$ and $c_2$, it holds:

$$Sim_{DICE}(c_1,c_2) \le Sim_{MIN}(c_1,c_2) \le Sim_{Base}(c_1,c_2) \text{ and}$$
$$Sim_{Kappa}(c_1,c_2) \le Sim_{Base}(c_1,c_2)$$

## 3.2 Evaluation Metrics

To evaluate the quality of a match result and thus the effectiveness of a match approach it is necessary to determine whether all real correspondences have been determined (completeness, high recall) and whether all determined correspondences are real correspondences (correctness, high precision). Exactly determining recall and precision thus requires the perfect match result to be known. Unfortunately, the perfect match result is generally unknown for large real-life match problems, especially for life science ontologies. A manual construction of a perfect match is also too laborious and extremely difficult for broad ontologies such as the Gene Ontology. For our evaluation we therefore focus on the relative quality of the differently obtained match results and use rough approximations for recall and precision.

With respect to recall or completeness we consider the so-called *match coverage,* i.e., the share of concepts that is covered by an ontology mapping, i.e., for which there is at least one correspondence in the match result. Let $C_{O1\text{-}Match}$ ($C_{O2\text{-}Match}$) be the set of matched concepts of ontology $O_1$ ($O_2$) and $C_{O1}$ ($C_{O2}$) the set of all concepts of ontology $O_1$ ($O_2$). We then define the match coverage of ontology $O_1$ ($O_2$) as follows:

$$MatchCoverage_{O_1} = \frac{|C_{O_1-Match}|}{|C_{O_1}|} \qquad MatchCoverage_{O_2} = \frac{|C_{O_2-Match}|}{|C_{O_2}|}$$

Match coverage can be determined for any match approach, in particular both metadata-based and instance-based schemes. For instance-based approaches the maximal coverage is limited by the number of concepts which have at least one associated instance (w.r.t. the considered instance data source). To take this into account we additionally determine the *instance match coverage* which is defined as the ratio of the matched concepts w.r.t. to the concepts having at least one associated instance. Let $C_{O1\text{-}Inst}$ ($C_{O2\text{-}Inst}$) be the set of concepts of ontology $O_1$ ($O_2$) having at least one associated instance. We then define the $O_1$-specific and $O_2$-specific instance match coverage as follows:

$$InstMatchCoverage_{O_1} = \frac{|C_{O_1-Match}|}{|C_{O_1-Inst}|} \qquad InstMatchCoverage_{O_2} = \frac{|C_{O_2-Match}|}{|C_{O_2-Inst}|}$$

In addition, we can define the combined instance match coverage for a match result:

$$InstMatchCoverage = \frac{|C_{O_1-Match}| + |C_{O2-Match}|}{|C_{O_1-Inst}| + |C_{O_2-Inst}|}$$

For estimating the precision of a match approach we determine the so-called *match ratio,* i.e., the ratio between the number of found correspondences and the number of matched concepts:

$$MatchRatio_{O1} = \frac{|Corr_{O1-O2}|}{|C_{O1-Match}|} \qquad MatchRatio_{O2} = \frac{|Corr_{O1-O2}|}{|C_{O2-Match}|}$$

Analogously we define the combined *match ratio.*

$$MatchRatio = \frac{2 \cdot |Corr_{O1-O2}|}{|C_{O1-Match}| + |C_{O2-Match}|}$$

In the above formulas, $Corr_{O1-O2}$ denotes the set of found correspondences in a match result. The intuition is that the precision (and thus value) of a match result is better if a concept is not loosely matched to many other concepts but only to fewer (preferably the most similar) ones. The match ratio for the baseline matcher is expected to provide a worst-case value for instance-based matching and can thus be used as a yardstick.

## 4   Instance-Based Match Results

We first analyze different instance-based match results using direct association. We then study the impact of combining different match results (mappings) and the use of indirect associations.

### 4.1   Match Results Using Direct Association

We applied the introduced instance-based similarity metrics to determine ontology mappings between the three GO ontologies on molecular functions (MF), biological processes (BP), cellular components (CC) and genetic disorders (GD) of OMIM. We thus solved six match tasks: three to match between the GO subontologies (MF-BP, MF-CC, BP-CC) and three GO-OMIM match tasks (MF-GD, BP-GD, CC-GD). As discussed in Section 2, we utilize the Ensembl protein associations for the three species Homo Sapiens, Mus Musculus and Rattus Norvegicus and first focus on direct associations. The three similarity metrics $Sim_{Base}$, $Sim_{Min}$, and $Sim_{Dice}$ are evaluated with a high similarity threshold of 1.0; for $Sim_{Kappa}$ we applied a significance level of 95%.

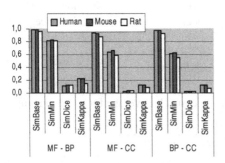

**Table 2.** Match Ratios of GO ontology mappings (direct associations; Homo Sapiens)

|  | MF – BP | | MF – CC | | BP – CC | |
|---|---|---|---|---|---|---|
|  | MF | BP | MF | CC | BP | CC |
| Base | 20.4 | 17.0 | 7.6 | 28.6 | 9.8 | 46.3 |
| Min | 4.4 | 4.0 | 2.2 | 7.8 | 2.4 | 8.6 |
| Dice | 1.3 | 1.2 | 1.0 | 1.3 | 1.0 | 1.3 |
| Kappa | 2.0 | 2.0 | 1.9 | 2.7 | 1.7 | 2.6 |

**Fig. 4.** Combined Instance Match Coverage of GO ontology mappings (direct associations)

Figure 4 illustrates the obtained values for combined *instance match coverage* for the three GO match tasks and the three considered species. Table 2 shows the corresponding *match ratios* for Homo Sapiens; the match ratios for the other species are similar and omitted due to space constraints. We observe that there are big differences between the considered similarity metrics while the match coverage results are very similar for the three species. The latter is because the species-specific proteins match the same concepts to a large degree (as noted in Section 2) so that the derived ontology mappings are also very similar for a given similarity metric and match task. As expected the baseline similarity metric $Sim_{Base}$ achieved the best coverage (recall) and worst match ratios (precision) for all match tasks. Its instance match coverage is up to 99% (for Homo Sapiens and the MF – BP match) so that almost every concept with an associated instance is matched. On the other hand, match ratios achieve values between about 8 and 46, i.e., concepts are mapped to many other concepts indicating a low precision. On the other hand, $Sim_{Dice}$ and $Sim_{Kappa}$ turn out to be very restrictive with match ratios close to 1.0. This is they focus on the best matching concepts. Unfortunately this is only achieved for very few correspondences so that the match coverage remains rather low (around 5-10% for $Sim_{Dice}$ and 10-20% for $Sim_{Kappa}$). For all match tasks the metric $Sim_{Min}$ achieves very promising precision/recall values which lie between the extreme cases discussed so far. In particular instance match coverage is as good as between 60-80% while match ratios are significantly lower than for $Sim_{Base}$. On average, a concept is matched with 2–9 concepts of another ontology which is still a reasonably low number, e.g., to be checked by a biologist.

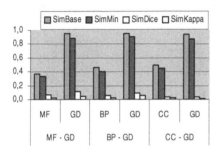

**Fig. 5.** Instance Match Coverage of GO-GD mappings (direct associations; Homo Sapiens)

**Table 3.** Match Ratios of GO-GD mappings (direct associations; Homo Sapiens)

|  | MF – GD | | BP – GD | | CC – GD | |
|---|---|---|---|---|---|---|
|  | MF | GD | BP | GD | CC | GD |
| Base | 7.1 | 4.3 | 2.5 | 6.3 | 2.5 | 3.4 |
| Min | 5.9 | 3.5 | 2.5 | 4.6 | 1.7 | 3.4 |
| Dice | 1.6 | 1.5 | 1.1 | 1.5 | 1.4 | 1.4 |
| Kappa | 1.4 | 1.2 | 1.1 | 1.2 | 1.1 | 1.2 |

The GO-OMIM match tasks are only performed for the species Homo Sapiens since there are no protein associations to OMIM for the other two species. Figure 5 shows the instance match coverage for the three match tasks; Table 3 illustrates the corresponding match ratios. For these experiments (and in contrast to the previous match tasks) we observed substantial coverage differences for the individual ontologies so that we indicate the ontology-specific coverage values in Figure 5. We observe that for both $Sim_{Base}$ and $Sim_{Min}$ the instance match coverage of the GO ontologies is only about half of the instance coverage of GD (40-50% vs. more than 88%). The reasons are twofold. On the one hand, many proteins are associated with concepts of the GO ontologies but have no correspondence to OMIM. On the other hand, if a protein is associated with OMIM then it is mostly also connected with a concept of the GO ontologies. For instance, there are 20,936 proteins of the Homo Sapiens that have at least one molecular function, but only 1,581 of these proteins are associated with a genetic disorder. Conversely, only 110 human proteins are described by a genetic disorder but not by a molecular function.

The relative outcome for the different similarity metrics is in agreement with the observations made for the previous match tasks. While $Sim_{Base}$ and $Sim_{Min}$ have a

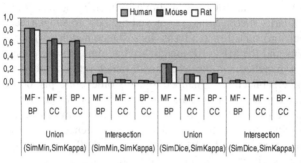

**Fig. 6.** Combined Instance Match Coverage of combined GO ontology mappings

**Table 4.** Match Ratios of combined GO ontology mappings (Homo Sapiens )

| | Match | | $\cup$ | $\cap$ |
|---|---|---|---|---|
| Min – Kappa | MF-BP | MF | 4.6 | 1.2 |
| | | BP | 4.2 | 1.3 |
| | MF-CC | MF | 2.4 | 1.0 |
| | | CC | 8.0 | 1.3 |
| | BP-CC | BP | 2.5 | 1.0 |
| | | CC | 8.8 | 1.2 |
| Dice – Kappa | MF-BP | MF | 1.8 | 1.1 |
| | | BP | 1.8 | 1.1 |
| | MF-CC | MF | 1.8 | 1.0 |
| | | CC | 2.6 | 1.3 |
| | BP-CC | BP | 1.6 | 1.0 |
| | | CC | 2.4 | 1.5 |

relatively high recall (instance match coverage), the metrics $Sim_{Dice}$ and $Sim_{Kappa}$ are very restrictive but precise (only about 1 to 2 correspondences per matched concept on average).

### 4.2 Combining Ontology Mappings

The match results discussed so far were each derived for a certain similarity metric and a species-specific set of instances. Combining several such ontology mappings for a given match task is a promising way to obtain an improved ontology mapping, e.g., with improved recall and/or precision. For example, taking the union of two independently derived ontology mappings is likely to improve recall (coverage) while building the intersection can improve precision. Other combination strategies are also conceivable (e.g., weighted or majority-based selection of correspondences) but are not further considered in this paper.

To illustrate the idea we analyze the combination of mappings obtained for different similarity metrics. This is not useful for all metrics since according to Section 3 all instance-based similarity measures generate subsets of correspondences of the base-line approach and $Sim_{Dice}$ produces a subset of correspondences of $Sim_{Min}$ . Therefore, we comparatively study the intersection and union of the ontology mappings generated by $Sim_{Dice}$ ($Sim_{Min}$) and $Sim_{Kappa}$.

Fig. 6 depicts the instance match coverage of the combined mappings between GO ontologies, while Table 4 shows the corresponding match ratios (for Homo Sapiens). We observe that the union mappings for $Sim_{Min}$ and $Sim_{Kappa}$ only slightly improve coverage (84%) compared to $Sim_{Min}$ (81%). The match ratios are also not significantly higher than for $Sim_{Min}$ alone (Table 2). This is because $Sim_{Min}$ alone achieved already a high coverage so that $Sim_{Kappa}$ could add only few new correspondences. On the other hand, the union mapping between $Sim_{Dice}$ and $Sim_{Kappa}$ is very effective and more than doubles coverage (30%) compared to $Sim_{Dice}$ alone (12%). The match ratio still remains low (1.8–2.6) indicating a high-quality ontology mapping.

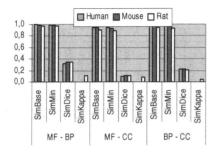

**Table 5.** Match Ratios of GO ontology mappings (indirect associations; Homo Sapiens)

| | MF – BP | | MF – CC | | BP – CC | |
|---|---|---|---|---|---|---|
| | MF | BP | MF | CC | BP | CC |
| Base | 90.2 | 60.4 | 27.2 | 96.2 | 38.8 | 210.2 |
| Min | 16.6 | 10.9 | 7.4 | 23.3 | 6.6 | 33.9 |
| Dice | 1.9 | 1.3 | 1.2 | 1.7 | 1.6 | 1.8 |
| Kappa | 6.7 | 5.1 | 5.5 | 7.6 | 6.3 | 11.7 |

**Fig. 7.** Combined Instance Match Coverage of GO ontology mappings (indirect associations)

### 4.3 Match Results Using Indirect Instance Associations

Another way to improve match coverage is to not only consider direct but also indirect object associations. As already discussed in Section 2 (Table 1), this increases the number of concepts for which instance-based matching can be applied (e.g., the number of GO processes with associated instances is increased by 45%). Although we restrict the propagation of object associations to two levels (parents, grandparents) the number of object associations is increased by almost a factor of 3 compared to direct associations.

Figure 7 shows the instance match coverage results for the GO match tasks using indirect associations; Table 5 illustrates the corresponding match ratios (for the species Homo Sapiens). The coverage for $Sim_{Base}$ was already high for direct associations; the use of indirect associations primarily is thus little helpful but leads to extremely high match ratios (27–210). For $Sim_{Min}$, on the other hand, the instance match coverage improvement is substantial, e.g., from 61% (direct) to 86% (indirect) for the match BP - CC. However, match ratios are also increased, e.g., from 6 (direct, MF) to almost 17 (indirect, MF) for matching MF with BP using $Sim_{Min}$ and human proteins.

The results suggest that the use of indirect associations can be helpful but also be harmful. Hence we see a need for more sophisticated approaches to intelligently make use of intra-ontology relationships in combination with instance-based matching. One idea is to restrict the use of indirect associations to concepts that remain otherwise unmatched. Another option is to use direct associations to determine instance-based concept similarities which are then propagated along intra-ontology relationships by a context matcher [18].

## 5   Metadata-Based Match Results

### 5.1   Metadata Match Results Using Concept Names

For comparison purposes we also use a simple metadata-based matcher to determine mappings between the considered ontologies. We apply a name matcher based on trigram similarity for comparing pairwise the concept names of different ontologies. Table 6 shows the name matcher results for the six match tasks by using the trigram similarity and different thresholds ($\geq 0.5$). Note that the match coverage values refer to all concepts not only to the ones with instances.

**Table 6.** Name matching results between selected ontologies

| Sim$_{Trigram}$ | | | 0.5 | | | 0.6 | | | 0.7 | | | 0.8 | | |
|---|---|---|---|---|---|---|---|---|---|---|---|---|---|---|
| Match | | # Correspondences | Match Coverage | Match Ratio | # Correspondences | Match Coverage | Match Ratio | # Correspondences | Match Coverage | Match Ratio | # Correspondences | Match Coverage | Match Ratio |
| MF – BP | MF | 15,415 | 47% | 4.4 | 2,770 | 15% | 2.4 | 602 | 6% | 1.4 | 69 | <1% | 1.1 |
| | BP | | 18% | 6.9 | | 8% | 2.9 | | 3% | 1.4 | | <1% | 1.1 |
| MF – CC | MF | 2,663 | 14% | 2.5 | 1,274 | 6% | 2.7 | 225 | 3% | 1.1 | 31 | <1% | 1.1 |
| | CC | | 23% | 6.3 | | 15% | 4.6 | | 8% | 1.5 | | 1% | 1.2 |
| BP – CC | BP | 2,563 | 8% | 2.5 | 693 | 3% | 1.7 | 175 | 1% | 1.4 | 32 | <1% | 1.1 |
| | CC | | 40% | 3.4 | | 17% | 2.0 | | 7% | 1.4 | | 2% | 1.2 |
| MF – GD | MF | 667 | 7% | 4.4 | 124 | 2% | 1.1 | 27 | <1% | 1.0 | 1 | <1% | 1.0 |
| | GD | | 2% | 6.9 | | 1% | 2.1 | | <1% | 1.1 | | <1% | 1.0 |
| BP – GD | BP | 1,400 | 9% | 1.3 | 174 | 1% | 1.0 | 11 | <1% | 1.0 | 2 | <1% | 1.0 |
| | GD | | 2% | 8.8 | | <1% | 5.0 | | <1% | 1.1 | | <1% | 1.0 |
| CC – GD | CC | 364 | 11% | 1.8 | 36 | 2% | 1.2 | 1 | <1% | 1.0 | 0 | 0% | 0.0 |
| | GD | | 1% | 5.9 | | <1% | 2.3 | | <1% | 1.0 | | 0% | 0.0 |

We observe a rather low number of correspondences especially for a similarity threshold of 0.7 or higher. This indicates a high diversity in the concept names so that name matching is not very effective. There are no correspondences with a threshold of 0.9 or greater (not shown in Table 6). The match coverage and match ratios grow for smaller similarity thresholds but probably due to many wrong correspondences.

Most correspondences are found between molecular functions and biological processes which are the largest ontologies considered (Table 1). As already indicated by the examples in Figure 1 many similar terms only differ in pre-/suffixes or an additional word, such as *activity* for naming a function. Moreover, in many cases concepts inherit their name from their parents and use an additional term representing the specialization, such as *transport, anion transport* (both BP), *transporter activity* and *anion transporter activity* (both MF). Hence, if the additional word is short enough then concepts from different levels are matched, e.g., *anion transport* with *transport activity*. Of course, a low threshold (e.g., 0.5) can lead to the generation of false correspondences, e.g., between the function *Inorganic anion transporter activity* (MF) and the process *Organic anion transport* (BP) due to a trigram similarity of 0.66.

Most correspondences for OMIM GD are found for the GO subontology on biological processes. The reason is that some genetic disorders refer to biological processes, such that their names only differentiate in modified suffixes or additional words. For instance, the concepts *vitamin A metabolism* (BP) and *vitamin A metabolic defect* (GD) are matched with a trigram similarity of 0.72. Of course, low threshold values also lead to false positives matches, such as *betaine transport* (BP) and *citrulline transport defect* (GD) with a trigram similarity of 0.5.

## 5.2  Comparison Between Metadata and Instance-Based Matching

To study the relationship between metadata- and instance-based matchers, we analyze the union and intersection (overlap) of the generated ontology mappings. For this purpose, we combine the name matcher results (threshold ≥ 0.5) with the instance-based results using the similarity metric Sim$_{Min}$ (similarity threshold = 1) and direct instance associations. Figure 8 shows the match coverage per ontology for the union results (species Homo Sapiens). The highest coverages are achieved for molecular functions (approx. 60%) in the combined MF–BP match result and for cellular components (54%) in the BP–CC result, both when using a trigram similarity of 0.5. These high coverage values are mainly due to the name matcher. According to Table 6, the name-based correspondences for threshold 0.5 cover already 47% of the functions (match MF-BP) and 40% of the components (match BP-CC). For trigram thresholds of 0.6 and higher, match coverage is primarily influenced by instance-based matching using $Sim_{Min}$. This is also the case for the unified match results between GO subontologies and OMIM; around 22% of the genetic disorders and between 11% (MF, BP) and 15% (CC) of GO subontology concepts are covered by using $Sim_{Min}$.

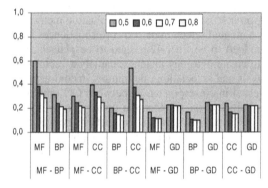

**Fig. 8.** Match Coverage per ontology for unified matches of $Sim_{Name}$ and $Sim_{Min}$ (Homo Sapiens)

**Table 7.** Match Ratios of combined ontology mappings between $Sim_{Name}$ (0.7) and $Sim_{Min}$ (Homo Sapiens)

| Match | | ∪ | ∩ |
|---|---|---|---|
| MF-BP | MF | 4.1 | 1.0 |
| | BP | 3.7 | 1.0 |
| MF-CC | MF | 2.2 | 1.0 |
| | CC | 6.7 | 1.0 |
| BP-CC | BP | 2.4 | 1.0 |
| | CC | 7.6 | 1.3 |
| MF-GD | MF | 5.8 | 1.0 |
| | GD | 3.4 | 1.0 |
| BP-GD | BP | 5.4 | 0.0 |
| | GD | 4.5 | 0.0 |
| CC-GD | CC | 12.9 | 0.0 |
| | GD | 2.5 | 0.0 |

The match coverage of the intersection results is in most cases only 1% or less (and therefore not shown in an extra plot). This is because the name-based and instance–based match results have only a very low number of correspondences in common. Especially for a lower trigram threshold (0.5) the vast majority of name correspondences has no corresponding instance similarity.

Table 7 illustrates the achieved match ratios for both, the union and intersection of the ontology mappings generated by the name matcher (similarity threshold = 0.7) and $Sim_{Min}$. We observe a moderate ratio (mostly less then 6) for the union results while the ratios for mapping intersection is seldom larger than 1.0. This is influenced by the fact that the number of correspondences is very low. The intersection of the mappings between genetic disorders and biological processes (cellular components) is even empty, therefore the match ratios also equal zero.

The experiment shows that simple name matching is not very effective and less promising than the proposed instance-based approaches. Still we believe that more sophisticated metadata-based matchers may be helpful to complement instance-based matching and leave the investigation of such combined approaches for future work.

# 6 Related Work

Overviews of approaches for ontology matching in general are given in [18, 10, 2, 20]. Typically, matching utilizes metadata, associated instances or both. The match approaches [13, 15, 12, 1] are based on metadata, such as concept names, synonyms and descriptions, and applied in different domains. More specific to bioinformatics, [4] utilizes a metadata matcher to link GO with ChEBI, an ontology of chemical entities for biological interest.

Instance-based ontology matching is investigated in [8, 9, 3, 11]. They follow statistical or machine learning approaches and apply them in different application domains. [8] focuses on integrating internet catalogs, represented by hierarchical collections of web links. Similar to our study, it applies the Kappa similarity metric including a significance test. [9] applies decision trees and Bayesian networks to create matches between GO subontologies that is different to our approach. It uses available annotations (instances) of two species (mouse and human) as training data and for cross validation to test the models. In contrast to our approach using the proposed evaluation metrics, the predicted match result is evaluated by a manual selection of 100 correspondences which are then validated by an expert (41 judged to be true, 42 judged to be plausible). [3] utilizes three non-lexical approaches to create ontology matches, namely a vector space model, a statistical co-occurrence analysis and association rule mining. In contrast to our match application where we are interested in correspondences between GO ontologies, they associate GO concepts without a distinction whether the concept is a function, process or component. Therefore, the result can also contain associations between concepts of the same GO subontology, e.g., between two functions. [11] applies association rule mining and formal ontological concepts to create mappings between the GO subontologies whereas we use simple and comprehensible metrics for ontology matching.

[14] is a mixed match approach, i.e., it follows lexicographic and instance-based approaches, with the goal to create a second ontology layer that maps the GO subontologies. Instead of using complete concepts names as we have applied they create specific patterns for the metadata-based matching such that ontology-specific words (e.g., *activity* for molecular functions) are ignored. Moreover, it applies association rule mining by using available gene annotations and reuses existing associations to metabolic pathways to create ontology matches. In contrast to our match scenario, the generated matches are validated by human experts.

# 7 Conclusions

We proposed the use of simple instance-based approaches for matching life science ontologies. The idea is to utilize the domain knowledge expressed in existing object-ontology associations for finding related concepts in different ontologies. The approach is extensible as ontology mappings obtained for different match approaches or different instance sources (e.g., different species) can be combined to improve overall recall or precision. We experimentally evaluated four alternatives for instance-based matching and one metadata-based approach for six match tasks involving the GO subontologies and OMIM. We observed that instance-based matching using the $Sim_{Min}$ metric achieves a high match coverage while limiting the number of correspondences per matched concept.

In future work, we will further study combined approaches for ontology matching and the interplay between instance-based and metadata-based matching in life sciences. We also plan applications that utilize the computed ontology mappings and gather user feedback to help validate the proposed match correspondences.

# References

1. Aumüller, D., Do, H.-H., Massmann, S., Rahm, E.: Schema and ontology matching with COMA++. In: Proc. ACM SIGMOD (2005)
2. Avesansi, A., Giunchiglia, F., Yatskevich, M.Y.: A large taxonomy mapping evaluation. In: Gil, Y., Motta, E., Benjamins, V.R., Musen, M.A. (eds.) ISWC 2005. LNCS, vol. 3729, Springer, Heidelberg (2005)
3. Bodenreider, O., Aubry, M., Bugrun, A.: Non-lexical approaches to identifying associative relations in the Gene Ontology. In: Proc. Pacific Symposium on Biocomputing (2005)
4. Bodenreider, O., Bugrun, A.: Linking the Gene Ontology to other biological ontologies. In: Proc. ISMB meeting on Bio-Ontologies (2005)
5. Hubbard, T., Andrews, D., Caccamo, M., et al.: Ensembl 2005. Nucleic Acid Research 33(Database Issue), D447–D453 (2005)
6. Cohen, J.: A coefficient of agreement for nominal scales. Educational and Psychological Measurement 20, 37–46 (1960)
7. The Gene Ontology Consortium: The Gene Ontology (GO) database and informatics resource. Nucleic Acids Research, vol. 32, pp. D258–D261 (2004)
8. Ichise, R., Takeda, H., Honiden, S.: Integrating multiple internet directories by instance-based learning. In: Proc. 18th Intl. Joint Conf. on Artificial Intelligence (IJCAI) (2003)
9. King, O.D., Fougler, R.E, Dwight, S.S., et al.: Predicting gene function from patterns of annotation. Genome research 13(5), 896–904 (2003)
10. Kalfoglou, Y., Schorlemmer, M.: Ontology mapping: The state of the art. The Knowledge Engineering Review Journal 18(1), 1–31 (2003)
11. Kumar, A., Smith, B., Borgelt, C.: Dependence relationships between Gene Ontology terms based on TIGR gene product annotations. In: Proc. 3rd Intl. Workshop on Computational Terminology (CompuTerm) (2004)
12. Mork, P., Bernstein, P.: Adapting a generic match algorithm to align ontologies of human anatomy. In: Proc. 20th Intl. Conf. on Data Engineering (ICDE) (2004)
13. Maedche, A., Staab, S.: Measuring similarity between ontologies. In: Proc. 13th Conf. on Knowledge Engineering and Management (2002)
14. Myhre, S., Tveit, H., Mollestad, T., Laegreid, A.: Additional Gene Ontology structure for improved biological reasoning. Bioinformatics 22(16), 2020–2037 (2006)
15. Noy, N., Musen, M.: The PROMPT suite: Interactive tools for ontology merging and mapping. Intl. Journal of Human-Computer Studies 59(6), 983–1024 (2003)
16. Ogren, P., Cohen, K., Acquaah-Mensah, G., et al.: The compositional structure of Gene Ontology terms. In: Proc. Pacific Symposium on Biocomputing (2004)
17. Online Mendelian Inheritance in Man, OMIM. McKusick-Nathans Institute for Genetic Medicine, Johns Hopkins University (Baltimore) and National Center for Biotechnology Information, National Library of Medicine (Bethesda) (2000)
18. Rahm, E., Bernstein, P.: A survey of approaches to automatic schema matching. The VLDB Journal 10(4), 334–350 (2001)
19. van Rijsbergen, C.J.: Information retrieval, 2nd edn. Butterworths, London (1979)
20. Shvaiko, P., Euzenat, J.: A survey of schema-based matching approaches. In: Spaccapietra, S. (ed.) Journal on Data Semantics IV. LNCS, vol. 3730, pp. 146–171. Springer, Heidelberg (2005)
21. Thor, A., Kirsten, T., Rahm, E.: Instance-based matching of hierarchical ontologies. In: Proc. 12th German Database Conf (BTW) (2007)

# Data Integration and Pattern-Finding in Biological Sequence with TESS's Annotation Grammar and Extraction Language (AnGEL)

Jonathan Schug[1], Max Mintz[2], and Christian J. Stoeckert Jr.[1]

[1] Department of Genetics in the School of Medicine
[2] Department of Computer and Information Science in the School of Engineering
University of Pennsylvania, Philadelphia PA, 19104, USA
jschug@pcbi.upenn.edu

**Abstract.** Decoding the functional elements in an organism's genome requires the integration of a wide variety of experimental and computational data from a wide range of sources. The location of this data, viewed as sequence features in the genome, must serve as one of the essential organizing principles for this integration. It is therefore important to have a data integration system that takes advantage of this fact. As part of the TESS project, we have developed a grammar-based data integration and pattern search tool, Annotation Grammar and Extraction Language (AnGEL), that follows this principle. AnGEL can represent most of the current work in cis-regulatory module (CRM) modelling in an intuitive way and can process data extracted from a variety of sources simultaneously. Here we describe AnGEL's capabilities and illustrate its use by querying for gene arrangements, CRMs, and protein domain structure.

## 1 Introduction

It is a common metaphor to speak of 'breaking the regulatory code' or 'understanding the language of gene regulation'. We took this metaphor seriously and have developed a grammar-based system that allows the user to search for patterns of interest occurring in genome sequence and/or in experimental or computational annotation of the genome. The tool, Annotation Grammar and Extraction Language (AnGEL), is part of the Transcription Element Search System (TESS) project[1] and so was originally aimed at modeling cis-regulatory modules (CRMs), but is more widely applicable. AnGEL adds a number of extensions to ordinary stochastic context free grammars (SCFGs) that allow it to represent most of the current work in CRM modelling as well as supporting *ad hoc* queries of biological interest such as gene and protein domain structure that we demonstrate here. It uses plug-in software modules that allow it to extract data from a variety of sources and then to integrate these data using a grammar to describe the positional relations between the sequence and data. For example, AnGEL can extract sequence from the UCSC DAS server[2], gene models from an Oracle database, and perform local positional weight matrix (PWM) matches to predict transcription factor binding sites (TFBS), in order to identify genes

S. Cohen-Boulakia and V. Tannen (Eds.): DILS 2007, LNBI 4544, pp. 188–203, 2007.
© Springer-Verlag Berlin Heidelberg 2007

that may be regulated by a particular CRM. We have also created the TESS relational sub-schema as part of the Genomic Unified Schema (GUS) project[3] to store AnGEL models so that they may form the basis for a knowledge warehouse. AnGEL's structured descriptions have the advantage of being interpretable by humans, and in fact can be translated automatically into natural language sentences.

Our goal in designing AnGEL was to create a tool that would allow a person to readily create many different queries to find patterns of interest but to do this in a common framework that allows for these, and more complicated, queries to be captured as well. In addition, we wanted to be able to integrate and expand upon existing techniques so that they could be compared on an equal footing. To accomplish this AnGEL uses a combination of ideas from two bioinformatics systems, K2/Kleisli[3,4] and GenLang[5,6,7,8]. Kleisli is a query system that allows the user to formulate queries that span multiple databases. The query author can specify that the query result be a list, set, or a multi-set. When the query returns a list, the order of the rows matters and the result may contain duplicate rows. For set and multi-set query results, the order does not matter and there may (multi-set) or may not (set) be duplicate rows. Kleisli connects to various data sources using drivers that create a suitable abstraction layer between the Kleisli query engine and the data source. GenLang is a tool based on a grammar formalism that adds extra capabilities on what is essentially a CFG. The additions raise the formal power of the system to mild context-sensitivity by allowing Genlang to recognize direct repeats and other constructs. Genlang also includes the notion of reversing complementing DNA strings.

The outline of this paper is as follows. First we briefly introduce our notation for stochastic context-free grammars (SCFG). Second, we describe the problems one encounters when applying SCFGs to recognizing patterns in biological sequences to motivate the enhancements we have developed. Third, we describe the enhancements and compare the resulting system to other grammar formalisms. Finally, we demonstrate AnGEL's usefullness by showing the queries for patterns that represent a variety of CRM models as well as gene and protein domain structure.

## 2    Background

A grammar is a set of rules that describe the structure of the legal sentences or strings of a language. In a stochastic context-free grammars (SCFG) the rules consist of a left-hand side (LHS) which is the name of a structure, and a right-hand side (RHS) which is a list of the subparts of that structure. Rules are usually hierarchical and often recursive. More formally, a SCFG $G$ consists four parts $(V, \Sigma, P, S)$: a set of variables $(V)$ which are names for the structures of the language, an alphabet $(\Sigma)$ of letters (or terminals) that make up the strings of the language, productions $(P)$ which are the rules defining the phrase structures, and the special start variable $(S)$ from which all legal strings are derived. Here we consider context-free grammars where the rules all have the form:

$$X \longrightarrow [p_{X,i}] \; Y_1, \; Y_2, \; \ldots \; Y_n; \tag{1}$$

The LHS, $X \in V$, is a variable name. In the RHS, $[p_{X,i}]$ is the probability of this expansion for the LHS and the terms $Y_i \in V \cup \Sigma$ are either variables or letters. Each rule may have a different length and can even have no terms on the right-hand side if the structure in question can consist of a phrase with zero length. There may be more than one rule with a given variable on the left-hand side; these represent alternative ways of forming a structure. The probabilities of such productions, $[p_{X,i}]$, are subject to the condition that $\forall X : \sum_i p_{X,i} = 1$. Hidden Markov models (HMMs) are a subclass of stochastic grammars that have been widely used to model protein domains[9,10,11], genes[12,13,14,15], and promoters[16]. Positional weight matrices (PWMs)[17] are an even simpler form of SCFG which are widely used to represent TFBSs and protein domains.

## 2.1    Applying Grammars to Gene Regulation

While GenLang formed the basis of a successful gene finder[6], experience suggested some improvements for identifying CRMs. To see this, consider a grammar rule for a CRM consisting of binding sites for three TFs that may occur in any order. Part of the grammar would look something like this.

$$CRM \longrightarrow A, B, C \mid A, C, B \mid B, A, C \mid B, C, A \mid C, A, B \mid C, B, A; \tag{2}$$

This is a cumbersome representation of what is a conceptually simple model. In addition, what is the spacing between the elements of RHS? Ordinarily, the tokens or strings for A, B, and C would have to appear immediately adjacent to each other. This rarely occurs in biological situations. On the other hand there must be some limit on the distance between each TFBS or at least on the entire size of the match. While there might prove to be differences in spacing preferences depending on the order of the sites, the simplest model is that the spacing is subject to some common upper limit: a constraint that is not succinctly expressible in this format. Furthermore, the first-encountered instance of a TFBS may be poor and may precede a much better site. Thus we would like the parser to be free to skip over TFBS in some cases. SCFGs do allow this, but at the cost of a representation that is fairly difficult to specify manually for *ad hoc* queries.

A limitation of a string-based grammatical approach is that some sequence features do not have a grammatical definition. Clearly features that are based on experimental data, e.g., mutations, chromatin configuration, actual (as opposed to predicted) TFBS, do not (initially) have a grammatical definition. There are also features that are defined computationally, but can not be represented in a SCFG. An example of this class are CpG islands defined as a region with at least 50% C-G content and at least 60% as many CpG dinucleotides as expected. Another example are conserved regions and BLAST hits. It is essential to include these kinds of data when analyzing and modeling genomic sequence. Thus, while ordinary SCFGs grammars offer a compelling scenario for intuitively describing and combining models of sequence features, they need some modification before being easily applicable. AnGEL is designed to address these issues.

# 3   Methods

In this section we give a description of each the extensions included in AnGEL. A more detailed description can be found elsewhere[18]. We envision this system having a usefulness beyond the applications described in this work. There are, therefore, features that we do not take advantage of or fully develop at this time. Or current parser supports only non-recursive grammars and returns all possible parses.

## 3.1   Sequence and Annotation Streams

AnGEL grammars describe patterns involving both sequence and sequence annotation. Sequence and annotation are organized in *streams* that each reference the same sequence ID and have the same coordinate system. The primary sequence is placed in the main stream and is loaded by a main sequence plug-in module. The AnGEL package includes several sequence plug-ins that read sequence from flat files using BioPerl[19], a DAS server, a GUS database instance, or from internal memory. The AnGEL parser expects to be able to extract one or more sequences from the sequence source and will parse each sequence in turn. For example a DAS server may provide one sequence for each chromosome in a genome, whereas a flat file may provide sequences consisting of just the proximal promoter region for thousands of genes.

If a grammar makes reference to annotation on a sequence, then it must specify a rule to define how the corresponding stream is loaded with annotation. The rule assigns a name to the stream and an annotation plug-in to populate the stream with annotation data. The rule can also provide arguments to control the behavior of the plug-in. During parsing, the parser will provide the annotation plug-in with the ID of the sequence currently being parsed as well as its actual sequence if need be. The plug-in can then access its data resource to extract or create annotation for the sequence and place it in the stream. Annotation is stored in the stream in a DAS-like hierarchical interval format. Each interval has a name, type, start and end position, score, strand, and other attributes. For example the GUS database plug-in that extracts gene models includes intervals with the types 'TranscriptionUnit', 'Tss', 'UpstreamRegion', 'Gene', 'FivePrimeRegion', 'Exon', and 'Intron'. Many different resources can be accessed via plug-ins in one grammar. The only limitation is that they must share the same set of sequence IDs so that the annotation can be correctly linked to the sequence being parsed. The AnGEL package currently includes annotation plug-ins that access GFF files, DAS servers, GUS databases, the EMBOSS[20] package program NEWCPGREPORT, a PWM scoring program for TFBS from TESS[1], and the TFBS predictor PSPAM[21]. Both sequence and annotation plug-ins are implemented as Perl packages that must live up to a simple programming interface standard. It is quite easy for a programmer to write new plug-ins as described below.

## 3.2 RHS Terminals

AnGEL uses an extended set of terminals for recognizing sequences and annotation. First, individual lowercase or quoted letters of the sequence alphabet, e.g., a or "A" for adenine or "Q" for glutamine, can be used to match sequences letter by letter. AnGEL also supports 'smart' alphabet matching such as the IUPAC-defined abbreviations for ambiguous bases, e.g., "W" matches either an "A" or "T". The AnGEL parser understands these abbreviations when matching DNA sequence but uses ordinary exact matching when parsing AA sequence. It is easy to create new smart alphabets if need be. Second, a sequence of letters can be specified as a single RHS term rather than a series of terms, e.g., "TATA" instead of "T", "A", "T", "A". Third, AnGEL includes a general *gap character*, '...', that matches any character of any alphabet. The range of possible lengths of a gap is indicated with a term count as illustrated in Table 1. Gap characters are also a signal to the parser that it can take shortcuts in identifying a match to the next RHS term in the production. Fourth, because positional weight matrices are such a big part of genomic sequence analysis, positional weight matrices are included as a terminal. Fifth, AnGEL includes positional terminals that match a particular absolute or relative position in a sequence. Sixth, AnGEL terminals can be a *path expression* that is matched against annotation. The simplest path expression is a single path term that consists of the name of the stream and the type of interval, e.g., Genes::Intron. Multiple path terms can be joined together to form a path expression using operators like 'part-of' ('.'), 'contained-in' ('/') or 'intersects' ('%'). Examples of each these terminals are shown in Table 1. Finally, a variable is an alphanumeric string beginning with an uppercase letter, e.g., Crm1. Matches to nonterminals are recorded in the main stream and can be included in a path expression as, for example, ::Crm1. Overlap between terms is controlled on a term-by-term basis; an '!' following a term prevents the subsequent term from overlapping it. The orientation or strand of a term is controlled by a leading '+', '-', or '*' or the sense attribute of an annotation.

## 3.3 Collection Productions

Ordinary grammars require that the strings that match the terms of the right-hand side of a production occur immediately adjacent to each other and in the order the RHS specifies. AnGEL adds new *collection productions* that relax this constraint. Table 2 contains examples of collection productions and their interpretation. The collection type is indicated by a pair of brackets in the production's arrow. *List* (-[]->) productions release the adjacency constraint. *Multiset* (-<>->) productions release the order constraint as well, but still require the number of each RHS term to be met. The number of instances is indicated with the term count. *Set* (-{}->) productions only require one instance of each RHS term. Finally, *linear* (-//->) productions are like set productions but they allow features to be missing. The parsing each type of collection production is accomplished by a specialized algorithm that avoids expanding the production into an exponential number of equivalent ordinary productions. For example,

**Table 1.** Examples of AnGEL terminals: Terminals are RHS terms that can be used to recognize substrings, positions, or annotation in any of the data streams

| AnGEL Syntax | Meaning |
|---|---|
| gatw or "GATW" | a consensus sequence for the GATA transcription factor |
| ...:{20,50} | a gap of between 20 and 50 characters |
| <acgt; 10,0,0,0; 0,9,1,0; 0,0,1,9; 0,0,8,2>[score>=10.2] | a match to a PWM with a consensus of 'ACTG' that has a log-odds score of at least 10.2 |
| @@1000 | position 1000 in a sequence |
| @-50 | a position 50 characters from from the end of the bounding interval |
| Annotation::CpgIsland | a CpG island feature loaded into the Annotation stream |
| Annotation::CpgIsland[length>=500] | a CpG island longer than 500bp |
| Annotation::CpgIsland / BindingSites::Sp1[score>=9] | an Sp1 site that is located inside a CpG island |
| Gene::Intron[index=1] | the first intron in a gene. |
| Gene::TranscriptStartSite % Annotation::CpGIsland=>center | the center of a CpG island that overlaps the start of transcription of a gene |

when parsing a linear collection production (which emulates a logistic regression model), AnGEL collects all of the feature instances in the bounds, then selects the best (up to the term count limits) of these and reports that group as the match. The score for a linear collection match amounts to the exponent in the logistic regression formula. In all production types, the parser is not required to match every letter or annotation in the current stream(s); it only has to find a group of letters and annotations that meet the pattern of the grammar. Some examples of each type are shown in Table 2.

### 3.4   Production Bounds

Faced with the freedom we have now given our grammars by relaxing the spacing, order, and count of the RHS terms, we need some compensating 'force' to allow the grammar to try to focus on statistically and biologically significant matches. To do this we introduce *production bounds*. If a production is bounded, then any match of the RHS of the production must satisfy the conditions of the bounds. A production may have a *numeric bound* and/or a *location bound*. Numeric bounds control the maximum size of a match to the production and are an easy way to ensure that, for example, TFBS's occur in close proximity. Location bounds force a match to occur inside annotation intervals or instances of matches to other productions. Contexts are specified using path expressions as described earlier. The specifications of the two bounds are placed inside the production's arrow.

**Table 2.** Examples of collection productions with various collection bounds

| AnGEL Syntax | Meaning |
|---|---|
| `S ----> A, B;` | This is an ordinary production requiring A and B to appear adjacent to each other. |
| `S -{1[kb];;}-> A, B;` | This is a set production that can be no longer than 1000bp. Since it is a set, A and B can appear in any order. |
| `S -<400[bp];;>-> A:2, B:3;` | This multi-set production requires two As and three Bs within 400bp of each other. |
| `S -(;Annotation::CpgIsland;)-> A;` | This ordinary production requires A to occur inside an annotated CpG island. |
| `S -[400[bp];Genes::Intron[1];]-> A, B, C;` | Here we require three features to appear in order within 400bp of each other and in the first intron of a gene. |

### 3.5 Scoring a Parse and Negated Features

AnGEL is based on a stochastic grammar formalism. Each terminal is expected to produce a score which is interpreted as a log probability or log odds score. Each RHS term can be preceded by a *weight* $w_i$ that is multiplied by the term's score to produce the effective term score. When a variable has more than one definition, each production is given a probability $p(P_j)$ which must sum to one. The default weight is one and default probability is $1/|X|$ where $|X|$ is the number of productions for a given variable X. The score for a match of a production as shown in Equation 1 is

$$\text{score}(\mathsf{X}) = \lg(p(P_j)) + \sum_{i=1}^{n} w_i \, \text{score}(\mathsf{Y}_i) \tag{3}$$

When a variable appears in a RHS its score is the score calculated as above. In this way the scores are passed up the parse tree to the top. The score for the start variable is the score for the parse.

Although RHS term weights are generally positive, they can be negative as well. A negative weight can either be used to convert the scores of an annotation source or to penalize the presence of a given feature. There is a special negative term weight NO. If a feature is found that matches a term that has a weight of NO then the entire match for that production is canceled. This feature allows AnGEL grammars to require the absence of, for example, a TFBS for an inhibitor.

### 3.6 The GUS: Tess Relational Schema

One of the goals of the TESS component of the GUS schema is to support a database of knowledge not just of facts. We want to store in our database not just instances of CRMs or other patterns, but also models of these patterns. Ideally the database representation would allow the models to be queried and to serve as a link between TFs, regulatory activity, and instances of binding sites in regulatory regions. We have designed a part of the TESS section of the GUS schema to store AnGEL grammars in a structured form, rather than as free text, to facilitates queries against models. In addition AnGEL contains a comment structuring convention that allows GUS primary keys to be attached to parts of an AnGEL grammar so that parsing results can be loaded back into GUS while maintaining a connection between the parsed feature and the grammar fragment that identified or defined it. Use of this mechanism is optional and a GUS database is not required to use AnGEL.

### 3.7   Data Plugin Specification

The programming specification for Perl plugins that generate or retrieve annotation is simple and consists of the methods listed in Table 3. The init method is called when the plugin object is created, but before its run-time parameters are known. The initialize method is called once these parameters are known and will typically connect to an external database if necessary. The prepareRegion method will retrieve the data for a portion of the current sequence, format them as AnGEL annotation features, and place the features in the plugin's stream. Plugins may implement region caching to improve performance. If the total amount of annotation for a sequence is known to be small, then all of the annotation can be retrieved during the beginEntry method. The init or initialize method should set the NeedsSequence attribute for the plugin to indicate whether the plugin will need the actual sequence to generate annotation. If none of the plugins or grammar rules require the sequence it is not loaded, saving time and memory. A similar set of methods is used for plugins that retrieve the main sequence entities to be parsed.

## 4   Results

Now we show a series of examples of the application of AnGEL grammars to some biological problems. We start with a few simple examples to cover the basics of

**Table 3.** Methods for annotation plugins. New plugins need only implement these methods to access data from sources not covered in the AnGEL distribution.

| Method | Description |
| --- | --- |
| init | set the attributes of the plugin object |
| getOptionDescriptions | define grammar-settable options for run-time behavior |
| initialize | prepare for a parsing run using option settings |
| beginEntry | prepare for a new sequence entry |
| prepareRegion | retrieve data for a region of the current sequence |
| discardRegion | discard data for the current region |
| endEntry | discard data for current sequence |
| finalize | called after parsing is complete |

the system. Next we discuss how to represent many of the recent CRM modeling techniques. Third, we consider an application for amino acid sequences.

### 4.1   Simple Examples

Our first example is a simple grammar to identify divergently-transcribed gene pairs.

```
S -[1[kb];;]-> *Genes::Tss[sense=-1], *Genes::Tss[sense=1];

@Genes -> 'DAS::Gene' --Types='refGene'
  --Categories = 'transcription'
  --Server     = 'http://genome.ucsc.edu/cgi-bin/das/hg18' ;
```

The grammar requires the two transcription start sites to be within 1000bp, on opposite strands, with the reverse-strand TSS on the 5' side of the pair. The example extracts gene annotation from the UCSC DAS server. AnGEL supports macro expansion in grammars, so the annotation stream could be defined as a macro that is expanded just prior to parsing. This helps separate the specification of the arrangement of interest from the specification of the data sources. Additional filtering could also be applied to, for example, limit the TSSs to only full-length clones.

We can look for a specialized subset of these genes by requiring that a CpG island cover both TSSs. In this case we have a grammar shown below which gets CpG island features from the mysql server at UCSC. This source of CpG islands can be combined with gene models from a different database.

```
S -[1[kb];CpgIsland::Feature;]-> *Genes::Tss[sense=-1],
                                 *Genes::Tss[sense=1];

@CpgIsland -> 'UCSC::MySql::CpgIsland'
  --mysqlServer='genome-mysql.cse.ucsc.edu'  --mysqlPort='3306'
  --mysqlUser='genome'  --mysqlPassword=''  --genomeRelease='hg17';
```

The next grammar looks for transversion SNPs located inside likely p53 binding sites and illustrates more data integration. Such SNPs may have phenotypic consequences. It uses experimental data generated by chromatin immunoprecipitation and paired end-tag sequencing (ChIP-PET) to identify regions of the human genome to which the transcription factor p53 binds[22]. However, the experimental data has limited resolution; the target regions are often a few hundred base pairs long. A SNP in such a region may have no regulatory impact if it is not located in the binding site. To increase resolution, we use a PWM for p53 to identify likely binding sites in the target regions and look for SNPs in these, much smaller, regions. To access the ChIP-PET locations, we cut them from the supplemental data from [22] and pasted them into a text file. The SNPs are extracted from the UCSC mysql database.

```
S -[;ChipTfbs::P53/PwmTfbs::BindingSite;]-> *Snps::Transversion;

@ChipTfbs -> TextFile  --tabFile='p53-spots.tab'
  --skipHeaderLinesN=0 --typeSpec='P53' --senseSpec='+'
  --ucscLocationSpec='%2' --nameSpec='%3' --scoreSpec='%1' ;

@PwmTfbs -> 'WMS' --mat='p53.pwm'  --id='V_P53_02  --mlo=10.0
  --mxd=100  --tpc=1.0 ;

@Snps -> 'UCSC::MySql::SNP' --genomeRelease='hg18' --preload=0
  --mysqlPort='3306'  --mysqlUser='genome'  --mysqlPassword=''
  --mysqlServer='genome-mysql.cse.ucsc.edu';
```

All of the grammars in this section can be run against the human genome in about 1 to 2 hours on a 1GHz/1GB laptop when the remote data is taken from a mysql or Oracle RDBMS. Using a DAS server for data slows the parsing down significantly.

## 4.2 Modeling CRMs

AnGEL can emulate many of the models used in previous work on modeling CRMs as well as express richer models. Here we review a few recent approaches to CRM modeling and illustrate how they are represented in AnGEL. We then show how these models can be combined and placed in larger contexts. In this section we use a more abstract syntax to illustrate the form of the solution, rather than an example of a particular model.

A simple CRM model is a cluster of a fixed set of TFBS. This is handled directly with a set collection production, e.g.,

$$ \text{CRM} \xrightarrow{\{n;loc;\}} F_1, \; F_2, \; \ldots, \; F_n; \tag{4} $$

where $n$ and $loc$ are the optional bounds and the $F_i$ are the TFBS. Instances of this model include work by Kreiman[23], Hannenhalli and Levy[24], and MSCAN[25]. The TOUCAN system [26] uses this model as well, though it can also penalize missing features.

Another popular model is a cluster with a fixed number $k$ of sites drawn from a set of $m$ possible TFs. There is no required order for the sites. This CRM is modeled in AnGEL as follows:

$$ \text{CRM} \xrightarrow{<n;loc;>} F : k; \tag{5} $$
$$ F \longrightarrow F_1 \mid F_2 \mid \cdots \mid F_m; $$

where $F_i$ is the $i$-th feature of interest. CISTER[27] and an earlier, related, system COMET[28], use this model but also include a Markov model of the background. The grammar fragment in Equation 5 will find instances with exactly $k$ TFBS, even if there are more nearby. A small modification to the term count (shown in

Equation 6) will find any cluster containing between $k_{min}$ and $k_{max}$ sites. If the upper count bound $k_{max}$ is left undefined, then as many sites as possible will be included in the parse.

$$\text{CRM} \xrightarrow{<n;loc;>} F : \{k_{min}, k_{max}\} \quad \text{e.g.} \quad \text{CRM -<400[bp];;>-> F:\{3,6\};} \quad (6)$$

Another way to make a site optional is to use a scoring formula such as logistic regression. This approach was used for models of muscle [29] and liver[30] regulatory regions. It is rendered in AnGEL using a linear collection production.

$$\text{CRM} \longrightarrow \text{CrmFeats}[\text{score} \geq c]; \quad (7)$$

$$\text{CrmFeats} \xrightarrow{/n;loc;/} w_1 F_1, \ w_2 F_2, \ \cdots, w_m F_m;$$

where $w_i$ are the weights and $c$ is a threshold needed to reject poor-scoring hits. Note that AnGEL allows one to use features that are more complex than individual binding sites as input to the logistic regression. For example, our assessment grammar shown at the end of this section uses a set of three TFBS as one of the logistic regression features.

A few CRM models consider the order of the sites. Older work by Frech [31,32,33] and more recent work by Dohr [34] and Terai [35] use this model. AnGEL can represent all of the models in these efforts using a list collection production, i.e.,

$$\text{CRM} \xrightarrow{[n;loc;]} F_1, F_2, \ldots, F_n; \quad (8)$$

Though it does not impose a fixed order, a Markov chain model developed in by Thompson[36] learns which TFBS prefer to follow each other. Because Markov chains are an instance of a stochastic grammar this can be rendered easily in AnGEL.

Note that each of these models can be combined with other CRMs or placed in a particular genomic context using AnGEL. For example the grammar fragment below places a CRM model in a conserved region even if it was not originally developed by considering conservation.

```
S -(;Conserved::Region;)-> CRM; @Conserved -> ...; @CRM -> ...;
```

Perhaps more interestingly, multiple CRMs can be placed in both proximal and distal regions relative to a gene as this grammar fragment suggests.

```
S      -[20[kb];Gene::Locus;]-> Enhancer, Proximal, Internal, Utr5;
Enhancer -(;Conserved::Feature;)-> Crm1;
Proximal -[400[bp];;]-> Crm2, TataBox::BindingSite, Gene::Tss;
Internal -[500[bp];Gene::Intron[1];]-> Crm3;
Utr5      -{;Gene::Exon[-1];}-> MiRnas;
Crm1  -> ... ;  Crm2-> ... ;  Crm3      -> ... ;
MiRnas-> ... ;  Gene-> ... ;  Conserved-> ... ;  TataBox-> ... ;
```

To illustrate the parsing time for CRM grammars described in Equations 4 to 8, we used a liver-specific CRM (shown below) based on earlier work [30,37] which incorporates both the basic liver-specific determining factors Hnf1, Hnf3, and Hnf4 as well as signal response factors GR, Oct1, and C/EBP. All of the TFBSs are computationally predicted, but would work just as well with experimentally identified sites.

```
S -{300[bp];Genes::UpstreamRegion;}-> *Hnf1::BindingSite,
   *Hnf3::BindingSite, *Hnf4::BindingSite, GrCrm;

GrCrm -{100[bp];;}-> *GR::BindingSite, *Oct1::BindingSite,
   *Cebp::BindingSite;
```

These simple and compound features were combined following the examples above. In most cases parsing, 3KB regions covering 10,000 promoters in the human genome (with repeat sequence masked) took less than 45 minutes on a 2.2GHz/4GB Intel-based server. The grammar patterned after Equation 4 took twice as along because it had so many matches. Though this is too slow for interactive use, it is perfectly acceptable for use in data analyses. The AnGEL parsing program supports sequence-level parallelization which can dramatically reduce parsing time when a compute cluster is available.

### 4.3   Protein Sequences

Here we show a simple example to indicate that AnGEL can also model problems relevant to proteins. The WNT inhibitory factor 1 protein (WIF1) has a signal peptide, WIF domain, and a Cripto growth factor domain. We show two simple grammars to recognize other proteins with these domains. The first version constrains the order of the domains to be the same as in WIF1. The second version keeps the signal protein at the N-terminus, but allows the other to domains to occur in either order. This shows a simple instance of the ability of AnGEL to apply constraints independently.

```
Wif1_like_Free  -[;;]-> SignalPeptide[end <= 30], Domains::WIF,
                        Domains::Cripto;

Wif1_like_Fixed -[;;]-> SignalPeptide[end <= 30], OtherDomains;
OtherDomains    -{;;}-> Domains::WIF, Domains::Cripto;
```

We note in passing that this same type of query can operate at a finer grained level to implement a protein model like PRINTS[38] which identifies protein domains using a fingerprint of short ungapped AA sequences.

### 4.4   Applications of AnGEL

As the following applications illustrate, AnGEL's capabilities are useful to practicing biologists. We have used the AnGEL system to provide a web query, available at the EPConDB website[39], for genes that are potentially regulated in pancreatic islet cells. An AnGEL grammar is generated automatically based on the user's input to a web form. The site uses DoTS transcripts[40] aligned to the mouse and human genome and stored in our GUS database as a source of gene models. The user-selectable location bound can be in a region defined relative to the TSS, or in one or any of the introns of DoTS genes. The user selects which TFBS models are of interest from a menu and inputs how many instances of each site are required, what the total spacing is, and whether the order matters. Predefined site models may be either literal matches, e.g., an E-box is CAnnTG, or PWMs with score thresholds based on estimated sensitivity. The user may also define consensus sequences for TFs not included in our list. (As we load the results of ChIP-chip and other similar experiments into EPConDB, we will augment the list of TFBS with confirmed sites from these experiments.) When the search is done, the user is presented with a list of the DoTS transcripts that contained their pattern as well as the details of the arrangement and scores of the sites. The transcript list can then be downloaded or used for further processing to help confirm the predictions with the other tools in EPConDB. Visualization of matches on the genome helps determine whether a rule has the intended behaviour. The AnGEL package includes a program that converts parsing results to GFF or BED format suitable for display on the UCSC and other genome browsers.

We have developed a machine learning algorithm that learns simple grammars for patterns that are enriched in a set of sequences. Given a set of possible features, the algorithm can learn the enriched component features, collection type, cut-off scores, and size bound of a collect production. We have applied this algorithm to identify possible partners of TFs that are the target of ChIP-chip experiments. Such an experiment helps identify which genes are likely targets of regulation by a particular transcription factor, but they do not explain why only these genes are targeted when other genes have good binding sites as well. One possible mechanism is that one or more TFs create the observed specificity by working with the ChIP-chip target TF. We used the AnGEL system to identify these TFs. First, in the case of C/EBP-beta binding during liver generation[41] we were able to identify a rule involving three other TFs as a good explanation of C/EBP-beta binding. In a second experiment[37] studying glucocorticoid receptor (GR) binding in fasting mice we were able to identify a number of potential partner TFs for GR, including AP-1, Gata-1, C/EBP-beta, and Oct-1. We also identified the spacing constraints on these potential partners. Many of the partners were supported by the literature.

## 5    Conclusion

We have given a brief introduction to our grammar-based query and data extraction tool AnGEL and have demonstrated how a grammar formalism makes it easy to express many of the current biosequence modeling techniques in a single framework. We have also showed how it allows simple models to be extended by combining them with other models and/or by placing them into a larger context of sequence features. AnGEL allows the grammar-based rules to operate on annotation extracted from local or remote resources, such as databases, DAS servers, flat files, and programs. Using a grammar allows a modeller to encapsulate local constraints and to build a complex model from simple pieces. The introduction of collection productions as a simple way of expressing otherwise complex models and the support for negation, provides an advance over other bioinformatics tools of this kind. Finally, by providing a relational schema we have a system capable of storing a wide variety of sequence models in a unified format. AnGEL is available by contacting the authors.

Although AnGEL can be used for a variety of biological problems, its development was driven by the problem of identifying CRMs. Essentially all of the attempts to model CRMs have tried to identify the combination and possibly the arrangement of TFBS that confer a particular regulatory function on a region of the genome, e.g., causing expression in a particular tissue or developmental stage. Early efforts considered mostly the position of TFBS relative to the transcription start site (TSS) or the order and orientation of a pair of TFBS. Some later work dropped order and orientation requirements and tried to identify the collection of TFs that were involved with only loose constraints on their positions, e.g., limiting the distance between consecutive sites. While these methods have achieved some success, the problem is still open. None of these efforts had any means of defining the context of a CRM (beyond sequence conservation), nor of easily combining CRMs to create larger complexes of CRMs to build up a more complex model of gene regulation. As work in sea urchin has shown[42], a full picture of a regulatory region can be very complex. Recently experiments are generating new types of genome-scale data, e.g., global assessment of TF binding[22], chromatin configuration[43], and histone modifications[44]. These data will help shed light on regulatory mechanisms, but also pose new challenges for analysis. AnGEL offers the ability to integrate this kind of data to build better models of gene regulation.

# References

1. Schug, J.: Using TESS to Predict Transcription Factor Binding Sites in DNA Sequence. In: Baxevanis, A.D. (ed.) Current Protocols in Bioinformatics, J. Wiley and Sons, New York (2003)
2. Karolchik, D., Hinrichs, A.S., Furey, T.S., Roskin, K.M., Sugnet, C.W., Haussler, D., Kent, W.J.: The UCSC Table Browser data retrieval tool. Nucleic Acids Res. 32 Database issue, D493–496 (2004)
3. Davidson, S., Crabtree, J., Brunk, B., Schug, J., Tannen, V., Overton, G., Stoeckert, C.: K2/Kleisli and GUS: Experiments in integrated access to genomic data sources. IBM Systems Journal 40(2), 512–531 (2001)
4. Buneman, P., Naqvi, S., Tannen, V., Wong, L.S.: Principles of Programming with Complex Objects and Collection Types. Theoretical Computer Science 149(1), 3–48 (1995)
5. Searls, D.B.: The Linguistics of DNA. American Scientist 80(6), 579–591 (1992)
6. Dong, S., Searls, D.B.: Gene Structure Prediction by Linguistic Methods. Genomics 23(3), 540–551 (1994)

7. Searls, D.B.: String Variable Grammar: A Logic Grammar Formalism for the Biological Language of DNA. Journal of Logic Programming, pp. 73–102 (1995)
8. Searls, D.B.: Languages, automata, and macromolecules. Biophysical Journal 76(1), A272–A272 (1999)
9. Grundy, W.N., Bailey, T.L., Elkan, C.P., Baker, M.E.: Meta-MEME: motif-based hidden Markov models of protein families. Comput Appl Biosci. 13(4), 397–406 (1997)
10. Eddy, S.R.: Profile hidden Markov models. Bioinformatics 14(9), 755–763 (1998)
11. Sonnhammer, E.L., Eddy, S.R., Birney, E., Bateman, A., Durbin, R.: Pfam: multiple sequence alignments and HMM-profiles of protein domains. Nucleic Acids Res. 26(1), 320–322 (1998)
12. Burge, C., Karlin, S.: Prediction of complete gene structures in human genomic DNA. J. Mol. Biol. 268(1), 78–94 (1997)
13. Reese, M.G., Eeckman, F.H., Kulp, D., Haussler, D.: Improved splice site detection in Genie. J. Comput. Biol. 4(3), 311–323 (1997)
14. Henderson, J., Salzberg, S., Fasman, K.H.: Finding genes in DNA with a Hidden Markov Model. J. Comput. Biol. 4(2), 127–141 (1997)
15. Yada, T., Nakao, M., Totoki, Y., Nakai, K.: Modeling and predicting transcriptional units of Escherichia coli genes using hidden Markov models. Bioinformatics 15(12), 987–993 (1999)
16. Pedersen, A.G., Baldi, P., Brunak, S., Chauvin, Y.: Characterization of prokaryotic and eukaryotic promoters using hidden Markov models. Proc. Int Conf. Intell. Syst. Mol. Biol. 4, 182–191 (1996)
17. Chen, Q.K., Hertz, G.Z., Stormo, G.D.: MATRIX SEARCH 1.0: a computer program that scans DNA sequences for transcriptional elements using a database of weight matrices. Comput. Appl. Biosci. 11(5), 563–566 (1995)
18. Schug, J.: Integrating Gene Expression Signals with Bounded Collection Grammars. PhD thesis, University of Pennsylvania (2005)
19. Stajich, J.E., Block, D., Boulez, K., Brenner, S.E., Chervitz, S.A., Dagdigian, C., Fuellen, G., Gilbert, J.G., Korf, I., Lapp, H., Lehvaslaiho, H., Matsalla, C., Mungall, C.J., Osborne, B.I., Pocock, M.R., Schattner, P., Senger, M., Stein, L.D., Stupka, E., Wilkinson, M.D., Birney, E.: The Bioperl Toolkit: Perl Modules for the Life Sciences. Genome Res. 12(10), 1611–1618 (2002)
20. Rice, P., Longden, I., Bleasby, A.: EMBOSS: the European Molecular Biology Open Software Suite. Trends Genet. 16(6), 276–277 (2000)
21. Wang, J., Hannenhalli, S.: Generalizations of markov model to characterize biological sequences. BMC Bioinformatics 6(1), 219 (2005)
22. Wei, C., Wu, Q., Vega, V., Chiu, K., Ng, P., Zhang, T., Shahab, A., Yong, H., Fu, Y., Weng, Z., et al.: A Global Map of p53 Transcription-Factor Binding Sites in the Human Genome. Cell 124(1), 207–219 (2006)
23. Kreiman, G.: Identification of sparsely distributed clusters of cis-regulatory elements in sets of co-expressed genes. Nucleic Acids Res. 32(9), 2889–2900 (2004)
24. Hannenhalli, S., Levy, S.: Transcriptional regulation of protein complexes and biological pathways. Mamm. Genome 14(9), 611–619 (2003)
25. Alkema, W.B., Johansson, O., Lagergren, J., Wasserman, W.W.: MSCAN: identification of functional clusters of transcription factor binding sites. Nucleic Acids Res. 32(Web Server issue), 195–198 (2004)
26. Aerts, S., Van Loo, P., Moreau, Y., De Moor, B.: A genetic algorithm for the detection of new cis-regulatory modules in sets of coregulated genes. Bioinformatics 20(12), 1974–1976 (2004)

27. Frith, M.C., Hansen, U., Weng, Z.: Detection of cis-element clusters in higher eukaryotic DNA. Bioinformatics 17(10), 878–889 (2001)
28. Frith, M.C., Spouge, J.L., Hansen, U., Weng, Z.: Statistical significance of clusters of motifs represented by position specific scoring matrices in nucleotide sequences. Nucleic Acids Res. 30(14), 3214–3224 (2002)
29. Wasserman, W.W., Fickett, J.W.: Identification of regulatory regions which confer muscle-specific gene expression. J. Mol. Biol. 278(1), 167–181 (1998)
30. Krivan, W., Wasserman, W.W.: A predictive model for regulatory sequences directing liver-specific transcription. Genome Res. 11(9), 1559–1566 (2001)
31. Frech, K., Werner, T.: Specific modelling of regulatory units in DNA sequences. Pac. Symp. Biocomput. pp. 151–62 (1997)
32. Klingenhoff, A., Frech, K., Quandt, K., Werner, T.: Functional promoter modules can be detected by formal models independent of overall nucleotide sequence similarity. Bioinformatics 15(3), 180–186 (1999)
33. Gailus-Durner, V., Scherf, M., Werner, T.: Experimental data of a single promoter can be used for in silico detection of genes with related regulation in the absence of sequence similarity. Mamm. Genome 12(1), 67–72 (2001)
34. Dohr, S., Klingenhoff, A., Maier, H., de Angelis, M.H., Werner, T., Schneider, R.: Linking disease-associated genes to regulatory networks via promoter organization. Nucleic Acids Res. 33(3), 864–872 (2005)
35. Terai, G., Takagi, T.: Predicting rules on organization of cis-regulatory elements, taking the order of elements into account. Bioinformatics 20(7), 1119–1128 (2004)
36. Thompson, W., Palumbo, M.J., Wasserman, W.W., Liu, J.S., Lawrence, C.E.: Decoding human regulatory circuits. Genome Res. 14(10A), 1967–1974 (2004)
37. Phuc, L.P., Friedman, J.R., Schug, J., Brestelli, J.E., Parker, J.B., Bochkis, I.M., Kaestner, K.H.: Glucocorticoid receptor-dependent gene regulatory networks. PLoS Genetics, vol. 1(2) (2005)
38. Attwood, T.K., Bradley, P., Flower, D.R., Gaulton, A., Maudling, N., Mitchell, A.L., Moulton, G., Nordle, A., Paine, K., Taylor, P., Uddin, A., Zygouri, C.: PRINTS and its automatic supplement, prePRINTS. Nucl. Acids Res. 31(1), 400–402 (2003)
39. Mazzarelli, J.M., Brestelli, J., Gorski, R.K., Liu, J., Manduchi, E., Pinney, D.F., Schug, J., White, P., Kaestner, K.H., Stoeckert, C.J.J.: EPConDB: a web resource for gene expression related to pancreatic development, beta-cell function and diabetes. Nucl. Acids Res. D751–D755 (2006)
40. CBIL: AllGenes: a web site providing access to an integrated database of known and predicted human (release 9.0, 2004) and mouse genes (release 9.0, 2004) (2004)
41. Friedman, J.R., Larris, B., Le, P.P., Peiris, T.H., Arsenlis, A., Schug, J., Tobias, J.W., Kaestner, K.H., Greenbaum, L.E.: Orthogonal analysis of C/EBPbeta targets in vivo during liver proliferation. In: Proc. Natl. Acad. Sci. vol. 101(35), pp. 12986–12991, USA (2004)
42. Yuh, C.H., Bolouri, H., Davidson, E.H.: Genomic cis-regulatory logic: experimental and computational analysis of a sea urchin gene. Science 279(5358), 1896–1902 (1998)
43. Crawford, G.E., Davis, S., Scacheri, P.C., Renaud, G., Halawi, M.J., Erdos, M.R., Green, R., Meltzer, P.S., Wolfsberg, T.G., Collins, F.S.: Dnase-chip: a high-resolution method to identify dnase i hypersensitive sites using tiled microarrays. Nat. Meth. 3(7), 503–509 (2006)
44. Kim, T.H., Barrera, L.O., Zheng, M., Qu, C., Singer, M.A., Richmond, T.A., Wu, Y., Green, R.D., Ren, B.: A high-resolution map of active promoters in the human genome. Nature 436(7052), 876–880 (2005)

# Inferring Gene Regulatory Networks from Multiple Data Sources Via a Dynamic Bayesian Network with Structural EM

Yu Zhang[1], Zhidong Deng[1], Hongshan Jiang[2], and Peifa Jia[1]

[1] State Key Laboratory of Intelligent Technology and System,
Computer Science and Technology Department,
Tsinghua University, Beijing, China, 100084
z-y02@mails.tsinghua.edu.cn, michael@mail.tsinghua.edu.cn,
dcsjpf@mail.tsinghua.edu.cn
[2] Department of Computer Science, Tsinghua University, Beijing, China, 100084
jhs03@mails.tsinghua.edu.cn

**Abstract.** Using our dynamic Bayesian network with structural Expectation Maximization (SEM-DBN), we develop a new framework to model gene regulatory network from both gene expression data and transcriptional factor binding site data. Only based on mRNA expression data, it is not enough to accurately estimate a gene network. It is difficult for us to estimate a gene network accurately only with the mRNA expression data. In this paper, we use the transcription factor binding location data in order to introduce the prior knowledge to SEM-DBN model. Gene expression data are also exploited specifically for likelihood. Meanwhile, we incorporate the prior knowledge into every learning step by SEM rather than only learning from the very beginning, which can compensate the attenuation of the effect with location data. The effectiveness of our proposed method is demonstrated through the analysis of *Saccharomyces cerevisiae* cell cycle data. The combination of heterogeneous data from multiple sources ensures that our results are more accurate than those recovered from only gene expression data alone.

**Keywords:** gene regulatory networks, dynamic Bayesian network, structural Expectation Maximization, microarray data, transcription factor binding location data.

## 1 Introduction

The establishment of gene regulatory network is critical to the understanding of the genetic regulation process. Several methodologies have been presented so far to learn gene network from microarray data, such as Boolean networks [2,3], differential equations [4,5], and Bayesian networks [6,7,8]. However, regulatory networks of the cell depend not only on the transcriptional regulation but also on the post-transcriptional and external signaling events. Recent studies show that using only gene expression data is not sufficient for estimating gene networks accurately [11], since microarray data is limited by its environment quality.

S. Cohen-Boulakia and V. Tannen (Eds.): DILS 2007, LNBI 4544, pp. 204–214, 2007.

Therefore, the use of other biological knowledge together with microarray data is important for extracting more reliable results.

Besides gene expression data, some other data sources are increasingly becoming available to aid in discovery of gene regulatory network. How to efficiently combine those different data sources together has become an important challenge in recent years [1]. Hartemink [7] developed a Bayesian network model to use location data corresponding to microarray data. Later, they improved to a dynamic Bayesian network to overcome the shortcoming of Bayesian[11]. Segal [9] used different types of data to identify sets of genes, which interact together in a cell or share common roles. In addition, there are some other methods that combine microarray data with more biological knowledge, such as DNA sequences of promoter elements [12,13] and protein-protein interaction [14,8].

In this paper, we propose a method for restructuring gene networks from microarray data and transcription factor binding location data. Our model embeds structural expectation maximization (SEM) to dynamic Bayesian network framework. We incorporate evidence from gene expression data through the likelihood, and from transcription factor binding location data through the prior. Because large-scale, genome-wide location analysis of DNA-binding regulators offers a second means for identifying regulatory relationships [10]. We expected that such an algorithm could get a more accurate learning regulatory network than methods based on gene expression data alone.

It is well-known that dynamic Bayesian network is a popular decision support model [17], which allows different strategies to integrate two data sources [26]. The unique point of our method is that we incorporate SEM algorithm, an efficient method to deal with missing data, to DBN framework. Our new dynamic Bayesian network is referred to SEM-DBN in this paper. Furthermore, we use the prior knowledge every learning step in our algorithm, in order to enhance the effect of location data.

We evaluate our method through the analysis of *Saccharomyces Cerevisiae* cell-cycle expression data. To evaluate our method, we conduct two experiments. We estimated one gene network by gene expression data alone, and by combining of expression and location data. The experimental results successfully show that the accuracy of the estimated gene network is improved after adding location data to gene expression data. The details of the experimental analyses are described in Section 4.

# 2   Methods for Estimating Gene Network

## 2.1   SEM-DBN Framework

As a graph model, a Bayesian network is defined by two parts. One is a graph structure $S$ , which is a directed acyclic graph (DAG) consisting of nodes and directed acyclic edges. The other is a parameter vector $\Theta$ comprising a set of conditional probability distributions. Given the parent $Pa(i)$ of one node $X(i)$, this node is conditionally independent of its non-descendants in a Bayesian network. Under the Markov assumption, the joint probability distribution of network can be written as:

$$P(X_1, X_2, \ldots, X_n) = \prod_{i=1}^{n} P(X_i|Pa_i) \tag{1}$$

Classical Bayesian network is unable to handle the cyclic edges [6]. Murphy and Mian [16] first employed a dynamic Bayesian network (DBN) to build such a gene expression model with cyclic edge, as shown in Fig.1. Apparently, the DBN is able to avoid the ambiguity of the edge directions [17].

The following two assumptions [18] are regarded to be a basis of our transition from a static Bayesian networks to a DBN: (1) the genetic regulation process is Markovian, that is, the expression state of one gene at one time point is dependent only on the expression state of other genes observed at the previous one time point; (2) the dynamic causal relationships among genes are invariable over all time slices, that is, the set of variables and conditionally probability definitions of the DBN are the same for each of time points.

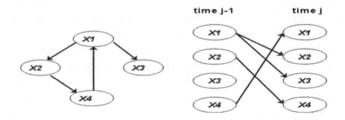

**Fig. 1.** Example of a cyclic network Bayesian network cannot handle the network (left) that contains a cycle . However, the DBN can build a cyclic structure by dividing states of a gene into different time slices (right).

In this case, the joint probability of network can be rewritten below

$$P(X_{11}, \ldots, X_{np}) = P(X_1)P(X_2|X_1) \ldots P(X_n|X_{n-1}) \tag{2}$$

where $(X_{i1}, \ldots, X_{ip})^T$ is a state vector of the $p$th gene at time $i$ , and

$$P(X_i|X_{i-1}) = P(X_{i1}|P_{i-1,1}) \times \ldots \times P(X_{ip}|P_{i-1,p}) \tag{3}$$

where $P_{i-1,j}$ denotes the state vector of the parent gene of the $j$th gene at $i-1$ time.

Our eventual goal is to learn the network from the data set $D$ generated by microarray experiments and other data sources, which requires finding the structure $S$ and parameters $\Theta$ that maximizing $P(S|D)$. To evaluate a network, we need to define a scoring function assigned to the graph. The Bayesian score is given below

$$Score(S|D) = \frac{P(D|S)P(S)}{P(D)} \qquad (4)$$

Ignoring structure prior $P(S)$ which will be discussed in Subsecsion 2.2 and the evidence $P(D)$, the problem becomes to find the best marginal likelihood given the data. It follows

$$P(D|S) = \int P(D|\Theta, S)P(\Theta|S)d\Theta \qquad (5)$$

Consequently, we will search for the DBN with highest score. Here we use an evaluation score with penalty as depicted in the previous work of Kim et al[19], $|\Theta|$ refers to the number of parameters used.

$$Score(S, \Theta|D) = logP(D|\Theta, S) - \frac{|\Theta|}{2}logp \qquad (6)$$

In the previous works [15,19], the dataset collected was assumed to be complete. But when the dataset has missing values, we cannot compute the marginal likelihood in a closed form. The Expectation Maximization (EM) algorithm is a commonly-used method to cope with missing data. In this article, we use the structural EM (SEM) [20] to learn the network from partially observable gene expression data. The concept is similar to that of the complete data problem, except that the score of the network is found using the expected sufficient statistics from the EM algorithm.

The EM algorithm has two steps. The E step assigns some random values to parameter $\Theta$ , and then the expected sufficient statistics for missing values are computed as:

$$E(N_{X_i=k,Pa_i=l}) = \Sigma_{j=1}^{p}P(X_i = k, Pa_i = l|d^j, \Theta, S) \qquad (7)$$

In the M step, the expected sufficient statistics are considered to be real sufficient statistics from a complete dataset $\hat{D}$. The next step is to estimate the value of $\Theta$ that maximizes the marginal likelihood $P(\hat{D}|\Theta, S)$ in formula (5),

$$\theta_{X_i=k,Pa_i=l} = \frac{E(N_{X_i=k,Pa_i=l})}{\Sigma_{X_i}E(N_{X_i=k,Pa_i=l})} \qquad (8)$$

In the structural EM,

$$E(N_{X_i=k,Pa_i=l})^{S'} \cong \sum_{j=1}^{p} P(X_i = k, Pa_i = l|d^j, \Theta, S) \qquad (9)$$

The resulting algorithm is shown in Fig. 2

## 2.2   Informative Structure Priors

The criterion score, introduced in the previous section, contains two quantities: the prior $P(S)$ of the network, and the marginal likelihood of the data. The

```
Choose an initial graph structure  S
While not converged
        For each  S' in neighborhood of  S
            Compute (9) using Bayesian inference algorithm [E step]
            Compute Score  S'
        End for

    S* := arg max_S' scoreS'

    If score(S*) > score(S), then improve parameters of  S* using EM

        S = S* [Structural M step]
    Else
        converged: =true
    End while
```

**Fig. 2.** Pseudo-code for structural EM

marginal likelihood shows the fitness of the model to the microarray data. The biological knowledge can then be added into the prior probability of the network $P(S)$.

Transcription factor binding location data provide evidence about regulatory relationship between a transcription factor and genes in the genome. This evidence is reported as a $p$-value, the $p$-value definition shows that: the smaller the $p$-value, the more likely the edge is to exist in the true structure[11].

We define random variable $P_i$ on the interval $[0,1]$, which is the $p$-value computed in experiment. $E_i$ is the edge, which is whether in the $S$ or not.

$$P(P_i = p|E_i \in S) = \frac{Ae^{-Ap}}{(1 - e^{-A})} \qquad (10)$$

$$P(P_i = p|E_i \notin S) = 1 \qquad (11)$$

In the first case, $E_i$ is present in structure $S$, $P_i$ is a exponential distribution, and $A$ is the parameter, it can be seen from formula (10) . In the second case, $E_i$ is not present in structure $S$, $P_i$ is uniformly distributed, as in formula (11).

We use $B$ denote $P(E_i \in S)$, according to Bayes rule, we can show that the probability edge $E_i$ is present after observing the corresponding $p$-value is:

$$P(E_j \in S|p_j = p) = \frac{ABe^{-Ap}}{ABe^{-Ap} + (1 - B)(1 - e^{-A})} \qquad (12)$$

$$P(E_k \notin S|p_k = p) = \frac{(1 - B)(1 - e^{-A})}{ABe^{-Ap} + (1 - B)(1 - e^{-A})} \qquad (13)$$

The following decomposition calculates the complete log prior probability over structures

$$logP(S) = \sum_{E_i \in S} logP(E_j \in S | p_j = p) + \sum_{E_k \notin S} logP(E_k \notin S | p_k = p) \quad (14)$$

Where formula (14) discards the normalizing constant term because it is the same for all structures. We compute prior when we search for the highest score in our SEM-DBN algorithm every time, which enhances the effect of prior information in our estimating. The values of the two parameters ($A$ and $B$)are critical to the final result. We simply use $P(E_i \in S) = B$ to indicate the probability that one edge is present in structure. The edges will be more likely present in the structure when $B > 0.5$ . Otherwise, prior is penalized if we make the selection of $B < 0.5$ . In the special case of $B = 0.5$ , the prior over structures is uniform.

As for $A$ , guessing is not a robust way for the results. We use a more robust Bayesian approach that can avoid selecting only one single value and instead marginalizes over an internal. We assume that $A$ is wih uniformly distribution over the interval $[a, b]$. So it follows

$$P(E_i \in S | p_i = p) = \frac{1}{a - b} \int_a^b \frac{ABe^{-Ap}}{ABe^{-Ap} + (1 - B)(1 - e^{-A})} dA \quad (15)$$

## 3 Results

### 3.1 Data Preparation

As a real data application, we applied our SEM-DBN method to uncover gene regulatory networks. We analyze *Saccharomyces Cerevisiae* cell cycle gene expression data given by Spellman [21]. These data were treated by four different methods: cdc15, cdc28, alpha-factor and elutriation. In the estimation of a gene network, we used the times series data from the four methods, the number of which was 24, 17, 18 and 14 respectively. We focus on a set of 14 genes, three of which are known as transcription factors with available location data. We can get their $p$-value from Lee's experiments [22].

Before feeding into our SEM-DBN model, the continuous gene expression data should be discretized. The level of data discretization is important for network inference. Yu et al. [24] stated that three categories seem to best balance the tradeoff between information loss and insufficient data for estimation. Therefore the expression values here were discretized. According to Friedman et al. [6], based on log-ratio cutoff of 0.5, we discretize the data in such a way that points less than negative 0.5 were considered "under-expressed (-1)", those between -0.5 and +0.5 were "normal (0)", while those above +0.5 were classed as "over-expressed (+1)".

To evaluate the quality of the recovery network, some criteria are useful (Hus-meier, et al., 2003): (1) the sensitivity, the proportion of recovered true edges

in the target network, and (2) the specificity, the proportion of erroneously recovered spurious edges. There are the four values needed to be calculated: true
positive (TP, a link that exists both in the true network and the recovered network), true negative (TN, a link that does not exist in either network), false
positive (FP, a link that exists only in the recovered network) and false negative
(FN, a link that exists only in the true network). Consequently, we can compute
the sensitivity and specificity from following two equations: sensitivity $= TP/$
$(TP + FN)$ and the specificity $= TN/ (TN + FP)$.

Our target dataset is stored in KEGG database [23] to be a cell-cycle sub-
network relative to G1 and S phases. The yeast cell cycle has many previously
established gene regulatory relationships from the KEGG, allowing confirmation
of the accuracy of our gene-gene relationship results. The experiments described
below are carried out with MATLAB Bayesian Network Toolbox [16] and Unscented Particle Filter Package [27]. Subsecsion 3.2 shows the results learning by
our new SEM-DBN method using expression data alone and jointly from both
expression data and location data.

### 3.2   The Experiments

We apply our SEM-DBN algorithm on the set of 14 genes, and there are three
transcription factors in our gene set, SWI4, SWI6 and MBP1. We first implement
our SEM-DBN to learn the network only by gene expression data and then learn
the network from both gene expression data and location data. In the second
experiment, when adjusting values of $A$ and $B$, we choose $A$ ranging from 2 to
5 as $B = 0.1$.

Fig.3 shows the pathway of the 14 genes in KEGG, which will be regarded as a
standard target to evaluate the results of our methods. Fig.4 is the reconstructed
gene network using only gene expression data via SEM-DBN method. Meanwhile,
Fig.5 shows learned network combining location data to expression data by our
SEM-DBN model. Both in Fig.4 and Fig.5, we used circle to represent the correct
estimation. Meanwhile, incorrectly inferred edges are marked with an 'X', and
the triangle indicated either a misdirected edge or an edge skipping at most one
node [25]. There are two correct edges related to transcription factor in Fig.5,
when comparing with Fig.4. They are SWI4 to CLB5 ($p$-value = 0.000079),
SWI6 to CLB5 ($p$-value = 0.0000013). The two edges are detected because the
evidence of location data, which are below threshold for inclusion. Consequently,
when learning from both location data and expression data, the power of location
data is enhanced. Specifically, our algorithm exploit the joint learning for each
learning step instead of learning only from the initiation.

Table 1 is the results from our two experiments. Comparing the results that
are obtained by incorporating the location information into expression data with
the result from expression data alone given in Table 1, we could observe that
adding prior knowledge was capable of improving the accuracy.

From the summary in Table 1, it is clear that we reduced the wrong and misdirected edges when use location data as supplement. Therefore, the specificity
de-creases from 42.1% to 21.1%. Table 1 also shows that the sensitivity improves

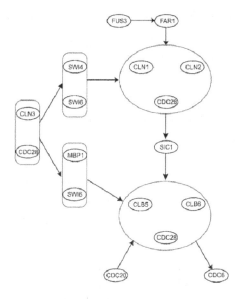

**Fig. 3.** This picture gives the target pathways from the KEGG

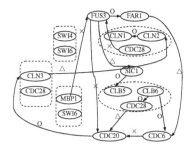
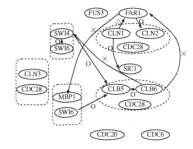

**Fig. 4.** The picture indicates the result of experiment using only expression data

**Fig. 5.** The picture one demonstrates the result of experiment 2 when combining expression data and location data

from 42.8% to 60.0%. It is apparent that the total number of edges decreased, which is because we add the informative prior in SEM-DBN model. When there are any inconsistent results between the prior and the posterior learning, the present of the edge will be penalized. That is why the total learning edges decreased, but the correct edges still high.

At the same time, the parameter impacted within a limit range. During the experiments, when adjust values of $A$ and $B$, the results indicates that it is more sensitive to the internal than the absolute values. $B = 0.1$ is proper for this case, because it will let the structure learn the edge in the following steps by SEM algorithm from a relative empty initiation.

**Table 1.** Comparison of results achieved by our two experiments

|  | Only expression data | Expression data and location data |
|---|---|---|
| correct estimation | 6 | 6 |
| wrong estimation | 4 | 3 |
| misdirected and skipping | 4 | 1 |
| sensitivity | 42.8% | 60.0% |
| specificity | 42.1% | 21.1% |

## 4    Conclusion

In this paper we developed a model for estimating gene regulatory networks by combining microarray gene expression data and transcription factor binding location data. Our method based on the framework of dynamic Bayesian network (DBN) with structural Expectation Maximization (SEM) learning algorithm. An advantage of our method is that we introduce SEM algorithm to improve the accuracy by handling the missing data. We introduce location data to the model by the usage of an informative prior. Owing to the fact that different data sources have different noise distributions, the integration can reduce the overall error present in the learned network.By adding transcription factor binding location data as a supplement with the real time series microarray data of *Saccharomyces Cerevisia* cell cycle, we estimated the gene network more accurately than using only microarray data. Specifically, the results indicated that adding prior can lower the number of false negative edge than without prior.

There are several research lines for the future work. First, our method is strongly dependent on the quality of the microarray data. Second, the discretization of data may lead to losing useful information and the data noise also has an impact on the result. An extension to solve this problem is to improve our method to deal with continuous data. In the future, one of our goals is to employ the framework reported here to deal with other data sources, such as protein-protein interaction, gene annotation, and promoter sequence. How to jointly incorporate all these additional data sources as prior knowledge together may be worth trying. Finally, our method can be used for either gene network modeling or many other problems of computational biology. The framework is a good platform to investigate biological process.

## Acknowledgement

This work was supported in part by the National Science Foundation of China under Grant No.60321002 and the Teaching and Research Award Program for Outstanding Young Teachers in Higher Education Institutions of MOE (TRAPOYT), China.

# References

1. Imoto, S., Higuchi, T., Goto, T., Miyano, S.: Error tolerant model for incorporating biological knowledge with expression data in estimating gene networks.Statistical Methodology, pp. 1–16 (2006)
2. Liang, S., Fuhrman, S., Somoyi, R.: REVEAL: a general reverse engineering algorithm for inference of genetic network architectures. In: Proc. Pacific Symposium on Biocomputing, pp. 18–29 (1998)
3. Akutsu, S., Miyano, S., Kuhara, S.: Algorithms for inferring qualitative models of biological networks. In: Proc. Pacific Symposium on Biocomputing, pp. 290–301 (2000)
4. Chen, T., He, H.L., Church, M.: Modeling gene expression with differential equation. In: Proc. Pacific Symposium on Biocomputing, pp. 29–40 (1999)
5. D'haeseleer, P., Wen, X., Fuhrman, S., Somogyi, R.: Linear modeling of mRNA expression levels during CNS development and injury. In: Proc.Pacific Symposium on Biocomputing, pp. 41–52 (1999)
6. Friedman, N., Linial, M., Nachman, I., Pe'er, D.: Using Bayesian networks to analyze expression data. Computational Biology, pp. 601–620 (2000)
7. Hartemink, A.J., Gifford, D.K., Jaakkola, T.S., Young, R.A.: Combing location and expression data for principled discovery of genetic regulatory network models. In: Proc. Pacific Symposium on Biocomputing, pp. 437–449 (2002)
8. Imoto, S., Goto, T., Miyano, S.: Estimation of genetic networks and functional structures between genes by using Bayesian networks and nonparametric regression. In: Proc. Pacific Symposium on Biocomputing, pp. 175–186 (2002)
9. Segal, E., Yelensky, R., Koller, D.: From Promoter Sequence to Expression: A Probabilistic Framework.Bioinformatics, pp. 273C282 (2003)
10. Bar-Joseph, Z., et al.: From promoter sequence to expression: A probabilistic framework. Nature Biotechnology, pp. 1337C1341 (2003)
11. Bernard, A., Hartemink, A.: Informative structure priors: Joint learning of dynamic regulatiory networks from multiple types of data. In: Proc. Pacific Symposium on Biocomputing, pp. 459–470 (2005)
12. Pilpel, Y., Sudarsanam, P., Church, G.M.: Regulatory networks by combinatorial analysis of promoter elements. Nature Genetics, pp. 153–162 (2001)
13. Tamada, Y., Kim, S., Bannai, H., Imoto, S., Tashiro, K., Kuhara, S., Miyano, S.: Estimating gene networks from gene expression data by combing Bayesian network model with promoter element detection. Bioinformatics, pp. 227–236 (2003)
14. Nariai, N., Kim, S., Imoto, S., Miyano, S.: Using protein-protein interactions for refining gene networks estimated from microarray data by Bayesian network. In: Proc. Pacific Symposium on Biocomputing, pp. 336–347 (2004)
15. Li, S.P., Tseng, J.J., Wang, S.C.: Reconstructing gene regulatory networks from time-series microarray data. Physica, pp. 63–69 (2005)
16. Murphy, K., Mian, S.: Modelling gene expression data using dynamic Bayesian networks. Technology Report, Computer Science Division, University of California Berkeley, CA (1999)
17. Husmeier, D.: Sensitivity and specificity of inferring genetic regulatory interactions from microarray experiments with dynamic Bayesian networks. Bioinformatics, pp. 2271–2282 (2003)
18. Wu, C.C., Huang, H.C., Juan, H.F., Chen, S.T.: GeneNetwork: An interactive tool for reconstruction of genetic network using microarray data. Supplementary information, Taiwan (2003)

19. Kim, S., Imoto, S., Miyano, S.: Dynamic Bayesian network and nonparametric regression for nonlinear modeling of gene networks from time series gene expression data. Biosystems, pp. 57–65 (2004)
20. Friedman, N., Murphy, K., Russell, S.: Learning the structure of dynamic probabilistic networks. In: Proc. Conf. Uncertainty in Aritif. Intell, pp. 139–147 (1998)
21. Spellman, P.T., Sherlock, G., Zhang, M.Q., Iyer, V.R., Aders, K., Eisen, M.B., Brown, P.O., Botstein, D., Futcher, B.: Comprehensive identification of cell cycle-regulated genes of the yeast Saccaromyces Cerevisiae by microarray hybridization. Mol.Biol.Cell, pp. 3273–3297 (1998)
22. Lee, T., et al.: Transcriptional Regulatory Networks in Saccharomyces cerevisiae. Science, pp. 799–804 (2002)
23. Home page of KEGG: http://www.genome.ad.jp/kegg
24. Yu, J., Smith, V.A., Wang, P.P., Hartemink, A.J., Jarvis, E.D.: Advances to Bayesian network inference for generating causal networks from observational biological data. Bioinformatics, pp. 3594–3603 (2004)
25. Zhang, Y., Deng, Z., Jiang, H., Jia, P.: Gene regulatory network constructiong using dynamic Bayesian network (DBN) with structure expectation maximization. In: Proc. International Conference on Bioinformatics and Computational Biology (BIOCOMP'06), pp. 41–47 (2006)
26. Gevaert, O., Smet, F.D., Timmerman, D., Moreau, Y., Moor, B.D.: Predicting the prognosis of breast cancer by integrating clinical and microarray data with bayesian networks Bioinformatics, pp. e184–e190 (2006)
27. Merwe, R., Doucet, A., Freitas, N., Wan, E.: The Unscented Particle Filter. Technical Report, Cambridge University Engineering Department (2000)

# Accelerating Disease Gene Identification Through Integrated SNP Data Analysis

Paolo Missier[1], Suzanne Embury[1], Conny Hedeler[1], Mark Greenwood[1],
Joanne Pennock[2], and Andy Brass[1,2]

[1] University of Manchester, School of Computer Science, Manchester, UK
[2] University of Manchester, School of Biological Sciences, Manchester, UK

**Abstract.** Information about small genetic variations in organisms, known as single nucleotide polymorphism (SNPs), is crucial to identify candidate genes that have a role in disease susceptibility, a long-standing research goal in biology. While a number of established public SNP databases are available, the specification of effective techniques for SNP analysis remains an open issue. We describe a secondary SNP database that integrates data from multiple public sources, designed to support various experimental ranking models for SNPs. By prioritizing SNPs within large regions of the genome, scientists are able to rapidly narrow their search for candidate genes. In the paper we describe the ranking models, the data integration architecture, and preliminary experimental results.

## 1 Introduction

The integration of scientific data sets can reveal opportunities for performing new forms of data analysis that cannot be supported by individual data sets, or which would otherwise lack sufficient coverage or depth. When the details of this analysis are known in advance, then we can design the integrated schema and the necessary data transformation steps with the needs of the intended application in mind. However, in many cases, converting the scientific ideas into concrete algorithms over the data is a non-obvious task. Different approaches must be prototyped and experimented with, before the most appropriate algorithm or model can be found. This requires a more flexible approach to data integration, since we cannot afford to lose information in the integration that may turn out to be critical to the implementation of the best analysis algorithm.

A problem in the life sciences that illustrates the need for experimentation, and consequent complication of the integration process, is the identification of the genes that are responsible for phenotypes in model organisms. A phenotypic trait is some observable behaviour or disease response and includes, for example, body size and susceptibility to some disease. Many phenotypes are typically the result of complex interactions among several genes, thus posing considerable challenges to the biologist wishing to understand their genetic origins.

Establishing the relationship between phenotype and one or more regions of the genome has been a research objective for quite some time [1]. The current methodology for establishing the genes which may be responsible for a quantitative trait uses elaborate breeding schemes to identify genomic regions where sequence differences among strains of the organism under study can be correlated to differences in the phenotype of

S. Cohen-Boulakia and V. Tannen (Eds.): DILS 2007, LNBI 4544, pp. 215–230, 2007.

interest. These regions are known as Quantitative Trait Loci (QTLs). They vary in size but inevitably contain many genes (100's to 1000's), all with the potential to influence the trait by some means. The challenge for biologists is then to narrow this down to a more manageable set of candidate genes, the roles of which can then be investigated using less expensive and time consuming experimental techniques.

As a result of recent research on this problem, a large number of studies identifying genetic variations within the mouse genome are now available, for many inbred strains with documented phenotypes. Each variation takes the form of a Single Nucleotide Polymorphism (SNP) — that is, a difference in a single base pair between one strain and the reference strain of the model organism. SNPs thus provide a key tool for scientists wishing to target likely candidate genes within a QTL. If a variation in phenotype (such as susceptibility to a particular disease) has a genetic cause, then there should be clear differences in the SNPs of the strains exhibiting this variation. Moreover, the locations of the SNPs within the genome can indicate the genes that play a role in determining whether an individual will exhibit the phenotype of interest or not. While it is clear that some SNPs found in QTLs are more informative than others, the precise criteria needed to isolate these SNPs are not completely clear, and their investigation is part of current research. At the same time, the sheer volume of SNPs under consideration, typically of the order of tens of thousands for a single QTL region, calls for an automation of the analysis process. Our goal is to support this exploration by providing biologists with a software environment for the semi-automated SNP analysis of SNP "informativeness".

Recognition of the value of SNPs in detecting the genes involved in specific phenotypes has fuelled the development of several publicly-accessible SNP databases. Notable among these are Ensembl [5], dbSNP [14], the Perlegen Sciences database[1], MGD [4], UCSC [8], and Wellcome-CTC Mouse Strain SNP Genotype Set[2]. Each of these resources allows the retrieval of SNPs from a given chromosome region, but they are also highly heterogeneous, in terms of access mechanisms, structure, content and quality. For example, Ensembl contains high-quality data that has been assessed by expert curators, while dbSNP contains more recent but more speculative SNPs that have not been subjected to such rigorous quality control.

In order to get a good coverage of both strains and chromosomal regions for SNP analysis, therefore, it is necessary to integrate data from several sources. Since data volumes are high (there are currently around 8 million confirmed SNPs in the mouse genome, for example), and since the various resources do not all provide suitable programmatic access to data, a materialised integration is necessary. However, at present, the main purpose of this integration is not to support a specific known application but to allow experimentation with a variety of hypothesised algorithms for assessing the likely role of a SNP in producing a given phenotypic response. We do not know at the outset what quality or coverage of SNPs will be required to provide reliable analyses of this kind. Therefore, rather than a conventional, tight integration to a fixed common schema, with "one-time" data cleaning steps, we have instead adopted a loose integration approach, which allows the user to experiment with different combinations of sources and integration approaches.

---

[1] Perlegen: http://www.perlegen.com/

[2] http://www.well.ox.ac.uk/mouse/INBREDS/

The first results of this experimentation have been implemented in a web-accessible database called *SNPit*. The *SNPit* database is populated with a loose integration of SNP and strain data covering the entire mouse genome. This paper describes our experiences in constructing *SNPit* and the loose integration approach that supports it. We begin, in Section 3, by describing the kinds of SNP scoring models that must be supported by a system such as *SNPit*. We then discuss the integration problems that arise and our solutions for them (Section 4), and present experimental evidence for the usefulness of the resulting scores (Section 5). Finally, Section 6 concludes and outlines our plans for further exploitation of the *SNPit* database through the discovery and implementation of additional SNP scoring models.

## 2   Related Work

While many examples of data integration projects can be found in bioinformatics, it is interesting to note the increased importance of automating SNP analysis, a sign that the role of SNPs in the discovery of genes responsible for particular phenotypes is widely recognized. It is no surprise, therefore, that a number of SNP searching tools are available in the public domain. A common goal of these tools is to perform large-scale searches through genome-wide collections of SNPs, in order to narrow the genotyping analysis to a small set of "optimal" SNPs. Where the tools differ is in the specific type of search filters, the analysis features offered, and the choice of primary data sources. *SNPHunter*, for example, retrieves SNPs that lie inside or around a given candidate gene [12]. The *SNPper* application described in [11] lets the user focus on highly poly-morphic regions, and filter SNPs based on their submitter (since users may attribute different reliability to SNPs coming from different submitters). Some systems, like *PolyDoms* [7] and the *SNP function portal* [13], integrate multiple data sources, but only one of these is a SNP database (dbSNP). The former provides filter options for predicted functional properties of SNPs, such as "Damaging non-synonymous SNPs", while in the latter search criteria can be expressed on a long list of annotations obtained from various other databases, e.g. at the genome, protein, pathway levels. Others, like *PupaSuite* [2] and *SNPeffect* [9], add functionality to predict the functional effect of SNPs on the structure and function of the affected protein.

We note two important differences between these tools and our *SNPit* database. Firstly, we integrate multiple sources of SNP data, allowing users to perform searches on specific sources, or to compare analysis results across sources. Secondly, since all the cited tools are specific to the human genome, SNP analysis cannot be based on observed phenotype differences among strains (because no collections of strains are available for humans). In contrast, by targeting the mouse (an important model organism), we are able to exploit the complete genome sequencing of different mouse strains, along with the growing number of available QTLs already identified for the mouse. One secondary mouse SNP database, called *Mouse SNP Miner*, is indeed described in the recent liter-ature [10]; but it is designed to perform batch analysis of potential damaging effect of SNPs, rather than for interactive search.

# 3   Capturing SNP "informativeness"

As mentioned, SNPs allow us to identify sets of candidate polymorphic genes within a QTL region which may be responsible for the disease response (or other behaviour) observed in various strains. The main intuition behind this process is the following: since different strains of the model organism exhibit the phenotype in different ways, if we can identify the regions of greatest genetic difference between those strains then we can prioritise the genes that are located in those regions for further investigation. In other words, we would like to rank the SNPs within a QTL in some way that indicates the likelihood that it contributes to the phenotypic differences observed between strains. From this, we can create a secondary ranking on the genes in which the SNPs appear.

In order to perform this ranking reliably, we need to gather together information about as many known SNPs in the QTL as possible. Since no one database, at present, can guarantee to provide this, we must collect data from several databases and integrate it. Unfortunately, there is no single way to translate the biologists' intuition regarding the informativeness of SNPs in identifying candidate genes into a procedure precise enough to be implemented in software. Therefore, we have proposed several different variants on the basic score model. The integrated data must be able to support experimentation with all these variants, so that their relative reliabilities can be explored.

The basic score model compares, for each SNP, the nucleotide base replacement observed in a single, user-selected strain, i.e., the strain that exhibits the phenotype under investigation, with those that occur in all other known strains. Each such alternative base is called an *allele*. A SNP in which the allele for the selected strain is different from that observed in all the others supports the hypothesis that the SNP plays a role in the phenotype associated with the selected strain; the SNP should therefore receive a high score.

To make this intuition precise, consider the set $S = \{S_1, ..., S_N\}$ of all known mouse strains (about 60) for which SNPs have been sequenced. Ideally, we would like to have allele information about each SNP in the entire genome for each of the strains, i.e., $A_{i,j} \in \{G, C, A, T\}$ for each SNP $i$ and strain $S_j$. In reality, sequencing efforts focus on particular genome regions and on particular strains, so that this information is missing for some strains on some SNPs – we indicate missing alleles with $A_{i,j} = N$. Note that the set $A_i = \{A_{i,j}, j : 1..N\}$ of all alleles for a SNP is a bag, rather than as set, because alleles from different strains may coincide, as shown in the example of Table 3(a).

In the basic score model, the user selects a single strain $S_{ref} \in S$ as the reference. For each SNP $i$, we compute a base score $s_{i,0}$ as the number of non-null alleles $A_{i,j} \neq N$ that are distinct from the reference, or $j \neq ref$ and $A_{i,j} \neq A_{i,ref}$. This is then normalized by the number $n'_i$ of non-null, distinct alleles $A_{i,j}$ for each $j \neq ref$, to yield the final score:

$$s_{i,ref} = s_{i,0}/n'_i$$

This model gives a high score to SNPs for which the selected strain has a unique allele but where all other alleles are the same. Consider the example of Table 3(a), available from Perlegen for SNP $\text{rs}61647296$ on chromosome 12. For selected strain $A/J$, the allele $G$ is indeed unique ($s_0 = 9$ because 9 non-reference strains have allele $T$), and furthermore, the only other known allele is $T$ ($n' = 1$). This yields a score $s_{A/J} = \frac{s_0}{n'} = 9$.

For comparison, in the SNP in Table 3(b) (rs61646963), the score for the reference strain (allele G) is only 0.5, because the allele appears among the non-selected strains, only one allele (A for strain BALB/cByJ) is different, and the non-selected strains contain two distinct values.

**Table 1.** Strains and alleles example 1

| (a) Strain | StrainAllele | | (b) Strain | StrainAllele |
|---|---|---|---|---|
| DBA/2J | T | | | |
| A/J | G | | (b) | |
| BALB/cByJ | T | | Strain | StrainAllele |
| C3H/HeJ | N | | DBA/2J | N |
| AKR/J | T | | A/J | G |
| FVB/NJ | T | | BALB/cByJ | A |
| 129S1/SvIm | N | | C3H/HeJ | G |
| NOD/LtJ | T | | AKR/J | N |
| WSB/EiJ | N | | FVB/NJ | N |
| PWD/PhJ | T | | 129S1/SvIm | G |
| BTBR T+ tf | N | | NOD/LtJ | N |
| CAST/EiJ | T | | WSB/EiJ | N |
| MOLF/EiJ | T | | PWD/PhJ | N |
| NZW/LacJ | N | | BTBR T+ tf | G |
| KK/HIJ | N | | CAST/EiJ | N |
| C57BL/6J | T | | MOLF/EiJ | N |
| | | | NZW/LacJ | G |
| | | | KK/HIJ | G |
| | | | C57BL/6J | G |

In summary, this simple model rewards SNPs where the selected strain is unique, and the alleles for all other strains are the same. Note that the score is 0 for SNPs where the allele is missing for the selected strain.

The second score model, called the *group score model*, generalises the first by allowing the comparison of two user-selected groups of strains, rather than comparing a single strain against all others. This is useful because it is often the case that a particular phenotype is observed in more than one strain. For example, it is common to want to compare strains which are known to be susceptible to a particular disease with those strains which are known to be resistant. There may be other strains for which we do not know the phenotype, and these should be excluded from the analysis. This score therefore rewards SNPs for which (i) the sets of alleles in the two selected groups are disjoint, and (ii) the alleles for each individual group are homogeneous — in the ideal case, the strains in one group will all exhibit one allele, while the strains in the other group all exhibit another allele.

Consider two disjoint sets of strains $\mathcal{S}_1 = \{S_1, ..., S_n\}$ and $\mathcal{S}_2 = \{S'_1, ..., S'_m\}$, and, for a given SNP, the corresponding bags of alleles $\mathcal{A}_1 = \{A_1, ..., A_n\}$ and $\mathcal{A}_2 = \{A'_1, ..., A'_m\}$ (the SNP index $i$ is omitted for simplicity). Let $\delta$ be the number of distinct, non-null alleles that are common to $\mathcal{A}_1$ and $\mathcal{A}_2$: $\delta = |\mathcal{A}_1 \cap \mathcal{A}_2|$, and $n'$, $m'$ the number of distinct alleles in $\mathcal{A}_1$ and $\mathcal{A}_2$, respectively. We define three variations for the group score. The simplest takes the form:

$$gs_0(\mathcal{A}_1, \mathcal{A}_2) = 1 - \frac{\delta}{n' + m'}$$

This model rewards disjoint sets of alleles, regardless of their internal homogeneity. Note however that, when using $gs_0$, one SNP for which one entire group of alleles is null gets a perfect score, because $\delta = 0$ in that case. This seems counter-intuitive, i.e., it would be misleading to give a high rank to SNPs for which a score simply cannot be computed. To counter this effect, the second variation of the model, $gs_1$, extends $gs_0$ by introducing penalty factors with values proportional to the number of null alleles in the groups under consideration:

$$p_1 = \frac{|\{A_j \in \mathcal{A}_1 | A_j = \text{N}\}|}{|\mathcal{A}_1|}$$

($p_2$ is defined similarly for $\mathcal{A}_2$). The resulting adjusted score is

$$gs_1(\mathcal{A}_1, \mathcal{A}_2) = gs_0(\mathcal{A}_1, \mathcal{A}_2) \cdot p_1 \cdot p_2$$

**Table 2.** Strains and alleles example 2

| (a) Strain | StrainAllele | | (b) Strain | StrainAllele |
|---|---|---|---|---|
| DBA/2J | G | | DBA/2J | A |
| A/J | A | | A/J | N |
| BALB/cByJ | G | | BALB/cByJ | A |
| C3H/HeJ | A | | C3H/HeJ | N |
| AKR/J | G | | AKR/J | N |
| FVB/NJ | G | | FVB/NJ | A |
| 129S1/SvIm | A | | 129S1/SvIm | N |
| NOD/LtJ | G | | NOD/LtJ | A |
| WSB/EiJ | G | | WSB/EiJ | N |
| PWD/PhJ | G | | PWD/PhJ | T |
| BTBR T+ tf | A | | BTBR T+ tf | N |
| CAST/EiJ | G | | CAST/EiJ | A |
| MOLF/EiJ | G | | MOLF/EiJ | N |
| NZW/LacJ | A | | NZW/LacJ | N |
| KK/HlJ | A | | KK/HlJ | N |
| C57BL/6J | G | | C57BL/6J | A |

Note that the values of penalties decrease as expected (because they are multiplying factors) when the number of null alleles increases. Consider the example in Table 3(a), and the two groups $\{\text{A/J}, \text{BALB/cByJ}\}$ and $\{\text{AKR/J}, \text{C57BL/6J}\}$, corresponding to allele groups $\mathcal{A}_1 = \{\text{A}, \text{G}\}$ and $\mathcal{A}_2 = \{\text{G}, \text{G}\}$. We have $\delta = |\{\text{G}, \text{G}\}| = 1$, $n' = n = 2$, $m' = 1$, and $gs_1(\mathcal{A}_1, \mathcal{A}_2) = \frac{2}{3}$, with no penalties since there are no missing alleles. The effect of penalties can be observed in the example of Table 3(b), where the alleles for A/J and AKR/J are missing. Here $p_1 = p_2 = \frac{1}{2}$, and $gs_1(\mathcal{A}_1, \mathcal{A}_2) = (1-\frac{1}{2}) \cdot p_1 \cdot p_2 = \frac{1}{8}$.

The third variation of this model accounts for the *heterogeneity* of each of the two groups, represented by the elements $h_1 = \frac{n'}{n}$ and $h_2 = \frac{m'}{m}$. The resulting score:

$$gs_2(\mathcal{A}_1, \mathcal{A}_2) = \frac{gs_0(\mathcal{A}_1, \mathcal{A}_2)}{h_1 + h_2}$$

is lower for highly heterogeneous groups. Using $gs_2()$, the score for the example of Table 3(b) would become $\frac{4}{9}$, because $h_1 = 1$, $h_2 = \frac{1}{2}$. It is possible, of course, to combine $gs_1()$ and $gs_2()$ to take into account both penalties and group heterogeneity. Note also that the scores do not take into account the number of strains in each group, which is typically very small.

Preliminary results on the performance of one of these models, $gs_1()$, are presented in Section 5.

# 4   Gathering and Integrating SNP Data

From the SNP analysis described in the previous session, we derive a number of requirements and design decisions for the management of SNP data. First of all, there is choice of publicly-accessible databases containing SNP data for specific organisms — including the mouse. These databases partially overlap in structure and content, depending on the submission policy and procedures of the controlling organization. The update frequency of the data, and thus its currency, also varies. Users tend to choose among the available data sources based on their prior confidence in its reliability, possibly cross-referencing the retrieved data with other sources for validation afterwards.

There is currently no single reference data source for SNP data, and therefore SNPs from multiple sources must be combined in order to achieve the coverage levels required by the score models described in the previous section. We have selected three of the most prominent public SNP databases, on the basis of their completeness, authoritativeness, and the complementarity of their respective content. First is Ensembl Mouse[3], a well-known source for the mouse genome, which is regarded as being of high quality thanks to the team of expert curators who make sure that only confirmed and established data is included. The second database is dbSNP, maintained by NCBI[4]; the quality of its data is known to be less consistent, since the submission process involves relatively little quality control. This, at the same time, makes dbSNP a good source for recently discovered SNPs. Thirdly, we have selected the database from Perlegen Sciences, the result of a project devoted specifically to sequencing the whole mouse genome across 15 mouse strains with high accuracy.

Programmatic access to these data sources is provided through a range of different mechanisms, including Web services (for instance, NCBI's eUtils), direct data layer access (Ensembl accepts public connections to its mySQL database) and through bulk data download. Since each region-wide SNP analysis involves retrieving and joining data sets of the order of tens of thousands of elements, followed by the execution of ad hoc algorithms, the performance of frequent bulk queries on the remote sources is likely to be poor. There is thus a need for some form of data localization, and the potential for developing further of analysis algorithms also requires the design of an integrated schema.

## 4.1   Data Integration Approach

These considerations led us to the design of a new database for SNPs, called *SNPit*, which consolidates data from the three data sources mentioned. Unlike typical OLAP

---

[3] Ensembl Mouse genome: http://www.ensembl.org/Mus_musculus/
[4] dbSNP: http://www.ncbi.nlm.nih.gov/SNP/

integration projects, the new schema is designed so that individual relations are very similar in structure to the corresponding relations in the source schemas. In practice, the database consists of a collection of materialized views on the sources, which can be pairwise joined through the use of common identifiers for the SNPs. A sketch of the data integration scenario appears in Figure 1, where the flow of SNP data across the sources is highlighted (top half). Perlegen SNPs are gradually being submitted to dbSNP, making this one of its major contributors (although the process is not yet complete).[5] In turn, data from dbSNP is gradually incorporated into Ensembl, through a slower curation process. Ensembl also includes SNP data that has been discovered by the ongoing sequencing work of the Sanger Institute in the UK.[6] As the figure shows, independent loading procedures processes have been setup for each of the three sources, using various offline data transformation techniques. As a result, we expect some of the Perlegen SNPs to appear in our dbSNP and Ensembl tables.

Maintaining separate sets of relations for each data sources has several advantages. Firstly, by directing their queries to views on a specific source, users may limit the scope of their analysis to familiar data. Secondly, overlapping SNPs from different sources are retained as separate data items, thus avoiding the problem of having to resolve all possible inconsistencies (eg different alleles detected for the same SNP and strain) upon loading. Also, tracking the correct propagation of the same SNP information from one database to the next can be done at the application level. Thirdly, both dbSNP and Ensembl are subject to ongoing revision and using separate relations makes the reloading of updated versions more manageable. Finally, since there is built-in redundancy in the loosely integrated schema, additional data sources with partially overlapping data may be added without disrupting the schema. One minor shortcoming of this approach is the need to create additional views for each useful combination of sources that are frequently queried together. The schema is designed to model the following main aspects of SNP data:

- the one-to-many relationship between a SNP and the strains in which it is known to occur. The number of strains alleles available for each SNP varies, in Ensembl, between 1 and over 60, depending on the sequencing effort carried out by the originating lab. In general, we expect that the more alleles are available, the better the chance of correlating the SNPs to phenotype differences among the strains;
- the position of the SNP, expressed as the number of bases from the start of a chromosome. This translates into a one-to-many relationship between a gene (whose position is identified by an interval of bases within a chromosome) and the SNPs that occur within its boundaries.[7]
- SNP provenance, i.e., the submitter institution along with the version of the genome used to specify the SNP position (called "build"), and other similar data;
- SNP *location*,i.e., whether the SNP occurs in a DNA fragment that is involved in the translation process for protein synthesis (a *coding* region), or in a non-coding region. This is relevant in assessing the potential consequences of a single-base mutation.

---

[5] Perlegen currently contributes about 44% of the dbSNP SNPs.

[6] http://www.sanger.ac.uk/

[7] SNPs may also occur in between genes. For this reason, we have complemented the collection of genes with a set of labels corresponding to the intergenic regions, for the purpose of our study.

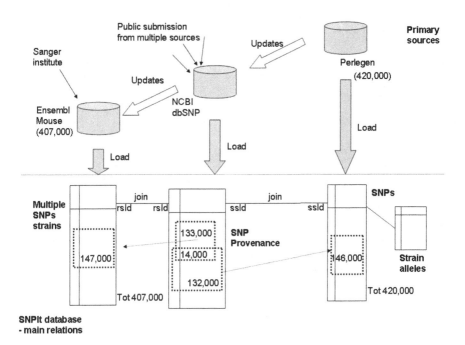

**Fig. 1.** Primary sources and main relations for the SNPit database. The number figures are data volumes for a single chromosome (12).

Both the Ensembl and the Perlegen views include SNP-to-strain and SNP-to-gene relationships, and are used to calculate the score models for data in these sources. Native Perlegen data does not include gene information, however, and we have had to add it to our database separately, as part of the loading process. This was done using the mouse genome in the Ensembl gene database. The provenance data is currently being used in a separate study concerning methods to assess the reliability of SNPs (as opposed to their "biological informativeness"), and is not discussed further in this paper. We plan to exploit location data to improve upon our current score models for SNPs, as explained in our conclusions section.

Successful joins in our schema rely upon the use of common SNP identifiers. Unfortunately, SNPs are given different types of identifier at different stages of their "acceptance" (they are also known by different names, as described in [3]). While a reference ID "rsId" (for instance rs61647296) is issued by Ensembl curators for accepted SNPs, Perlegen uses its own private naming scheme. To complicate matters still further, dbSNP makes a distinction between the SNP reference ID (when available) and the *submitter ID* ssID, issued by dbSNP at the time the SNP is entered into the database. The purpose of using reference IDs is to represent SNPs that have been identified by more than one lab, using a submitter-independent numbering scheme. This complicates the task of tracking multiple occurrences of the same SNP in our schema, since, for example, only the Perlegen SNPs that have already reached Ensembl will have an rsID. The bottom part of Figure 1 shows how rsID and ssID are used in

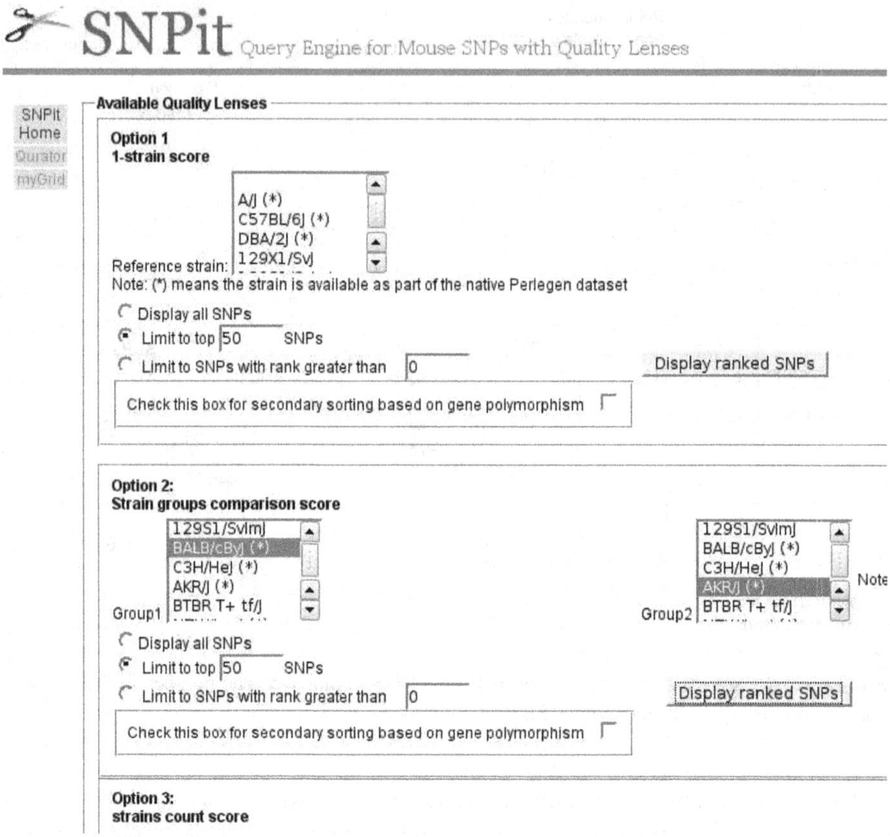

**Fig. 2.** Score model selection in the SNPit application

combination with the dbSNP view, to mediate between the Ensembl and Perlegen views. This means that the scope of a comparative analysis over the SNPs that occur in both views is limited to the subset indicated by the dotted box. The numbers in the figure provide an example, for one sample chromosome, of the amount of overlapping SNPs among the sources.

### 4.2   The *SNPit* Web Application

The *SNPit* MySQL database contains the entire set of known mouse SNPs from the three sources. A Web application (written using JSP technology) makes the score models available to end users. The application allows the biologist to (i) select SNPs for a region of interest, eg. an entire QTL, or for a set of individual genes, with some filtering capability, for example by selecting SNPs that belong to highly polymorphic regions; and (ii) repeatedly apply various score models on this selection. The available scoring options are shown in Figure 2. At this stage, the application has already fetched about 38,000 SNPs from both Perlegen and Ensembl, for a user-specified region[8]. Next, users may select a score model (three of them are available in the Web form), along with the strain or strain groups they wish to analyse (Figure 2).

Once the user has selected their preferred ranking method, the SNPs are retrieved, scored and displayed, as illustrated in Figure 3. The ranked SNPs are shown in the table on the left (with the name of each associated gene shown in the leftmost column). On the right of the main SNP result table, we also show histograms of the score distribution for the returned SNPs. Ideally, we would hope to see a highly skewed distribution, with most of the SNPs receiving low scores, but with a long thin tail showing a small number of high scoring SNPs. In order to assist the user in understanding the characteristics of this tail, we also display the histogram using a log-linear scale, which amplifies the results in the tail. The application is scheduled to be released for public access in

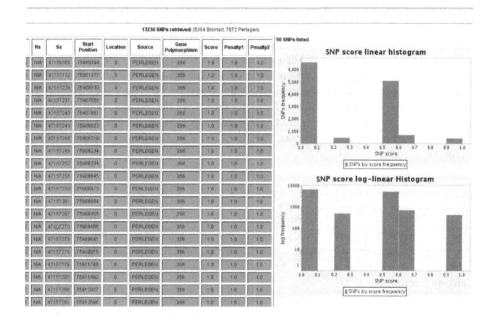

**Fig. 3.** Ranked SNPs in the SNPit application

the near future. In addition, Web Service access to the analysis functionality is also being implemented. This will make the score models available to scientific workflow applications, i.e., as part of the myGrid suite of services, which includes the Taverna workflow management system [6].

---

[8] The Ensembl data source is known in the application as "Biomart", since the Biomart version of the data has been used to populate the view. Biomart (http://www.biomart.org) is an open source project that makes data available as a data mart for analysis purposes.

## 5    Experimental Evaluation of the Score Models

The integrated views of SNP data that we have created for the *SNPit* application are only of value if they can support experimentation with different score models. In this section, we describe how we have evaluated the $gs_1()$ model over the integrated views.

### 5.1    Experiment Design

The goal of the experiments was twofold: firstly, to assess the performance of the $gs_1()$ model, i.e., to determine how well the resulting SNP ranking reflects an expert's judgment of their informativeness. More importantly, we also wanted to test the hypothesis that the SNP ranking induces a meaningful ranking on the genes themselves, by placing the best candidates at the top with sufficiently high accuracy.

The $gs_1()$ model was evaluated using three independent, manually selected test data sets consisting of SNPs from three separate, highly polymorphic QTL regions on the mouse genome, two on chromosome 12, with a size of 23 (denoted as Chr12-A) and 6 Mbases (Chr12-B), respectively, and one on chromosome 17 (8.2 Mbases), denoted as Chr17. In the experiment, the biologist selected a limited number of SNPs from a few genes that are known to be good candidates for a particular phenotype. The selection was made based on the known difference in phenotype between two groups of strains; the same two groups were then used to assign a $gs_1()$ score to all the SNPs in the selected regions. The main limitation factor for the size of the test sets, as is usually the case, is the amount of effort required to manually sift through the SNPs (the number of SNPs found in each of these regions ranges in the tens of thousands, as shown in Table 3).

### 5.2    SNP-Level Performance

A common way of assessing the performance of a score model is to compare the computed ranking with a correct binary classification (i.e., interesting vs. non-interesting) for a test data set. The performance can then be expressed in a standard way using a ROC curve, in which the ratios of false positives to true positives are plotted for various ranking thresholds.

In our case, two problems complicate this procedure. Firstly, biologists find that providing positive examples, i.e. for "definitely interesting SNPs", is much easier that providing negative examples. This reflects the nature of the experimental process, whereby the initial, large set of SNPs are all considered potentially interesting, and experimental

**Table 3.** SNPs and genes volume for the experiment QTL regions

| Data set | | SNPs in region | threshold criteria | SNPs above threshold | above threshold / total SNPs (selectivity) | genes count |
|---|---|---|---|---|---|---|
| Chr12A | Ensembl | 73242 | score > 0 | 231 | 0.3% | 81 |
| | Perlegen | 82281 | score >= 2/3 | 2656 | 3% | 145 |
| Chr12B | Ensembl | 13044 | score > 0 | 189 | 1.4% | 57 |
| | Perlegen | 25408 | score = 1 | 1471 | 5.8% | 111 |
| Chr17 | Ensembl | 40572 | score > 0 | 2849 | 7% | 64 |
| | Perlegen | 18916 | score = 1 | 2062 | 11% | 323 |

evidence as well as prior experience is applied to make some of them stand out as genuinely important. Thus, while it is natural for the expert to indicate that a data element is of interest, ruling it out completely seems harder. The second problem is the high cost of manual SNP analysis, which results in a small test set (less than 100 SNPs for each of three experiments).

Given these limitations, we decided to perform only an informal SNP analysis, and instead invest additional expert time into higher-level gene-level analysis. Thus, we only count the user-selected SNPs which are found towards the top of our ranking (the true positives), normalized by the total number of user-selected positives. These rates are greater than 95% throughout (details are omitted due to space constraints), with the exception of one of the three experiments. In that case, SNP information was simply missing from the Ensembl database at the entire gene level. The ability to perform the same analysis on alternate data sources for the same region proved important in this case; indeed, the corresponding rate for the Perlegen SNPs, our second source, is unsurprisingly high.

## 5.3   Gene-Level Performance

In the second part of the performance assessment, the genes corresponding to the test SNPs were compared to the genes for the top-ranked SNPs. As we have mentioned, not all SNPs occur within genes – many occur in between genes, and indeed, these SNPs may be among the most important, since some of these inter-genic regions are responsible for controlling the transcription rates of the neighbouring genes. We use labels of the form "between X and Y" to record the location of each such SNP; these labels count as actual genes for the purposes of our study.

The comparison of the automatically and manually ordered genes was performed as follows. For each of the three test data sets (i.e. regions), the entire set of genes for that region was ranked according to the underlying ranking of their corresponding SNPs, using a novel metric that we call *density of interesting SNPs*. Specifically, suppose that $X$ SNPs are known for gene $G$, and that $x$ SNPs out of the $X$ are above a given threshold $t$, applied to the computed ranking. We say that $G$ has a density $x/X$ of interesting genes at threshold $t$. This choice of ranking metric follows the intuition that, from a biology perspective, a gene whose SNPs are considered for the most part informative, according to our definition based of strain differences, has a higher chance of explaining the phenotype than genes with only few interesting SNPs.

As in the case of the SNPs, we were again only given positive examples of strong candidate genes by the biologist, making it difficult to estimate the number of false negatives. In this case, however, the number of genes is much smaller than the number of SNPs (less than one hundred for each experiment). Thus we can afford to have our biologist manually analyse the top-ranked genes, in order to identify *additional positives* that may have escaped attention due to the size of the original list of genes. These represent the real added-value information to the biologist: interesting genes that have been spotted only thanks to the ranking model.

Thus, our performance model is based on a two-step process, whereby the expert first provides an initial list of positive examples, which is used to plot a ROC curve where all the non-selected genes are assumed to be negatives. This is a pessimistic estimate,

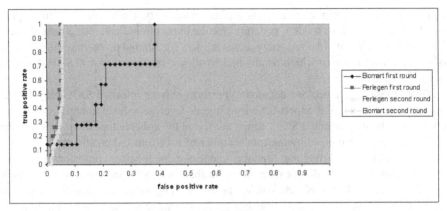

(a) Chr17

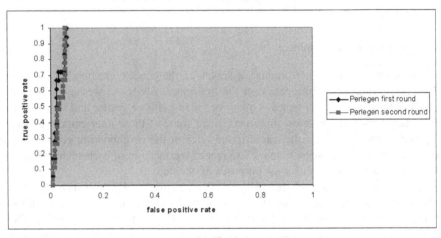

(b) Chr12-A

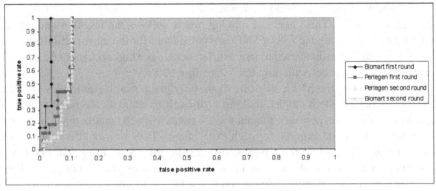

(c) Chr12-B

**Fig. 4.** ROC curves for gene-level scores

because each non-selected gene in the top ranks counts as a false positive. Then, the expert identifies additional positives from the list. These count as true positives if they lie above the threshold, and as false negatives otherwise. Although non-selected genes are again considered negatives, the new curve obtained from this list is more realistic.

Concerning the choice of threshold $t$ used to compute the gene SNP density, we observe that the model only assigns a handful of scores from the available [0,1] interval, namely 0, .25. .5, .67, and 1, effectively creating a discrete classification. This is due to the very small size (2) of the strain groups selected by the analyst for the comparison[9], which limits the possible overlaps among the alleles. By observing the frequency of occurrence of each score over all SNPs, we may select a suitable threshold that captures the majority of them – this is typically score = 1 for Perlegen, and score > 0 for Ensembl. With this assumption, we compute interesting SNP density as the ratio of SNPs that are above the threshold, to the total SNPs for the gene.

The resulting curves for each of the three experiments and for the two data sources are shown in Figure 4. In this type of chart, good results are represented by curves that rapidly reach the upper left corner, representing a region of many true positives and few false positives. Although not conclusive, our preliminary results are promising. The first chart shows the improvement of the additional expert selection (indicated as "second round"). This effect seems to be reverted in the last chart; this may be due to the relatively large number of false negatives, i.e., interesting genes with low ranking. The initial, subjective reaction from our users is that this level of accuracy may already be sufficient to significantly accelerate the search for candidate genes.

We are now experimenting with further score models that exploit some of the additional information associated to the SNPs, notably whether the SNP occurs in a coding region of the gene, and whether the base substitution actually causes a change in the corresponding amino acid. This additional knowledge can be used to improve upon our models, for example by adding weight factors to SNPs. Most of the required information for this study is already captured in our schema.

# 6   Conclusions

The problem of correlating phenotype with genotype information is important to determine the genetic cause of diseases. SNPs play an important role in current methodology, but their high volume limits the potential for their exploitation.

In this paper we have described an approach to partially automate SNP analysis, based on a data integration architecture that makes it easy to implement ranking models on large collections of SNPs, using multiple data sources. In our loose integration approach, we begin by capturing the essential attributes of SNPs as views on the primary sources, and then materialize the views into our new *SNPit* database.

We have shown encouraging experimental results for the initial SNP ranking models implemented using the database. We are now experimenting with more elaborate models,

---

[9] This could be due to the complexity involved in manual analysis when larger groups are chosen, and we expect that automated support will encourage the biologists to investigate analyses involving more strains.

that take into account the relative importance of individual SNPs, e.g. based on their location in the genome, as well as provenance information to assess their trustworthiness.

# References

1. Chakravarti, A.: Population genetics – making sense out of sequence. Nature Genetics, 21(Suppl. 1) (January 1999)
2. Conde, L., Vaquerizas, J.M., Dopazo, H., Arbiza, L., et al.: PupaSuite: finding functional single nucleotide polymorphisms for large-scale genotyping purposes. Nucleic Acids Res. 34,W621–W625 (2006)
3. Coulet, A., Smaïl-Tabbone, M., Benlian, P., Napoli, A., Devignes, M.: SNP-Converter: An ontology-based solution to reconcile heterogeneous SNP descriptions for pharmacogenomic studies. In: Leser, U., Naumann, F., Eckman, B.A. (eds.) DILS 2006. LNCS (LNBI), vol. 4075, pp. 82–93. Springer, Heidelberg (2006)
4. Eppig, J.T., Blake, J.A., Bult, C.J., Kadin, J.A., Richardson, J.E.: The Mouse Genome Database Group. The mouse genome database (MGD): new features facilitating a model system. Nucl. Acids Res. 35(Database issue), D630–D637 (2007)
5. Hubbard, T.J.P., Aken, B.L., Beal, K., et al.: Ensembl 2007. Nucl. Acids Res 35(suppl_1), D610–D617 (2007)
6. Hull, D., Wolstencroft, K., Stevens, R., Goble, C., Pocock, M., Li, P., Oinn, T.: Taverna: a tool for building and running workflows of services. Nucl. Acids Res. 34(Web Server issue), W729–W732 (2006)
7. Jegga, A.G., Gowrisankar, S., Chen, J., Aronow, B.J.: PolyDoms: a whole genome database for the identification of non-synonymous coding SNPs with the potential to impact disease. Nucleic Acids Res. 35,D700–D706 (2007)
8. Kuhn, R.M., Karolchik, D., Zweig, A.S., Trumbower, H., et al.: The UCSC genome browser database: update 2007. Nucl. Acids Res. 35(Database issue), D668–D673 (2007)
9. Reumers, J., Maurer-Stroh, S., Schymkowitz, J., Rousseau, F.: SNPeffect v2.0: a new step in investigating the molecular phenotypic effects of human non-synonymous SNPs. Bioinformatics 22(17), 2183–2185 (2006)
10. Reuveni, E., Ramensky, V.E., Gross, C.: Mouse SNP miner: An annotated database of mouse functional single nucleotide polymorphism. BMC Genomics, 8(24) (2007)
11. Riva, A., Kohane, I.S.: SNPper: retrieval and analysis of human SNPs. Bioinformatics 18(12), 1681–1685 (2002)
12. Wang, L., Liu, S., Niu, T., Xu, X.: SNPHunter: a bioinformatic software for single nucleotide polymorphism data acquisition and management. BMC Bioinformatics, 6(60), (2005) doi:10.1186/1471-2105-6-60
13. Wang, P., Dai, M., Xuan, W., McEachin, R.C., Jackson, A.U., Scott, L.J., Athey, B., et al.: SNP function portal: a web database for exploring the function implication of SNP alleles. Bioinformatics 22(14), e523–e529 (2006)
14. Wheeler, D.L., Barrett, T., Benson, D.A., Bryant, S.H., et al.: Database resources of the national center for biotechnology information. Nucl. Acids Res. 35(Database issue), D5–D12 (2007)

# What's New? What's Certain? – Scoring Search Results in the Presence of Overlapping Data Sources

Philipp Hussels, Silke Trißl, and Ulf Leser

Humboldt-Universität zu Berlin, Institute of Computer Sciences, D-10099 Berlin,
Germany
{hussels,trissl,leser}@informatik.hu-berlin.de

**Abstract.** Data integration projects in the life sciences often gather data on a particular subject from multiple sources. Some of these sources overlap to a certain degree. Therefore, integrated search results may be supported by one, few, or all data sources. To reflect these differences, results should be ranked according to the number of data sources that support them. How such a ranking should look like is not clear per se. Either, results supported by only few sources are ranked high because this information is potentially new, or such results are ranked low because the strength of evidence supporting them is limited.

We present two scoring schemes to rank search results in the integrated protein annotation database Columba. We define a *surprisingness* score, preferring results supported by few sources, and a *confidence* score, preferring frequently encountered information. Unlike many other scoring schemes our proposal is purely data-driven and does not require users to specify preferences among sources. Both scores take the concrete overlaps of data sources into account and do not presume statistical independence. We show how our schemes have been implemented efficiently using SQL.

## 1 Introduction

In research on molecular biology, very often knowledge from different domains is needed to answer practical questions. Imagine a researcher asking for the three-dimensional structure of a protein that participates in a certain metabolic pathway and is associated with a certain disease. This researcher has to query multiple data sources. For instance, she could access the Protein Data Bank (PDB) [1] for the protein structure, KEGG [7] for pathway information, and PubMed to find information about protein-disease associations. However, for the latter two aspects many other data sources could be used as well.

We call those different aspects of biomedical objects a *domain*. For a protein such domains are 3D structure, sequence, fold, functional classification, other proteins it interacts with, processes it is involved in, diseases it is associated with, etc. For many domains there exist multiple sources. For example, information about pathways can be found in KEGG [7], aMAZE [9], Reactome [6],

S. Cohen-Boulakia and V. Tannen (Eds.): DILS 2007, LNBI 4544, pp. 231–246, 2007.

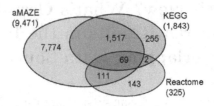

**Fig. 1.** The figures in brackets state the total number of enzyme-enzyme pairs that are connected through a chain of reaction → substrate → reaction. The figures in the different partitions show the number of pairs that occur in only one, two, or all three data sources.

and several other resources. These resources usually overlap extensionally, i.e., store the same information (and not only the same kind of information) because they partially draw their content from the same data sets. For three sources on pathways the overlap can be seen in Figure 1. But the sources also contain different and potentially unique information, due to different methodology used to curate, integrate, select, or produce the data. Thus, when a researcher looks for the pathways a given protein is involved in, the results may vary considerably depending on the chosen source. We want to provide users with a ranking of search results depending on the particular set of data sources that support it.

## 1.1   Data Model

We assume that a user is interested in information about a particular class of biological entities, called the *primary domain* $P$. Objects in $P$ are described by objects in other data sources. A group of data sources that contain information about the same type of entities or even the same entities is called *secondary domain* $D_i$. The content of secondary domains is comprised of data from various data sources $S_{i1}, \ldots, S_{im}$ and link sources $R_{i1}, \ldots, R_{il}$, where $i$ is the secondary domain. If the domain is clear $i$ can be omitted.

The link sources $R_1, \ldots, R_l$ contain entries $(s, p)$ with $s \in S_i$ and $p \in P$, i.e., they provide *links* between objects in data sources of a secondary domain and objects in the primary domain. Thus, every object in a data source of domain $D_i$ *is linked* through link sources to one or more objects in $P$ and vice versa. We also say that an object in $P$ is *annotated* by objects in data sources of $D_i$. This situation is depicted in Figure 2.

A query selects entries from $P$ by setting conditions on annotations in different domains. The result of a query, written as $res(q)$ is the set of objects in $P$ that comply with these conditions through at least one data and one link source for every domain mentioned in $q$. For a single result $p \in res(q)$ we say that the result *is supported* by at least one qualified annotation in every secondary domain. As the data and link sources in a domain overlap, an annotation supporting a result may stem from different data sources and may be linked by different link sources. According to the degree of dependence between the data and link sources of a

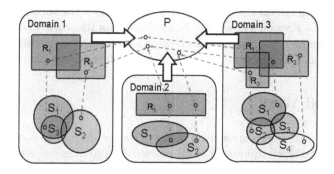

**Fig. 2.** The objects in the primary domain $P$ are annotated by objects from three secondary domains ($D_1 \ldots D_3$). Every domain contains several secondary data and link sources.

domain, certain combinations of sources frequently support query results, while other combination of sources rarely do. We make use of this fact to assign query results scores for *confidence* and *surprisingness*.

## 1.2 Scoring of Results

This paper is about ranking results in a setting described above. Integrating many sources instead of manually selecting some (the 'best' ones) comes at the risk of large result sets. Therefore, ranking of results becomes important. However, ranking is not a one-dimensional problem. Clearly, a user is most *confident* in results supported by all data sources. In the previous example a result that is supported by KEGG, aMAZE, and Reactome is one where users will be most confident that it is biologically true. However, such results are sort-of common place and thus potentially boring. Some researcher might be more interested in the contrary, i.e., *surprising* results supported by only few sources. For example, a result supported by Reactome but not by aMAZE is rather unexpected, because Reactome is much smaller. Thus, a-priori chances to find a result supported only by Reactome are small. If this occurs it makes a good starting point for a more thorough investigation with a higher chance to produce some new findings.

Both scores, confidence and surprisingness, are important. It depends on the concrete application which ranking scheme should be used for a search. In this paper we present a method to compute both scores for integrated search results over multiple domains where each domain is formed from multiple data and link sources. In contrast to much of the related work, our method does not require expert knowledge, but is merely based on the properties of the data sources themselves, i.e., the overlaps between them.

## 1.3 Paper Outline

The paper is structured as follows. We discuss related work in the next section. The surprisingness score is defined in Section 3 and the confidence score in Section 4. In Section 5 we show how to expand both measures to multi-domain queries. Section 6 describes the application of the scores in the integrated database Columba. Section 7 shows how to implement the scoring scheme and also gives some experimental results. Section 8 concludes the paper.

## 2   Related Work

One option to use information from different data sources is to provide the user only with information supported by all selected data sources, i.e., the information a user is most confident in. Marcotte et al. [10] proposed such a method for the reconstruction of metabolic pathways from protein-protein interaction data. Clearly, their results are highly trustworthy, but a biologically correct protein-protein interaction supported only by some data sources will not be considered. In contrast Yanai & DeLisi [17] used a union of different interaction data sources. This leads to good coverage, as all known interactions are listed, but possibly also many incorrect protein-protein interactions are included.

The problem of giving the user all possible information ordered according to some criteria is addressed by many projects. Internet search engines rank hits according to their expected usefulness for the query. The protein-protein database STRING [12] integrates information on protein-protein interactions from different data sources such as high-throughput experiments, literature search, or sequence comparison. A confidence score for every object is created. This score is either uniform within a data source, e.g., for an integrated source without further knowledge, or individual for every object, e.g., when text mining methods are used to extract protein-protein interactions from publications. Similar methods have been described in the area of functional analysis of microarray experiments [5]. A general framework for specifying and using such quality scores for query optimization and result ranking has been proposed in [13]. All these methods build on expert knowledge about the data sources. Such ratings are highly subjective and not easy to obtain.

In this paper we propose a method that ranks results without the need for expert knowledge. A similar idea was proposed by Florescu et al. in [4] for the purpose of query optimization. Given a query they want to optimize the ratio between the execution cost and the size of the result set. To answer the query they first estimate which sources will return most results and then choose $k$ sources, based on the selectivity of the source and the overlap with other sources.

A different approach is described by Lacroix et al. in [8] for estimating the size of the result set. They assume a network of interlinked sources and data objects. A query poses conditions on a start source and returns results from a primary source by analyzing all paths from the start to the primary source. To estimate the size of the result set they pre-compute overlap statistics for different paths using sampling. In Bleiholder et al. [2] these overlap statistics are used to optimize queries over multiple data sources to solve the Budgeted Maximum Coverage problem. In contrast to this work, we use a simpler model (primary and secondary sources) and focus on ranking of results in result sets, not on query optimization.

## 3    Surprisingness of Results

We now present a framework for measuring the surprisingness of a search result. Confidence will be defined in Section 4. We develop our model starting from a single domain with a single data and link source and then extend it to multiple data and link sources. The extension to multi-domain queries is given in Section 5.

We assume that the result set contains objects from the primary source. The user can restrict this set by setting conditions on objects in secondary domains. An object in the primary source is contained in the result set if it is supported by at least one qualified annotation in every queried domain.

### 3.1    Single Data Source

We start with the simple scenario of a single domain $D$, a single data source $S$, and a single link source $R$ as shown in Figure 3(a). Without loss of generality we assume that every annotation $s \in S$ is linked to at least one object $p \in P$ through at least on link $r \in R$ (we can safely delete all other annotations and links since they can never select entries in $P$). A query selects objects in $D$ and determines the set of objects in $P$ that are linked by at least one link in $R$.

For a given query $q$ we calculate the probability that a randomly chosen object $p \in P$ is part of the result set of $q$. We first derive the a-priori probability that a randomly chosen annotation $s \in S$ is linked to a randomly chosen object $p \in P$:

$$P((s,p) \in R) = \frac{|R|}{|P \times S|} = \frac{|R|}{|P| * |S|} \tag{1}$$

A randomly chosen $p \in P$ takes part in the query result if it is linked to at least one qualified annotation $s \in S$ by at least one link $r \in R$. If we assume that $q$ selects $k$ annotations and take into account that a single object in $P$ can be selected by multiple annotations in $S$, then the probability that a concrete $p \in res(q)$ is selected is precisely the probability that not none of the $k$ selected annotations is linked to $p$, which gives:

$$P(p \in res(q)) = 1 - \left(1 - \frac{|R|}{|P| * |S|}\right)^k \tag{2}$$

Clearly, we could also estimate the value of $k$ a-priori using attribute selectivities. Note that this formula ranks all objects in a result set of a query equal. This is expected, as we want to rank a result by the subset of sources that supports it in any domain. Therefore, differences in the computed score only appear when more than one source is present.

### 3.2    Multiple Sources in a Single Domain

We now extend our framework to the case of $m$ data sources $S_i$ and $l$ link sources $R_j$, $1 \le i \le m$ and $1 \le j \le l$ for a single domain $D$ as shown in Figure 3(b).

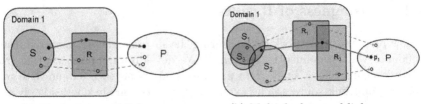

(a) Single data and link source.          (b) Multiple data and link sources.

**Fig. 3.** An object $p \in P$ is supported by objects $s \in S_i$ in a domain

An object $p \in P$ is in the result set of a given query $q$ if it is supported by at least one qualified annotation $s$ linked through at least one link $r$. However, $s$ as well as $r$ can be contained in various combination of sources. Consider the situation shown in Figure 3(b) with three overlapping data sources $S_1$, $S_2$, and $S_3$ and two overlapping link sources $R_1$ and $R_2$. $S_1$ and $S_3$ strongly overlap, while $S_2$ mostly contains divergent data. In this situation it is likely that a query result is linked to a qualified annotation contained in $(S_1 \cap S_3)\backslash S_2$ or $S_2\backslash(S_1 \cup S_3)$. Such query results shall be assigned a low score for *surprisingness*. We would rate a result more unlikely and therefore more surprising that is supported by a qualified annotation in $S_3\backslash(S_1 \cup S_2)$. Clearly, to compute the score we also have to consider over which combination of link sources $s$ is linked to $p$. Note that according to our understanding of surprisingness, a high score might also be assigned to results with incorrect annotations. This is in the line of our argument, since errors can be considered surprising and certainly require user attention.

The space of all annotations in $D$ is partitioned into disjoint subsets according to the overlaps of data and link sources. Some of these subsets are represented by different colors in Figure 3(b). We call these partitions in data sources $Z_1, \ldots, Z_n$. The assignment of annotations to partitions can be represented by a *domain-vector* $v$ of size $m * l$ for a domain with $m$ data and $l$ link sources. If annotation $s \in S_i$ and $(s,p) \in R_j$ we set $v_{i,j} = 1$, and $v_{i,j} = 0$ otherwise. In Figure 3(b) an annotation contained in $S_1 \cap S_3\backslash S_2$ that is linked over $R_1$ corresponds to the domain-vector $v_{i,1} = 101$ and $v_{i,2} = 000$. It follows that $2^{m*l}$ different domain vectors are possible. Now consider a single annotation $s$ selected by $q$. Intuitively, a $p$ linked to $s$ is the more surprising, the smaller the partition $Z_k$ is in which $s$ lies.

However, we need some more work to derive a suitable definition for surprisingness. We compute the surprisingness for each annotation selected by a query which might later be aggregated into a score for an object $p$ linked to multiple annotations. Let $Z_k$ be the partition in which an annotation $s$ lies that is selected by a query $q$. We estimate the probability that $p$ is verified by all sources that contain $Z_k$ and no others, which depends on the a-priori overlaps of sources. That means, we want to know how likely it is that a result for a given query is verified by a certain combination of available sources. The less likely, the more surprising is the result.

To answer this we first estimate the probability that for a given query a result is verified by a particular data source $S_x$ provided that it is verified by at least one source in $D$. This is different from Equation 2 because $p$ can be selected by other sources than $S_x$. Let $q_{Sx}$ denote the subset of $res(q)$ that is verified by $S_x$. Using Bayes's Theorem we get:

$$P(p \in q_{Sx} | p \in res(q)) = \frac{P(p \in res(q) | p \in q_{Sx}) * P(p \in q_{Sx})}{P(p \in res(q))} \tag{3}$$

Clearly, the probability that an object is verified by at least one data source provided that it is verified by a particular $S_x$, $P(p \in res(q) | p \in q_{Sx})$, is 1 because the first event logically implies the second one. The a-priori probability $P(p \in res(q))$ is given by Equation 2, where $|S|$ now denotes the set of all unique annotations in $D$. We only miss the a-priori probability $P(p \in q_{Sx})$. For this probability we must take into account that not every object in $S$ is contained in $S_x$ and not every link in $R$ links annotations $s \in S_x$ to a $p \in P$. We therefore can identify a subset of $R$, denoted as $R_x$, that only contains links from $s \in S_x$. Analogously, we can distinguish a subset of $P$, called $P_x$ that contains entries $p$ that are supported by an annotation $s \in S_x$.

$$P(p \in q_{Sx}) = 1 - \left(1 - \frac{|P_x|}{|P|} * \frac{|S_x|}{|S|} * \frac{|R_x|}{|P_x| * |S_x|}\right)^k$$
$$= 1 - \left(1 - \frac{|R_x|}{|P| * |S|}\right)^k \tag{4}$$

Thus, Equation 3 can be rewritten as:

$$P(p \in q_{Sx} | p \in res(q)) = \frac{1 - \left(1 - \frac{|R_x|}{|P| * |S|}\right)^k}{1 - \left(1 - \frac{|R|}{|P| * |S|}\right)^k} \tag{5}$$

We now determine the probability that a particular $p$ is supported by qualified annotations in a partition $Z_k$. Here as well we denote the subset of $res(q)$ verified by annotations in $Z_k$ as $q_{Zk}$. This gives:

$$P(p \in q_{Zk} | p \in res(q)) = \frac{1 - \left(1 - \frac{\left| \bigcap_{S_i \supseteq Z_k} R_i \setminus \bigcup_{S_i \not\supseteq Z_k} R_i \right|}{|P| * |S|}\right)^k}{1 - \left(1 - \frac{|R|}{|P| * |S|}\right)^k} \tag{6}$$

### 3.3   Surprisingness Score of a Single Annotation

To value the surprisingness of a single annotation we use the measure of self-information as defined by Shannon. Consider a domain-vector $v$ as a symbol in a message, the self-information of $v$, $I(v)$, depends on the probability of its occurrence and is defined as $I(v) = -\log_2(P(v))$. Accordingly, we calculate the surprisingness score for $p$ that is contained in the result set of a given query by applying the probability that $p$ is supported by an $s \in Z_k$ using Equation 7. Definition 1 formalizes this approach:

**Definition 1 (Surprisingness for a single annotation).** *Let $q$ be a query selecting an annotation $s$, let $p$ be linked to $s$, and let $s$ lie in partition $Z_k$ of $D$. The surprisingness $S(p,s)$ of $p$ with respect to $s$ is defined as:*

$$S(p,s) = -\log_2 P(p \in q_{Zk}|p \in res(q)) \tag{7}$$

### 3.4   Surprisingness Score for a Single Domain

Equation 7 only gives the probability that a given $p$ is linked to a given annotation $s$ selected by a query $q$. But we want a score for $p$ given all its linked annotations selected by $q$, as shown in Figure 4. Therefore, we will need to aggregate scores of multiple $s$.

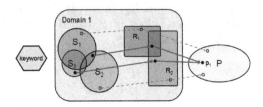

**Fig. 4.** An object linked to two qualified annotations

Suppose, the primary domain contains protein structures and the secondary domain multiple data sources about scientific publications. Assume we query with a keyword and receive a structure $p$ that is linked to multiple qualified publications. If all publications are contained in the same combination of data sources, intuitively the number of publications that verify $p$ does not influence its surprisingness. In this case $p$ shall be assigned the same surprisingness score as assigned to a single publication. Now imagine $p$ is linked to multiple qualified publications contained in different combinations of data sources as depicted in Figure 4. Clearly, if most selected publications linked to $p$ are highly surprising we also want to assign $p$ a high surprisingness score. We therefore define the surprisingness of $p$ as the average of the surprisingness scores for every qualified publication that supports $p$.

**Definition 2 (Surprisingness for multiple annotations).** *Let $q$ be a query and $p \in res(q)$ be linked to a set $T$ of annotations selected by $q$. The surprisingness $S(p,T)$ of $p$ is defined as:*

$$S(p,T) = \frac{1}{|T|} \sum_{s \in T} S(p,s) \tag{8}$$

## 4   Confidence of Results

As explained in Section 1, researches are not solely interested in highly surprising query results but also in trustworthy results. A researcher might want to rank those results high that are likely to be correct. Having multiple data sources in a domain, intuitively every data source that verifies a query result $p$ increases the confidence in the correctness of $p$. Thus, a straightforward method to value the confidence of a query result would be to count the number of sources verifying the result. But here too we have to consider that the different data sources within a domain are not independent. If, for example, a query result $p$ is verified by two data sources, the confidence in $p$ being correct is the higher the lower the degree of dependence between those data sources is, because then it is more likely that information contained in both sources is the outcome of independent experiments rather than information stemming from the same resource.

Consider again the situation shown in Figure 3(b). We are most confident in annotations that are contained in $S_1 \cap S_2 \cap S_3$ and linked through both link sources $R_1$ and $R_2$. If we consider annotations $s_1 \in (S_1 \cap S_3) \backslash S_2$ and $s_2 \in (S_1 \cap S_2) \backslash S_3$, both linked only over $R_1$, we intuitively assign $s_2$ a higher confidence score because $S_1$ and $S_3$ strongly overlap, while $S_1$ and $S_2$ do not. More generally, for the confidence score we want to use the probability that an annotation is contained in a combination of sources given that the annotation is contained in at least one source.

**Definition 3 (Confidence for a single annotation).** *Let $q$ be a query selecting an annotation $s$, let $p$ be linked to $s$ through $r$, and let $s$ be contained in the the partition $Z_k$. The confidence $C(p, s)$ of $p$ with respect to $s$ is defined as:*

$$C(p, s) = 1 - \log_2 \frac{\left| \bigcap_{S_i \supseteq Z_k} S_i \right|}{\left| \bigcup_{S_i \supseteq Z_k} S_i \right|} \tag{9}$$

Resulting from Definition 3 the score for a $p$ linked to an annotation $s$ that is contained in only one source is 1. Note, the confidence score for $p_1$ that is annotated by $s_1 \in Z_k$ is always lower or equal to the score for $p_2$ annotated by $s_2 \in Z_l$, with $Z_l$ being the intersection between data sources in $Z_k$ and an additional data source $S_i$.

So far, we considered the confidence for a result supported by only a single annotation. We shall now show how to aggregate confidence scores for multiple annotations. While the number of qualified annotations linked to a query result $p$ does not influence its score for surprisingness, it clearly enhances the trust in the correctness of $p$. As we consider every single annotation as an evidence that $p$ is an answer to a given query we sum up the confidence scores of all qualified annotations linked to $p$ to calculate the confidence score of $p$.

**Definition 4 (Confidence for multiple annotations).** *Let $q$ be a query and $p \in res(q)$ be linked to a set $T$ of annotations selected by $q$. The confidence $C(p, T)$ of $p$ is defined as:*

$$C(p, T) = \sum_{s \in T} C(p, s) \tag{10}$$

## 5   Multi-domain Query Results

In the last two sections we defined scores for surprisingness and confidence for single domain queries. In this section we explain how to use these values to compute a surprisingness and confidence score in a multi-domain setting.

We assume that different secondary domains are statistically independent. We can make that assumption as according to our model we group data sources that contain information about the same type of biological entities in one domain. To compute an overall surprisingness score we add up scores from all secondary domains given in a query. We do this as we consider a result surprising when it is surprising for at least one domain. In contrast, for the confidence, only those results of multi-domain queries shall be ranked high that have high confidence scores in many domains. To ensure this, we normalize single domain scores resulting from Equation 10 before multiplying all scores for the multi-domain confidence score.

## 6   Multi-domain Setting: Columba

In this section we introduce our real world example, where the scores presented in this paper are beneficial. We developed the integrated database Columba [14]. This database focuses on protein structures from the Protein Data Bank (PDB) [1] that are annotated by objects of different domains, such as fold, sequence, function, publication, metabolic pathway, or taxonomic classification.

We apply our scoring methods for ranking query results to parts of the Columba database. We use as primary domain the protein structures given by the PDB. Objects in the PDB are annotated by the secondary domains *sequence*, *publication*, and *metabolic pathway* as shown in Figure 5.

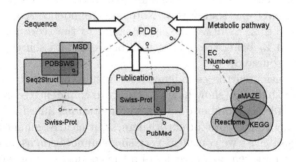

**Fig. 5.** The PDB annotated by secondary domains that contain multiple data sources. Note, the size and overlap of data and link sources does not necessarily reflect reality.

The domain *sequence* contains the data source Swiss-Prot [3] and has three different link sources linking entries from Swiss-Prot to entries in the PDB, namely PDBSWS [11], Seq2Struct [16], and MSD [15]. The overlap of link sources is given by identical entries in the sources.

Data sources for the domain *metabolic pathway* are KEGG [7], aMAZE [9], and Reactome [6]. To compare these heterogeneous data sources we extract information on the level of reactions. We store enzyme-enzyme pairs that are connected through a path enzyme → reaction → substrate → reaction → enzyme. The overlap of data sources is given by identical enzyme-enzyme pairs. If we consider enzymes as nodes and a pair of enzymes as edges in a graph we can compute paths between enzymes. We therefore can answer queries such as "Which PDB entries are less than 3 steps away from an enzyme with EC number 2.7.1.1 (Hexokinase)". The data sources are linked to the PDB through EC numbers given in the PDB as well as in the three data sources.

In the third domain *publications* we use PubMed as data source. The articles referenced in PubMed can be linked directly to the PDB using the references given in the PDB. But articles in PubMed can also be linked to the PDB via Swiss-Prot.

A multi-domain query for this setting is for example "Give me all protein structures that are up to 7 steps away from an enzyme with EC number 1.14.16.1 (Phenylalanine hydroxylase), linked to entries in Swiss-Prot that contain the keyword *Phenylalanine catabolism*, and linked to publications that mention the disease *Phenylketonuria*". This query returns in total 17 PDB chains. Using our scoring scheme we can rank the results according to their surprisingness and their confidence.

# 7    Implementation and Evaluation

In this section we show which values can be precomputed and how to implement this computation inside a relational database environment. We additionally show for some exemplary queries the impact of surprisingness and confidence scores.

## 7.1    Precomputation of Values

To compute the surprisingness of an object $S(p, T)$ we must first compute for every object in a domain the probability $P(p \in q_{Zk} | p \in res(q))$ as given by Equation 6. We therefore require information about the size of the primary domain and the size of data and link sources in secondary domains. To gather $|P|$ we simply count the number of objects in the primary domain. To gather $|S|$ and $|R|$ we first have to integrate all objects of the data and link sources of $D_i$ and then count the number of unique objects in both integrated sources. Knowing these values we can compute the value $c_1 = 1 - \frac{|R|}{|P|*|S|}$ for domain $D_i$. But we also require the size of different partitions of data and link sources. To gather these data we precompute and store the domain-vector $v$ for every unique object in $S$ of dimension $D_i$. We can determine the size for every partition $Z_k$ by determining the frequency of different patterns in $v$, denoted as $freq(v)$. But to solve Equation 6 we require the value for the size of partitions in the link sources, denoted as $link\_size(v)$. We can determine $link\_size(v)$ by summing up for every $s \in Z_k$ the number of $(s, p) \in R$. Knowing this value we can compute the value $c_2 = 1 - \frac{link\_size(v)}{|P| * |S|}$. The only value in Equation 6 that can not be precomputed is $k$, the number of qualified objects in $D_i$. But we can substitute variables in Equation 6 to gain the following equation for $S(p, s)$:

$$S(p, s) = \frac{1 - (c_2)^k}{1 - (c_1)^k} \tag{11}$$

To compute the confidence score for a result $p$ we must compute $C(p, s)$ as given by Equation 9. Here we require the size of all unions and intersections of data sources $S_i$ that contain $s$. Both values are independent of a particular query and therefore can be precomputed using $freq(v)$. For a given domain-vector $v$ of length $n$ the sum of $freq(v')$ with $v' : v' \wedge v = v$ is the size of the intersection and the sum of $freq(v'')$ with $v'' : v'' \wedge v \neq 0^n$ is the size of the union for a combination of sources. We can thus write Equation 9 as:

$$C(p, s) = 1 - \frac{\sum\limits_{v' \wedge v = v} freq(v')}{\sum\limits_{v'' \wedge v \neq 0^n} freq(v'')} \tag{12}$$

We can precompute the confidence score $C(p, s)$ for all possible intersections of sources in $D_i$, but we have to store $2^{m*l}$ confidence values for one domain with $m$ data and $l$ link sources. This means, we can precompute the size of partitions and unions only for a limited number of data and link sources. But we expect that in real world applications such as Columba this will not be a problem. If the problem arises, some heuristics for precomputation must be introduced.

## 7.2   Implementation

**Precomputation.** The integrated database Columba is implemented on Post-GreSQL 8.2. In Columba every data and every link source is stored in its own table. For every domain we store the domain-vectors $v$ in a separate table. We precompute and store all sizes and frequencies mentioned in the last section in statistics tables. To compute $freq(v')$ and $freq(v'')$ we use the provided functions `bit_and()` and `bit_or()` of PostGreSQL.

**Execution of Queries.** We now describe how to use the precomputed values to compile a ranked result set for a given query. The compilation of the result set with scores is done in four steps. For every domain given in the query we first select all annotations $s$ fulfilling the conditions posed in the query and link them to entries in $P$. In this step we also return the precomputed values for every pair $(p, s)$, including $C(p, s)$. In the second step we determine $k$ and calculate $S(p, s)$. In the third step we aggregate the surprisingness and confidence scores of a single domain for an object $p$. In the last step – if the query poses conditions on multiple domains – we aggregate the scores for an object $p$ over all domains.

We will explain this by a simple example that selects chains of protein structures from the PDB that are supported by entries in Swiss-Prot, which contain the keyword *Phenylalanine catabolism*. Figure 6 shows the SQL query to find all combinations of PDB chains and qualified entries in Swiss-Prot. For every combination we return the values for $c_1$ and $c_2$ and the confidence score $C(p, s)$. In the next step we determine $k$ by counting all unique Swiss-Prot ids and then compute $S(p, s)$. In the last step we aggregate the scores for every PDB chain by averaging over the surprisingness scores and sum over the confidence scores.

```
SELECT seq_int_links.pdb_chain, swissprot.id,
       stats.c1, stats.c2, stats.confidence_ps
FROM   swissprot, seq_int_data, seq_int_links, stats
WHERE  swissprot.keyword = 'Phenylalanine catabolism'
AND    swissprot.id = seq_int_data.swissprot_id
AND    seq_int_data.vector = stats.vector
AND    swissprot.id = seq_int_links.swissprot_id
```

**Fig. 6.** SQL query to return all PDB chain - Swiss-Prot id combinations given the keyword *Phenylalanine catabolism* and some constants

### 7.3 Evaluation

**Overlap of Sources in Columba.** The three data sources that link Swiss-Prot entries to chains in the PDB have an overlap of 51,051, i.e., most of the links of MSD (total 69,785) and PDBSWS (total 69,303) are contained in that overlap (data not shown). Seq2Struct contains in total 216,539 links, i.e., most links between the PDB and Swiss-Prot are only contained in that source. The overlap for the data sources of metabolic networks is given in Figure 1. aMAZE contains the highest number of enzyme-enzyme combinations, mainly due to the fact that reactions in aMAZE are always bi-directional. In the publications domain we have 73,945 links from PDB chains directly to PubMed, most of which are contained in the 223,156 links over Swiss-Prot to PubMed.

**Queries on Columba.** To evaluate our approach we queried the Columba database using keywords on a single domain, e.g., "Give me all PDB chains annotated by Swiss-Prot entries that contain the keyword *Phenylalanine catabolism*". We used all distinct keywords from Swiss-Prot (in total 881) to query the sequence domain and 1,000 randomly chosen MeSH terms to query the publication domain. For evaluation we excluded empty result sets and result sets in which all entries had the same confidence or surprisingness score. This results in 727 result sets for the sequence domain and 695 for the publication domain.

For every result set we normalized the confidence and surprisingness scores to gain values between 0 and 1. We sorted entries in the result set into 11 buckets ([0, 0.1), [0,1, 0.2), ..., and an own bucket for [1]) according to their confidence or surprisingness scores. Figures 7(a)-(d) show the average frequencies of entries in a bucket for the result sets of the sequence and publication domain.

Figures 7(a) and (c) show that on average only 14 % of the entries in a result set for the sequence and 12 % for the publication domain have a normalized

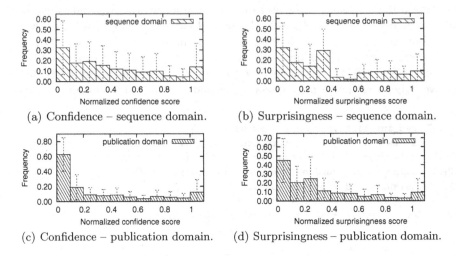

(a) Confidence – sequence domain.     (b) Surprisingness – sequence domain.

(c) Confidence – publication domain.     (d) Surprisingness – publication domain.

**Fig. 7.** Average frequency and standard deviation of normalized confidence and surprisingness scores for queries on the sequence and the publication domain

confidence value of 1. Most entries in the result set (50 % for sequence and 82 % for publication domain) have normalized confidence values between 0 and 0.2. Figures 7(b) and (d) show the data for the surprisingness score. In both domains on average about 9 % of entries in result sets have a normalized surprisingness score of 1. Here as well the largest bucket is the bucket that contains entries with scores between 0 and 0.1. The high standard deviation for all buckets can be explained by varying distributions of scores within the result sets. Consider a result set in which the entries only have two different scores, which is typical for small result sets. Clearly, a subset of entries will be in the bucket with value 1, while the other subset is in one of the remaining 10 buckets. This subset can contain one entry or all but one entry of the result set. The figures for the metabolic pathway domain are not displayed. Concluding, the figures indicate that both scores will nicely rank entries in the result set for the given domains.

We now present an exemplary query on multiple domains and parts of its result set. The query "Give me all PDB chains that are annotated by Swiss-Prot entries with the keyword *Glycolysis*, that are linked to PubMed articles containing the word *Glycolysis*, and that are at most three steps away from the protein with EC number 2.7.1.1 (Hexokinase)" returns 109 chains from the PDB. Table 1(a) and 1(b) show the top 5 results sorted either by confidence or surprisingness.

The most confident results are structures for the protein phosphoglucose isomerase. This is expected as the protein is only one reaction away from the hexokinase in the glycolysis pathway. Note, the top 5 most surprising results contain completely different chains in the PDB, including a pyruvate kinase (1pky) that is also in the glycolysis pathway, but further away from hexokinase than phosphoglucose isomerase.

**Table 1.** The top 5 query results for different sorting

| (a) Sorted by Confidence | | | | (b) Sorted by Surprisingness | | | |
|---|---|---|---|---|---|---|---|
| PDB id | chain | Confidence | Surprising-ness | PDB id | chain | Confidence | Surprising-ness |
| 1dqr | A | 1.0 | 20.6 | 1pky | C | 0.1 | 22.8 |
| 1dqr | B | 1.0 | 20.6 | 2pgi | - | 0.5 | 22.2 |
| 1g98 | A | 0.6 | 18.3 | 1c7q | A | 0.5 | 22.2 |
| 1g98 | B | 0.6 | 18.3 | 1c7r | A | 0.5 | 22.2 |
| 1xtb | A | 0.6 | 18.3 | 1i33 | D | 0.1 | 22.2 |

# 8 Conclusion

In this paper we defined the surprisingness and the confidence score for an object in a result set that is annotated by multiple, possibly overlapping data sources. Both scores can be used to rank objects in a result set. We showed its applicability to biological data using parts of the integrated database Columba. In the future we plan to integrate both scoring schemes in the Columba web interface. In addition we will further investigate the possibility to extend both score definitions to also account for the distribution of source combinations within a single result set.

# References

1. Berman, H., Westbrook, J., Feng, Z., Gilliland, G., et al.: The Protein Data Bank. Nucleic Acids Research 28(1), 235–242 (2000)
2. Bleiholder, J., Khuller, S., Naumann, F., Raschid, L., Wu, Y.: Query Planning in the Presence of Overlapping Sources. In: Ioannidis, Y., Scholl, M.H., Schmidt, J.W., Matthes, F., Hatzopoulos, M., Boehm, K., Kemper, A., Grust, T., Boehm, C. (eds.) EDBT 2006. LNCS, vol. 3896, pp. 811–828. Springer, Heidelberg (2006)
3. Boeckmann, B., Bairoch, A., Apweiler, R., Blatter, M.-C., et al.: The SWISS-PROT protein knowledgebase and its supplement TrEMBL in 2003. Nucleic Acids Research 31(1), 365–370 (2003)
4. Florescu, D., Koller, D., Levy, A.: Using Probabilistic Information in Data Integration.. In: Proceedings of the VLDB, pp. 216–225. Morgan Kaufmann, San Francisco (1997)
5. Huttenhower, C., Hibbs, M., Myers, C., Troyanskaya, O.: A scalable method for integration and functional analysis of multiple microarray datasets. Bioinformatics 22(23), 2890–2897 (2006)
6. Joshi-Tope, G., Gillespie, M., Vastrik, I., D'Eustachio, P., et al.: Reactome: a knowledgebase of biological pathways. Nucleic Acids Research 33, D428–D432 (2005)
7. Kanehisa, M., Goto, S., Kawashima, S., Nakaya, A.: The KEGG databases at GenomeNet. Nucleic Acids Research 30, 42–46 (2002)
8. Lacroix, Z., Murthy, H., Naumann, F., Raschid, L.: Links and Paths through Life Sciences Data Sources. In: Rahm, E. (ed.) DILS 2004. LNCS (LNBI), vol. 2994, pp. 203–211. Springer, Heidelberg (2004)

9. Lemer, C., Antezana, E., Couche, F., Fays, F., et al.: The aMAZE LightBench: a web interface to a relational database of cellular processes. Nucleic Acids Research 32, D443–D448 (2004)

10. Marcotte, E., Pellegrini, M., Ng, H., Rice, D., et al.: Detecting protein function and protein-protein interactions from genome sequences. Science 285(5428), 751–753 (1999)

11. Martin, A.C.: Mapping PDB chains to UniProtKB entries. Bioinformatics 21(23), 4297–4301 (2005)

12. von Mering, C., Jensen, L., Snel, B., Hooper, S., et al.: STRING: known and predicted protein-protein associations, integrated and transferred across organisms. Nucleic Acids Research 33, D433–D437 (2005)

13. Naumann, F., Leser, U., Freytag, J.-C.: Quality-driven Integration of Heterogenous Information Systems. In: Proceedings of the VLDB, pp. 447–458. Morgan Kaufmann, San Francisco (1999)

14. Rother, K., Müller, H., Trissl, S., Koch, I., et al.: Columba: Multidimensional Data Integration of Protein Annotations. In: Rahm, E. (ed.) DILS 2004. LNCS (LNBI), vol. 2994, pp. 156–171. Springer, Heidelberg (2004)

15. Velankar, S., Mcneil, P., Mittard-Runte, V., Suarez, A., et al.: E-MSD: an integrated data resource for bioinformatics. Nucleic Acids Research 33, D262+ (2005)

16. Via, A., Zanzoni, A., Helmer-Citterich, M.: Seq2Struct: a resource for establishing sequence-structure links. Bioinformatics 21(4), 551–553 (2004)

17. Yanai, I., DeLisi, C.: The society of genes: networks of functional links between genes from comparative genomics. Genome Biology, 3(11) : (research0064) (October 2002)

# Using Annotations from Controlled Vocabularies to Find Meaningful Associations

Woei-Jyh Lee[1], Louiqa Raschid[1], Padmini Srinivasan[2],
Nigam Shah[3], Daniel Rubin[3], and Natasha Noy[3]

[1] University of Maryland, College Park, MD 20742, USA
{adamlee,louiqa}@umiacs.umd.edu
[2] The University of Iowa, Iowa City, IA 52242, USA
psriniva@iowa.uiowa.edu
[3] Stanford University, Stanford, CA 94305, USA
{nigam,rubin,noy}@stanford.edu

**Abstract.** This paper presents the *LSLink* (or Life Science Link) methodology that provides users with a set of tools to explore the rich Web of interconnected and annotated objects in multiple repositories, and to identify meaningful associations. Consider a physical link between objects in two repositories, where each of the objects is annotated with controlled vocabulary (CV) terms from two ontologies. Using a set of *LSLink* instances generated from a background dataset of knowledge we identify associations between pairs of CV terms that are potentially significant and may lead to new knowledge. We develop an approach based on the logarithm of the odds (LOD) to determine a *confidence* and *support* in the associations between pairs of CV terms. Using a case study of Entrez Gene objects annotated with GO terms linked to PubMed objects annotated with MeSH terms, we describe a user validation and analysis task to explore potentially significant associations.

**Keywords:** links between data objects, annotations, associations, controlled vocabularies, LOD, confidence and support scores, life science link (*LSLink*).

## 1 Introduction

The vast amounts of knowledge that is being generated by the biological enterprise is captured and represented in a variety of disparate resources. This data is typically annotated with links to concepts from different ontologies. Data objects in one repository are also physically linked to objects in other repositories. The semantics of these physicals links is typically not explicit and not accessible to the scientists.

Biologists spend countless hours navigating this Web of inter-connected resources, following physical links from objects in one repository to objects in another, then following links from the data to annotations and back to the data, trying to aggregate the information that they need. While the annotated

S. Cohen-Boulakia and V. Tannen (Eds.): DILS 2007, LNBI 4544, pp. 247–263, 2007.
© Springer-Verlag Berlin Heidelberg 2007

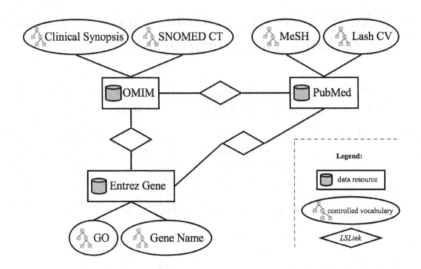

**Fig. 1.** Interconnected Web of Entrez Gene, OMIM and PubMed Resources

data objects and their physical links form a rich knowledge base, few tools allow users to explore the knowledge captured in these richly annotated graphs, and to find meaningful associations. This paper presents the Life Science Link (*LSLinks*) methodology [19] that will provide users with tools to explore the Web of interconnected and annotated objects in multiple repositories, and identify meaningful patterns.

Consider a simplified Web of interconnected resources shown in Figure 1. It includes three resources, Entrez Gene [20], OMIM [13] and PubMed [34], represented by rectangles. Objects in each data resource are annotated with terms from multiple controlled vocabularies (CVs); they are represented by ovals. A physical link between two data resources is represented by the relationship (association) diamond. The physical link occurs at the level of the data objects; for example, there is a many-to-many relationship between objects in Entrez Gene and OMIM, between objects in Entrez Gene and PubMed, etc. Thus, an object in Entrez Gene, annotated with Gene Ontology (GO) terms [11] can be linked to an object in PubMed, annotated with Medical Subject Headings (MeSH) terms [22]. Similarly, OMIM objects may be annotated with terms from the Systematized Nomenclature of Medicine Clinical Terms (SNOMED CT) [30], and they may be linked to objects in Entrez Gene and PubMed.

The *LSLinks* methodology provides a language and data model that allows a user to specify an experiment protocol or workflow to collect a background dataset of data objects, physical links between data objects, and the annotations (controlled vocabulary or CV terms) on the data objects. Next, *LSLink* instances are generated to represent the knowledge in the background dataset. An *LSLink* instance associates a pair of CV terms, where each CV term annotates one of the data objects that are connected by a physical link. The pairs of CV terms occur in two different ontologies. In the dataset used in this paper, *LSLink* instances associate a GO term with a MeSH term, where the GO term annotates an entry in Entrez Gene that has a physical link to an entry in PubMed that is annotated with the MeSH term. An example *LSLink* instance is provided in a later section.

Given some user query dataset, e.g., the *LSLinks* from a background dataset relevant to some specific gene, there may be many thousands of associations among the CV terms annotating the data objects. Among them, we must identify those that are potentially significant, so they can be used to annotate the link or analyzed to obtain meaningful knowledge. We develop an approach based on the logarithm of the odds (LOD) [3], to determine a *confidence* and *support* [1,2] in the associations between the pairs of CV terms. Users may then analyze those associations that score high in both confidence and support, to explore significant *LSLinks* that may lead to new knowledge. The contributions of this paper are as follows:

- The definition of a background dataset (and user dataset) of *LSLink* instances that associate pairs of CV terms from two ontologies annotating two data objects connected by a physical link.
- The definition of an LOD based confidence and support score, to identify potentially significant pairs of associations of CV terms in the dataset(s).
- A tool to assist users to identify queries and to analyze and evaluate the importance and meaningful nature of associations that are uncovered by the *LSLink* instances.
- A preliminary validation using *LSLink* instances generated from the physical links between Entrez Gene records annotated with GO terms and PubMed records annotated with MeSH terms.

The paper is organized as follows: Section 2 presents related work and Section 3 illustrates the *LSLink* methodology. In Section 4, we define the LOD based confidence and support. In Section 5, we use a case to illustrate how background and user datasets are generated. Section 6 presents a user interface, user analysis and validation. Section 7 concludes.

## 2    Related Work

There has been much research and development on interconnecting knowledge sources. The three major repositories NCBI, DDBJ and EBI have made significant efforts recently to provide integrated access, e.g., Links, LinkOut, and Entrez Gene at NCBI [34], LinkDB [10] at DDBJ, and Integr8 [16] at EBI. However, beyond providing ease of access to *related* material in allied databases, these typically do not attempt to enhance the representation and the semantics of individual links. Observe that navigational links are useful only to the extent that their semantics is readily visible to the user. Unfortunately this semantics remains in many cases unspecified. With a vast and growing network of links (and therefore paths between objects) it becomes urgent to remedy this situation by specifying as clearly as possible the semantics connecting linked pairs of objects. This situation is further complicated as the same pair of objects may be directly and indirectly connected in numerous ways.

However, research on link semantics is slowly evolving especially given recent examples of projects enhancing specific links [21,29,14,23]. For example, links in PDBSProtEC [21] identify SwissProt codes and Enzyme Commission numbers (EC numbers) for chains in the Protein Databank (PDB). The mapping identified by the links are useful to understand structure-function relationships. Protein-Interaction Map (PIMtool) [29] provides links from proteins to various kinds of interactions reported in multiple datasets. The relationships observed in these links are the protein-protein interactions, which do not connect the knowledge to genes or other data resources. In Information Hyperlinked over Proteins (iHOP) [14], there are links that connect genes and proteins to articles. It is an online service that provides a gene-guided network to access PubMed abstracts. By using genes and proteins as hyperlinks between sentences and abstracts, the information in PubMed is converted into navigable links. Sentences in a PubMed abstract are ranked with respect to the experimental evidence of the interaction between the proteins that appear in the sentence. Recently, Bio-DASH [23] is a semantic Web prototype of a drug development dashboard that generates links to associate disease, compounds, molecular biology, and pathway knowledge Unfortunately, while all of these projects enhance specific links, the enhancements are typically hardcoded to a specific dataset or task. In other words these efforts do not provide a general methodology for using the knowledge captured by these links to query and analyze across multiple independent datasets, to use multiple ontologies, and to be used by multiple applications or tools. The design of such a methodology is in fact a distinctive feature in our research.

Our research on link semantics also has knowledge discovery and text mining implications. The key goal in text mining is to come up with novel and interesting hypotheses typically involving a pair of objects such as a disease and a drug or a gene and a disease. A variety of approaches have been explored as for example those that focus on the free-text of MEDLINE records [9,31], those that exploit the MeSH terms associated with records [28,15,33,24], those that exploit the full text of published documents [17] etc. Our effort is similar to that of [33,24] that exploit interconnections between terms belonging to different vocabularies. In addition to labeling links with linked terms, our method has the potential to suggest novel connections through uncommon yet meaningfully paired terms.

Finally, our research contributes to annotation research. A key activity in bioinformatics is the annotation of genes/proteins from different species with terms from Gene Ontology [5,12]. While manual annotation is most common, there are several automatic or semi-automatic annotation efforts. This includes the the design of automatic annotation methods in the BioCreAtIvE I initiative [4], supervised machine learning based approaches [26], unsupervised methods [6], and n-gram based statistical models built using full text [25]. Research by Couto et al. substantiates uncurated annotations using a text similarity based method which also identifies novel annotations [7]. Note that our research is distinctive in that we are focussed on the semantic annotation of links between pairs of objects.

## 3    *LSLink* Methodology

Figure 2 presents the *LSLink* methodology by discussing its application to the task of generating *LSLink* instances between Entrez Gene and PubMed. The first step is to specify a protocol to navigate the objects in the resources and the physical links between the objects. In this example, the background dataset includes all entries in Entrez Gene that are human genes and annotated with GO terms, and all the entries that they reach in PubMed, following four types of physical links. The details of the experiment protocol to create the dataset is in Section 5. The next step is to specify the CV terms that must be extracted. In addition to identifying the sets of terms, one can also identify semantic concepts that are to be used to create the background dataset; an example is in Section 5. The next two steps are to generate the *LSLink* instances and calculate the confidence and support in associations of CV terms.

The left part of Figure 3 is a graphical representation of the physical link between Entrez Gene and PubMed. There are links between the objects $e_1$ and $e_2$ of Entrez Gene and objects $p_1$ and $p_2$ of PubMed. The terms $g_a$, $g_b$, $m_a$ and $m_b$ annotate these objects. Each object is associated with two terms. The physical links are between $e_1$ and $p_1$, $e_1$ and $p_2$, and $e_2$ and $p_2$. The right part of Figure 3 shows the corresponding *LSLink* instances. If we consider the physical link between $e_1$ and $p_1$, the two CV terms $g_a$ and $g_b$ annotating $e_1$ and the two CV terms $m_a$ and $m_b$ annotating $p_1$, then there will be four *LSLink* instances. An example instance is the following: $(g_a, e_1, m_a, p_1)$.

## 4    Calculating Confidence and Support in *LSLinks*

We regard the *LSLinks* derived from the physical link between Entrez Gene and PubMed as a background dataset that generates associations between terms from the GO and MeSH controlled vocabularies. This dataset may produce thousands of associations and many may not be not meaningful (false positives). Our task is to identify those associations that are significant, i.e., biologically meaningfully, and hopefully of interest to users.

A natural approach for doing this is by considering the perspective of a user, represented by a query of interest. For example, assume that a user is interested in a specific gene X. Then we may define a query dataset of *LSLink* instances that contains a subset of the background dataset, and are associated with this gene object. Our overall strategy is to use the query dataset to identify interesting *LSLinks*. We note that a particular association that is significant in one query dataset may not be significant for another query dataset.

Since we expect the vast majority of associations in the user query dataset to be irrelevant (false positives), we determine the confidence and support estimates

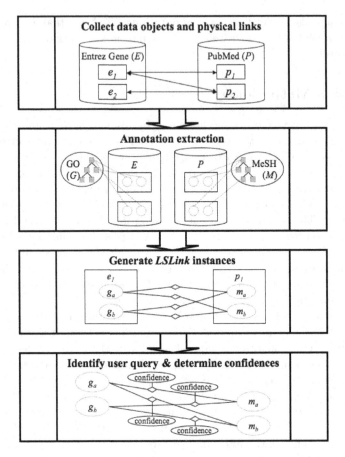

**Fig. 2.** Methodology to generate *LSLink* instances between Entrez Gene (annotated with GO) and PubMed (annotated with MeSH)

for each association. We rank the associations based on these scores and only select those that exceed some user defined threshold. The confidence and support calculations will involve statistics from the background dataset to ensure that we are identifying non random phenomenon. Our calculations are based on the Logarithm of the Odds (LOD) score - a score that has been used frequently, along with its variant forms, in text mining research [18,27]. We note that there are alternate techniques to determine confidence and support, e.g., association rule mining, [1,2]. The problem of mining association rules is to generate association rules between sets of items in a large database of transactions, and to find all significant association rules.

The log odds (LOD) ratio score used here measures the extent to which the association deviates from one resulting from chance alone (a random association). We note that support reflects the relative ratio of *LSLinks* instances that associate the two CV terms with respect to all *LSLinks* instances in the dataset.

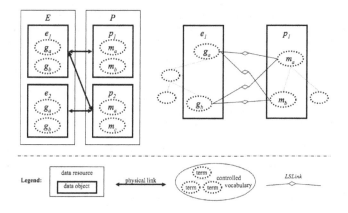

**Fig. 3.** Graphical representation of *LSLink* instances between Entrez Gene and PubMed

Confidence reflects the relative ratio of *LSLinks* instances that associate the two CV terms with respect to those *LSLinks* instances that are associated with one of the CV terms. Users may then analyze those associations that score high in both confidence and support since they are potentially significant associations that could be used to annotate the links and also lead to new knowledge. We further note that give the universe of terms, data objects and physical links between data objects, there are many possible approaches to obtain expressions for support and confidence. We made what we consider to be reasonable choices. The notation and definitions that we use are as follows:

- Two data resources:
  - Entrez Gene ($E$)
  - PubMed ($P$)
- Data objects:
  - $e_1$, $e_2$, ... in $E$
  - $p_1$, $p_2$, ... in $P$
- Two controlled vocabularies (CVs), one for each data resource:
  - Gene Ontology (GO, $G$)
  - Medical Subject Headings (MeSH, $M$)
- CV terms annotating data objects:
  - CV terms $g_a$, $g_b$, ... in $G$ annotate objects in $E$.
  - CV terms $m_a$, $m_b$, $m_c$, ... in $M$ annotate objects in $P$.
- Term probability:
  - Estimated using data object level term frequencies
    * $Pr(g_a, E) = \dfrac{number\ of\ objects\ containing\ term\ g_a\ in\ E}{total\ number\ of\ objects\ in\ E}$
    * $Pr(m_c, P) = \dfrac{number\ of\ objects\ containing\ term\ m_c\ in\ P}{total\ number\ of\ objects\ in\ P}$
  - Estimated using annotation level term frequencies
    * $Pr'(g_a, E) = \dfrac{number\ of\ annotations\ that\ are\ g_a\ in\ E}{total\ number\ of\ annotations\ in\ E}$
    * $Pr'(m_c, P) = \dfrac{number\ of\ annotations\ that\ are\ m_c\ in\ P}{total\ number\ of\ annotations\ in\ P}$

- If we assume that an object receives a specific annotation only once, then the value of the numerator for the two alternative term probability expressions will be the same. Hence the relative rankings of term probabilities will be the same. The assumption holds for both the GO annotations for Entrez genes and the MeSH terms for PubMed records. We choose the first expression for simplicity; the number of data objects is much smaller than the number of annotations.

- Link annotation probabilities estimated from the user query dataset:

  - Assumption: Given $\#(g_a|g_a \in e_i)$, $\#(m_c|m_c \in p_k)$ distinct CV terms annotating two objects $e_i$, $p_k$, respectively, and a physical link between $e_i$ and $p_k$, there are $\#(g_a|g_a \in e_i) \times \#(m_c|m_c \in p_k)$ *LSLink* instances, where a pair of CV terms is specified in each *LSLink*.

    * $Pr'(g_a, E', m_c, P') =$
    $$\frac{number\ of\ LSLinks\ in\ user\ query\ results\ containing\ terms\ g_a\ and\ m_c}{total\ number\ of\ LSLinks\ in\ user\ query\ results}$$

  - Conditional link annotation probability conditioned on either CV term appearing in the *LSLink* instances.

    * $Pr'(g_a, E', m_c, P'|g_a, m_c) =$
    $$\frac{number\ of\ LSLinks\ in\ user\ query\ results\ containing\ terms\ g_a\ and\ m_c}{number\ of\ LSLinks\ in\ user\ query\ results\ containing\ terms\ g_a\ or\ m_c}$$

- LOD based confidence and support:

  - LOD based confidence equals to the logarithm of the conditional link annotation probability given the appearance of either CV term divided by the corresponding term probabilities.

    * $Conf(g_a, E', m_c, P') = log(\frac{Pr'(g_a, E', m_c, P'|g_a, m_c)}{Pr(g_a, E)Pr(m_c, P)})$

  - LOD based support equals to the logarithm of the link annotation probability divided by the corresponding term probabilities.

    * $Supp(g_a, E', m_c, P') = log(\frac{Pr'(g_a, E', m_c, P')}{Pr(g_a, E)Pr(m_c, P)})$

## 5    Data Collection and Analysis

We discuss a case study where the background dataset includes *LSLinks* generated from Entrez Gene entries representing human genes with GO annotation and the PubMed entries that they reach along with their MeSH annotations. Note that this is one of the three links in Figure 1.

### 5.1    Background Dataset

We construct the background *LSLink* dataset as follows:

1. Retrieve all human gene objects in Entrez Gene and extract their GO annotations.
2. Follow all links from these objects to PubMed objects. There are four types of links. We do not use this knowledge in this study, but will distinguish them in future work.

**Table 1.** Background dataset

| | |
|---|---:|
| Number of active human gene objects in Entrez Gene | 38,608 |
| Number of distinct GO terms extracted | 6,038 |
| Number of distinct PubMed objects that are reached via four link types | 141,745 |
| Number of distinct MeSH descriptors extracted | 14,387 |
| Number of distinct MeSH qualifiers extracted | 82 |
| Number of distinct MeSH descriptors that are major topics | 11,103 |
| Number of *LSLinks* generated | 12,461,601 |
| Number of distinct association pairs of GO Term and MeSH descriptors | 1,742,325 |

(a) Gene References Into Function (GeneRIF) provided by the National Library of Medicine (NLM). These links are produced through user submissions in an Entrez Gene record or through manual curation from the published literature by staff of the NLM.

(b) Human Immunodeficiency Virus Type 1 (HIV-1) links provided by the National Institute of Allergy and Infectious Diseases (NIAID). These interactions are reported in the Human Protein Interaction Database, and there are links to PubMed publications that support the described interaction.

(c) General Interactions submitted by scientists with links to PubMed publications that support the described interaction.

(d) GO annotations provided by GOA. These links are generated by a combination of electronic mapping and manual curation.

3. Extract all MeSH annotations for the PubMed objects reached in step 2. We limit our protocol to use only the most relevant MeSH terms identified as topic headings in the PubMed publications.

The statistics for the background dataset as of January 18th, 2007 is reported in Table 1. There are 162,637 records for human in Entrez Gene, but we do not use the ones that were discontinued or replaced by other records.

## 5.2 User Query Dataset

We support multiple user scenarios for querying the background dataset. The input can be a simple set of gene symbols, object identifiers or medical terms. The scenarios include the following:

1. To find highly related articles associated with a human gene, we retrieve gene objects that are associated with a human gene symbol and follow all *LSLinks* to PubMed objects (in the background dataset).

2. A scientist wants to know all human genes associated with some set of articles. We retrieve these objects in PubMed and follow all links to human gene objects (in the background dataset).

3. A scientist is interested in specific medical terms and would like to retrieve highly related human genes. We retrieve objects in PubMed associated with these MeSH terms and follow all links to human gene objects (in the background dataset).

**Table 2.** 11 user queries on human genes

| Human gene name (official symbol) | Number of GO terms in the gene | Number of distinct directly linked PubMed objects | Number of distinct MeSH descriptors w/ major topic in the PubMed objects | Number of LSLinks |
|---|---|---|---|---|
| APOE | 27 | 407 | 452 | 41,985 |
| BRCA1 | 45 | 444 | 372 | 83,286 |
| BRCA2 | 16 | 175 | 186 | 13,056 |
| CFTR | 17 | 275 | 306 | 17,459 |
| DMD | 18 | 160 | 157 | 11,142 |
| HEMA (F8) | 7 | 115 | 140 | 2,667 |
| IFNG | 26 | 286 | 549 | 36,504 |
| P53 (TP53) | 52 | 1,615 | 1,243 | 398,268 |
| PSEN1 | 40 | 230 | 251 | 36,040 |
| PSEN2 | 22 | 84 | 114 | 6,952 |
| TNF | 33 | 884 | 1,155 | 140,943 |

Given a user query, we first retrieve the dataset corresponding to the query. Table 2 reports on the user query dataset for 11 human gene symbols. The second column reports on the number of GO terms annotating the gene objects. The third column reports on the number of PubMed objects that are directly linked from the corresponding gene object. The final column reports on the number of distinct MeSH terms extracted from the linked PubMed objects.

## 5.3   Confidence and Support of Associations

We process a specific user query dataset as follows to determine the confidence and support of the associations in that dataset:

- Determine the term probabilities for the corresponding GO and MeSH terms, $e_a$ and $m_c$, respectively, using the background dataset.
- Determine the link annotation probabilities for associations of pairs of terms, $e_a$ and $m_c$, using all relevant *LSLink* instances in the user query dataset.
- Determine the LOD score for confidence and support in all pairs of associations of CV terms $(e_a, m_c)$.
- Select a cutoff (threshold) for the confidence and support in the associations based on the distribution of LOD scores for this dataset.
- Finally, we applied some additional filtering steps. First, we limited our dataset to MeSH terms that were identified as topic headings in the PubMed object. Further, we identified the semantic type of the MeSH terms using a resource [8] that provided a many-to-many mapping between MeSH terms and semantic types. The semantic types that are of interest to the evaluation task could be selected by the user.

We made a number of assumptions and simplifications in our analysis and we discuss these issues next. First, multiple sources provide annotations and it is well understood that the confidence in the association is not identical. We assume equal confidence in all annotations. Determining if significant associations of terms also correlate with high confidence in the annotation will be addressed in future work. Second, terms within an ontology have relationships with other terms, e.g., their parent and child terms. These relationships may impact the knowledge obtained from our associations. Our current approach to compute LOD scores does not consider such relationships and their possible impact. This will be future research.

# 6    User Interface and Evaluation

In this section, we discuss the tools provided to users and we discuss the outcome of a preliminary validation study. Each user query dataset may yield hundreds or thousands of associations and users need the support of analysis tools to visualize the associations and assist in their exploration. The following are example features of an analysis tool:

- Given some GO term (or MeSH term) present all the associations of that term that are significant with respect to a threshold selected by the user.
- Group the significant associations based on semantic knowledge. An example is the semantic type associated with the MeSH terms.
- Group associations using either a GO term or MeSH term, so that users can analyze groups of associations rather than individual associations.

We aim for an interactive interface where the user can browse some results and then specify particular terms of interests in either vocabulary. This type of "relevance feedback" may be used to further refine information that is presented in an iterative manner. For example, the initial query provided by the user may be refined after the user has had an opportunity to look at the kinds of links presented that were found to be significant.

## 6.1    Sample User Interface

Figures 4 and 5 describe the interface that the scientist can use to analyze associations for some user dataset of *LSLinks*. We consider a simple query where the scientist identifies a gene symbol. Based on this dataset of *LSLinks* that are associated with the Entrez Gene record, the LOD based confidence and support scores are determined. We show only the LOD based confidence scores.

A threshold for significance can be determined by the scientist based on the range of scores for this dataset. A histogram of the distribution of confidence scores is in Figure 6. The left side of the figure presents the range of confidence scores for two human genes, *APOE* and *CFTR*, in the form of a histogram. For *APOE*, there are 12,204 associations and the rating ranges from 0.17 to 7.16 with a mean 4.24 and median 4.28. The variance of confidence scores appears to be much greater for *APOE*.

On the right Figure 6, we report on the range of confidence scores for associations that involve two GO terms, *apolipoprotein E receptor binding* and *cytoplasm*. The associations of the GO term *apolipoprotein E receptor binding* yields higher confidence scores compared to associations of the GO term *cytoplasm*. To

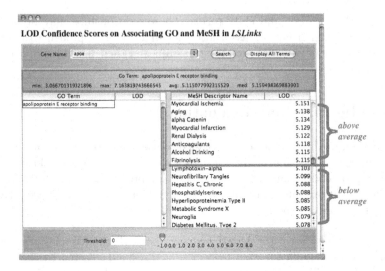

**Fig. 4.** User interface showing human gene *APOE* with GO term *apolipoprotein E receptor binding*

explain, the former term appears only once in the human gene *APOE* object, but the later term annotates 1,541 gene objects in the background dataset.

The scientist can then select a threshold LOD score. The system will use this threshold to identify all associations that exceed the score. The user can then either select a GO term (Figure 4) or a MeSH term (Figure 5). The interface will order all associations for this GO or MeSH term, based on the LOD score, and display these associations. Not that here we ordered the associations based on the confidence score. Figure 5 illustrates the result when the user selected associations of the MeSH term *Central Nervous System Infections* with a threshold of 6.50 on the confidence score for the *APOE* dataset.

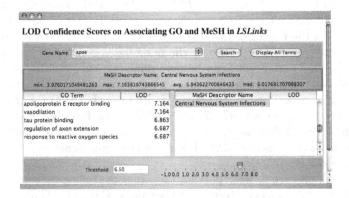

**Fig. 5.** User interface showing human gene *APOE* with MeSH descriptor name *Central Nervous System Infections* with confidence threshold over 6.50

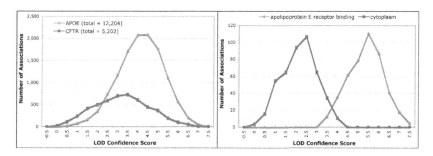

**Fig. 6.** Histograms on numbers of associations in two human genes (*APOE* and *CFTR*) and two GO terms in the *APOE* dataset

## 6.2   User Validation

A preliminary validation task was conducted to explore methods for assessing the *LSLinks* strategy. We identified associations that exceed a threshold on confidence and had them rated along two independent dimensions. The first dimension is to assign a score for a meaningful association; the rating is as follows: (Meaningful, Maybe Meaningful, Not Meaningful). The second dimension is to assign a rating based on whether the association is already known; this score is as follows: (Widely Known, Somewhat Known, Maybe Known, Unknown/Surprising).

The evaluation team (2 scientists) chose six genes and classified the top 20 associations. The associations that they examined for human gene *CFTR* are shown in Table 3. A majority of these associations were identified as meaningful and well known. Some associations were identified as possibly meaningful / not well known. Two associations were not semantically meaningful; we note that we have not determined the reason for these false positives. To complete the evaluation, we also examined a random sampling of associations with medium or low scores for confidence. The association of the GO term *chloride ion binding* with the MeSH term *Phosphoprotein Phosphatase (Enzyme)* had a medium score of 3.12. The association is not meaningful. The association of the GO term *membrane* with the MeSH term *Cloning, Molecular (Laboratory Procedure)* had a low score of 0.65. Both terms are generic and the association is not meaningful.

To summarize, the validation was successful since it identified that the significant associations are also meaningful.

While the validation did not immediately identify new interesting knowledge this is not unexpected. First, these genes are well studied so many associations are already known. Second, many MeSH terms are general terms used to classify the content of the paper rather than identifying specific results reported in the paper. Consequently, we do not expect that these general terms will lead to interesting results and the evaluation team planned to identify more specific MeSH terms using the semantic types of these terms.

The evaluation team further determined that more meaningful results would be obtained by combining these associations with additional knowledge about

**Table 3.** User evaluation of the human *CFTR* gene dataset

| LOD based Conf. | LOD based Supp. | GO term | MeSH descriptor w/ major topic (semantic type) | Meaningful | Known |
|---|---|---|---|---|---|
| 7.34 | 6.12 | ATP-binding and phosphorylation-dependent chloride channel activity | Mucociliary Clearance (Organ or Tissue Function) | yes | well known |
| 7.34 | 6.12 | channel-conductance-controlling ATPase activity | Mucociliary Clearance (Organ or Tissue Function) | yes | well known |
| 7.34 | 6.12 | ATP-binding and phosphorylation-dependent chloride channel activity | Salmonella typhi (Bacterium) | yes | well known |
| 7.34 | 6.12 | channel-conductance-controlling ATPase activity | Salmonella typhi (Bacterium) | yes | well known |
| ... | ... | ... | ... | ... | ... |
| 6.85 | 5.64 | ATP-binding and phosphorylation-dependent chloride channel activity | Pancreatitis, Alcoholic (Disease or Syndrome) | not | well known |
| 6.85 | 5.64 | channel-conductance-controlling ATPase activity | Pancreatitis, Alcoholic (Disease or Syndrome) | not | well known |
| 6.74 | 5.52 | ATP-binding and phosphorylation-dependent chloride channel activity | Fimbriae Proteins (Amino Acid, Peptide, or Protein) | maybe | not known |
| 6.74 | 5.52 | channel-conductance-controlling ATPase activity | Fimbriae Proteins (Amino Acid, Peptide, or Protein) | maybe | not known |
| ... | ... | ... | ... | ... | ... |
| 6.72 | 5.87 | ATP-binding and phosphorylation-dependent chloride channel activity | Cystic Fibrosis (Disease or Syndrome) | yes | well known |
| ... | ... | ... | ... | ... | ... |
| 6.62 | 6.08 | channel-conductance-controlling ATPase activity | Cystic Fibrosis Transmembrane Conductance Regulator (Amino Acid, Peptide, or Protein) | yes | well known |
| ... | ... | ... | ... | ... | ... |
| 6.60 | 5.38 | ATP-binding and phosphorylation-dependent chloride channel activity | Pseudomonas Infections (Disease or Syndrome) | yes | somewhat |
| 6.60 | 5.38 | channel-conductance-controlling ATPase activity | Pseudomonas Infections (Disease or Syndrome) | yes | somewhat |
| 6.57 | 5.34 | ATP-binding and phosphorylation-dependent chloride channel activity | Fallopian Tube Diseases (Disease or Syndrome) | yes | somewhat |
| 6.57 | 5.34 | channel-conductance-controlling ATPase activity | Fallopian Tube Diseases (Disease or Syndrome) | yes | somewhat |

the genes. They suggested exploring the associations between GO terms and phenotypes using the link from Entrez Gene to OMIM and the link from Entrez Gene to PharmGKB [32]. We note that the link from Entrez Gene to OMIM was identified in our initial study and we plan to extend to the second link.

## 6.3  Frequency Analysis

A further conclusion of the validation task was that some meta-level analysis is needed. One suggestion was to examine groups of associations rather than individual associations and the group frequency of occurrence. The rationale for the frequency analysis is that the GO terms associated with the gene record were determined a priori based on known knowledge about the gene. On the other hand, scientists may not have studied all the knowledge in the PubMed articles linked to the gene record and annotated the PubMed record with this knowledge. Hence, grouping the associations by MeSH terms may help to uncover hidden but possibly significant patterns. The higher frequency reflects those MeSH terms that are associated with many GO terms for some user query (gene). For those

**Table 4.** Highest frequencies of associations in the human gene *APOE* dataset with threshold 6.50 on LOD based confidence score

| MeSH descriptor w/ major topic | Number of associated GO terms |
|---|---|
| Akathisia, Drug-Induced (Disease or Syndrome) | 5 |
| Apolipoprotein E4 (Amino Acid, Peptide, or Protein) | 5 |
| Candidiasis, Cutaneous (Disease or Syndrome) | 5 |
| Central Nervous System Infections (Disease or Syndrome) | 5 |
| Hyperlipoproteinemia Type V (Disease or Syndrome) | 5 |
| Tinea Versicolor (Disease or Syndrome) | 5 |
| Hyperlipoproteinemia Type III (Disease or Syndrome) | 5 |
| Dyslipidemias (Disease or Syndrome) | 5 |

terms of interest, the distribution of these GO terms in the GO hierarchy may also be relevant in identifying meaning.

We consider those associations that are above a user specific threshold for the LOD score. We then *group* these associations by the MeSH terms. Table 4 identifies the results for the user query on gene *APOE* with 6.50 as the threshold on the confidence score. The first column identifies the MeSH term, and the second column identifies the cardinality of GO terms associated with the MeSH term. The highest cardinality is *5* in *APOE*, and the five GO terms are *apolipoprotein E receptor binding, vasodilation, tau protein binding, regulation of axon extension* and *response to reactive oxygen species*. The corresponding LOD scores are descending from the top row to the bottom row.

# 7    Conclusion

We presented the *LSLinks* methodology to explore the rich Web of interconnected and annotated objects in multiple repositories, and to identify meaningful patterns. We generate a set of *LSLink* instances to represent a background dataset of knowledge. We then identify those associations of pairs of CV terms that are potentially significant and may lead to new knowledge. We develop an approach based on the logarithm of the odds (LOD) to determine a *confidence* and *support* in the associations between the pairs of CV terms. We created an initial dataset of *LSLinks* from Entrez Gene objects annotated with GO terms linked to PubMed objects annotated with MeSH terms. We reported on the results of a preliminary user validation.

In future work, we will extend the dataset to include additional links so that associations across multiple resources can be analyzed. We will also extend the methodology to include more semantic knowledge associated with the CV terms, e.g., semantic types or other knowledge within an ontology. We will investigate how relationships within an ontology may impact the significance of some associations among CV terms. We also plan to further study cases where the associations are judged to be not meaningful. We will also further analyze techniques to identify significant associations, e.g., association rule mining techniques and also consider modifications to our approach to determine support and confidence.

**Acknowledgments.** This research has been partially supported by the National Science Foundation under grants IIS0222847, IIS0430915 and 0312356.

# References

1. Agrawal, R., et al.: Mining association rules between sets of items in large databases. SIGMOD Record 22(2), 207–216 (1993)
2. Agrawal, R., Srikant, R.: Fast Algorithms for Mining Association Rules in Large Databases. In: Proceeding of the 20th International Conference on Very Large Data Bases, pp. 487–499, San Francisco, CA, USA (September 1994)
3. Barnard, G.A.: Statistical inference. Journal of the Royal Statistical Society. Series B (Methodological) 11(2), 115–149 (1949)
4. Blaschke, C., et al.: Evaluation of BioCreAtIvE assessment of task 2. BMC Bioinformatics 6(Suppl 1), S16 (2005)
5. Camon, E., et al.: The Gene Ontology Annotation (GOA) Database: sharing knowledge in UniProt with Gene Ontology. Nucleic Acids Research 32(Database issue), D262–D266 (2004)
6. Couto, F.M., et al.: Finding genomic ontology terms in text using evidence content. BMC Bioinformatics 6(Suppl 1), S21 (2005)
7. Couto, F.M., et al.: GOAnnotator: linking protein GO annotations to evidence text. Journal of Biomedical Discovery and Collaboration, 1(19) (December 20, 2006)
8. Current Semantic Types in the Unified Medical Language System (UMLS) http://www.nlm.nih.gov/research/umls/META3_current_semantic_types.html
9. Fiszman, M., et al.: Integrating a hypernymic proposition interpreter into a semantic processor for biomedical text. In: AMIA 2003 Annual Symposium, pp. 239–243, Washington, DC, USA (November 8-12, 2003)
10. Fujibuchi, W., et al.: DBGET/LinkDB: an integrated database retrieval system. In: Third Pacific Symposium on Biocomputing (PSB 1998), pp. 683–694, Maui, Hawaii, USA, (January 4-9, 1998)
11. Gene Ontology (GO)., http://www.geneontology.org/
12. Gene Ontology Annotation (GOA). http://www.ebi.ac.uk/GOA/
13. Hamosh, A., et al.: Online Mendelian Inheritance in Man (OMIM), a knowledgebase of human genes and genetic disorders. Nucleic Acids Research 33(Database issue), D514–D517 (2005)
14. Hoffmann, R., Valencia, A.: A gene network for navigating the literature. Nature Genetics 36(7), 664 (2004)
15. Hristovski, D., et al.: Improving literature based discovery support by genetic knowledge integration. Studies in health technology and informatics 95, 68–73 (2003)
16. Kersey, P.J., et al.: Integr8: enhanced inter-operability of european molecular biology databases. Methods of Information in Medicine 42(2), 154–160 (2003)
17. Koike, A., Takagi, T.: Knowledge discovery based on an implicit and explicit conceptual network. Journal of the American Society for Information Science and Technology 58(1), 51–65 (2007)
18. Korbel, J.O., et al.: Systematic association of genes to phenotypes by genome and literature mining. PLoS Biology, 3(5) (April 5, 2005)
19. Lee, W.-J., Raschid, L., Vidal, M.-E.: A Generic, Flexible and Scalable Methodology to Enhance the Semantics of Links in Life Science Data Resources. Technical Report CS-TR-4809 (UMIACS-TR-2006-29), Univeristy of Maryland, (June 2006)

20. Maglott, D., et al.: Entrez Gene: gene-centered information at NCBI. Nucleic Acids Research 35(Database issue),D26–D31 (2007)
21. Martin, A.C.: PDBSprotEC: a Web-accessible database linking PDB chains to EC numbers via SwissProt. Bioinformatics 20(6), 986–988 (2004)
22. Medical Subject Headings (MeSH).
    http://www.nlm.nih.gov/mesh/meshhome.html
23. Neumann, E.K., Quan, D.: Biodash: A semantic web dashboard for drug development. In: Eleventh Pacific Symposium on Biocomputing (PSB 2006), pp. 140–151, Maui, Hawaii, USA, (January 3-7, 2006)
24. Perez-Iratxeta, C., Bork, P., Andrade, M.A.: Association of genes to genetically inherited diseases using data mining. Nature Genetics 31(3), 316–319 (2002)
25. Ray, S., Craven, M.: Learning statistical models for annotating proteins with function information using biomedical text. BMC Bioinformatics 6(Suppl 1), S18 (2005)
26. Rice, S.B., Nenadic, G., Stapley, B.J.: Mining protein function from text using term-based support vector machines. BMC Bioinformatics 6(Suppl 1), S22 (2005)
27. Siadaty, M.S., Knausg, W.A.: Locating previously unknown patterns in data-mining results: a dual data- and knowledge- mining method. BMC Medical Informatics and Decision Making, 6(13) (March 7, 2006)
28. Srinivasan, P., Libbus, B.: Mining MEDLINE for implicit links between dietary substances and diseases. Bioinformatics 20(Supplement 1), i290–i296 (2004)
29. Stanyon, C.A., et al.: A Drosophila protein-interaction map centered on cell-cycle regulators. Genome Biology 5(12), R96 (2004)
30. Systematized Nomenclature of Medicine Clinical Terms (SNOMED CT).
    http://www.snomed.org/snomedct/
31. Thomas, J., et al.: Automatic extraction of protein interactions from scientific abstracts. In: Fifth Pacific Symposium on Biocomputing (PSB 2000), pp. 538–549. Oahu, Hawaii, USA (2000)
32. Thorn, C.F., et al.: PharmGKB: the pharmacogenetics and pharmacogenomics knowledge base. Methods in Molecular Biology 311, 179–191 (2005)
33. Tiffin, N., et al.: Integration of text- and data-mining using ontologies successfully selects disease gene candidates. Nucleic Acids Research 33(5), 1544–1552 (2005)
34. Wheeler, D.L., et al.: Database resources of the National Center for Biotechnology Information. Nucleic Acids Research 35(Database issue), D5–D12 (2007)

# CONANN: An Online Biomedical Concept Annotator

Lawrence H. Reeve and Hyoil Han

College of Information Science and Technology, Drexel University
3141 Chestnut Street, Philadelphia, PA 19104, USA
lhr24@drexel.edu, hyoil.han@acm.org

**Abstract.** We describe our biomedical concept annotator designed for online environments, CONANN, which takes a biomedical source phrase and finds the best-matching biomedical concept from a domain resource. Domain concepts are defined in resources such as the United States National Library of Medicine's Unified Medical Language System Metathesaurus. CONANN uses an incremental filtering approach to narrow down a list of candidate phrases before deciding on a best match. We show that this approach has the advantage of improving annotation speed over an existing state-of-the-art concept annotator, facilitating the use of concept annotation in online environments. Our main contributions are 1) the design of a phrase-unit concept annotator more readily usable in online environments than existing systems, 2) the introduction of a model which uses semantically focused words in a given ontology (e.g., UMLS) to measure coverage, called *Inverse Phrase Frequency*, and 3) the use of two different filters to measure coverage and coherence between a source phrase and a domain-specific candidate phrase. An intrinsic evaluation comparing CONANN's concept output to a state-of-the-art concept annotator shows our system has an annotation precision ranging from 90% for exact match concept to 95% for relaxed concept matching while average phrase annotation time is eighteen times faster. In addition, an extrinsic evaluation using the generated concepts in a text summarization task shows no significant degradation when using CONANN.

**Keywords:** Biomedical semantic annotation, biomedical concept mapping, concept annotation.

## 1  Introduction

The biomedicine community maintains large and continuously-updated information sources. For example, United States National Library of Medicine's PubMed service contains in excess of 16 million publications from over 5,000 worldwide biomedicine-related journals [1]. The PubMed service consists of publication abstracts which can link with full-texts. For physicians and biomedical researchers, finding and using relevant texts within these large resources can be challenging. To address this challenge, annotation systems using domain-specific concepts, rather than terms, have been developed. Examples of such systems include MetaMap Transfer

S. Cohen-Boulakia and V. Tannen (Eds.): DILS 2007, LNBI 4544, pp. 264–279, 2007.

(MetaMap) [2], SAPHIRE [3], and KnowledgeMap [4]. Among the benefits of using concepts, rather than terms, is 1) synonym merging, where synonymous phrases are merged to a single concept, and 2) the use of a domain-specific language for querying. Biomedical concept annotations have been used in applications for indexing and retrieval, data mining, decision support, patient records, medical curriculum searching, and text summarization [2] [4] [5].

The task of a concept annotator is to map each text unit (typically a phrase) of a source text into one or more domain-specific concepts. In some systems, such as MetaMap [2], efforts are made to find a best-matching concept, while in other systems, such as IndexFinder [6], all possible concepts are found. Almost all existing concept annotators are slow performing, precluding their use in online applications, where the text is annotated dynamically, rather than statically. In typical search and retrieval applications, static annotation is fine since neither the text nor the concept resource is expected to change. However, in some applications, dynamic annotation is needed to allow for changing concept resources (such as UMLS and NCI Thesaurus) or unseen texts. An annotation system designed for online use can avoid concept annotation maintenance issues by providing annotations dynamically (at runtime), but require a level of acceptable end-user response time.

In this paper, we describe our biomedical concept annotator, CONANN, which supports both dynamic and static concept annotation. The current concept resource is UMLS, but can support other concept resources as well. We chose UMLS because a) it is an actively developed resource and is paired with a state-of-the-art annotator called MetaMap; and b) UMLS has been used for indexing and data mining work, which is most closely related to our work [2]. Its design is intended to achieve faster annotation time per phrase while maintaining annotation accuracy competitive with existing biomedical annotation systems. Such an online biomedical annotation concept system has the advantages of supporting texts unknown to the system ahead of time, as well as providing for constantly changing concept resources, which is common in a field such as biomedicine. These advantages overcome the limitations of purely static biomedical concept annotators, which form the majority of existing systems.

Our main contributions are 1) the design of a phrase-unit concept annotator more readily usable in online environments than existing systems, 2) the introduction of a model which uses semantically focused words in a given ontology (e.g., UMLS) to measure coverage, called *Inverse Phrase Frequency*, and 3) the use of two different filters to measure coverage and coherence between a source phrase and a domain-specific candidate phrase.

This paper is organized as follows. Section 2 provides background on the biomedical resource we use for concept annotation of texts. Section 3 discusses the architecture of our concept annotator system, as well as the algorithms for two important parts of the mapping process. Section 4 describes previous work in this area. Section 5 describes our evaluation methods, and Section 6 discusses the evaluation results. Section 7 concludes.

## 2  Background

### 2.1  Biomedical Concept Resource

Automated semantic annotation is the process of mapping data instances to an ontology [7], [8]. In the biomedical domain, the National Library of Medicine (http://www.nlm.nih.gov/) provides resources for identifying concepts and their relationships under the framework of the Unified Medical Language System (UMLS). The UMLS Metathesaurus contains concepts and real-world instances of the concepts, including a concept name and its synonyms, lexical variants, and translations [9], known as concept instances. A concept name is the name given to a particular UMLS concept.

### 2.2  Text-to-Concept Mapping Process

The task of a biomedical concept annotator is to map small text units in the source text to concept instances which in turn, determine the concept name the text unit should have. For example, assuming our ontology consists of the single concept *Multiple Myeloma* and the five concept instances {Multiple Myeloma, Myeloma, Plasma Cell Myeloma, Myelomatosis, Plasmacytic myeloma}, the phrase "Plasmacytic myeloma" is mapped to the concept Multiple Myeloma. Some systems focus on finding all possible matches, while other systems find the best possible match.

## 3  Concept Annotator - CONANN

In this section, we discuss the general design of our CONANN concept annotator and describe how the UMLS domain resource is pre-processed for use within CONANN. In addition, we detail CONANN's coverage and coherence filter algorithms.

### 3.1  Architectural Overview

CONANN finds the best matching concept for a source phrase and uses an incremental approach, as shown in Figure 1. There are several phrase types used by CONANN. A *source phrase* is a phrase from the source text which the system will attempt to annotate with a biomedical concept. A *concept instance* is phrase belonging to a UMLS concept (each UMLS concept is associated with one or more synonymous phrases). *Candidate phrases* are concept instances having words in common with the source phrase. A *candidate concept* identifies the UMLS concept a candidate phrase belongs to. A *concept name* is the name given to a particular UMLS concept. The idea of CONANN is to successively filter out candidate concepts using basic techniques, and compute more complex candidate phrase scores for a small subset of possible candidate phrase matches. This approach is different than existing approaches, which typically score a candidate phrase completely in one pass and then rank the set of resulting concepts [10], [2].

In CONANN, a list of candidate phrases is generated based on the overlap of words from the source phrase and all concept instances. If only a single candidate phrase exists, its associated concept name is returned. If there is more than one candidate phrase generated, filtering process to remove unlikely candidate phrases begins. The candidate phrase filters are based on n-gram co-occurrences between the source phrase and the candidate phrases. The incremental filtering is done to improve computational efficiency by applying successively more computationally complex filters.

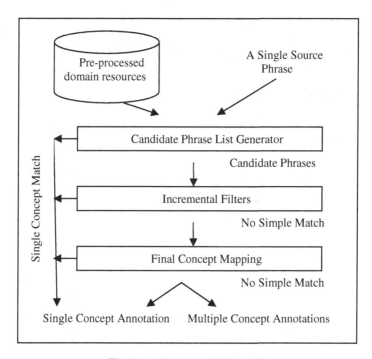

**Fig. 1.** Architecture of CONANN

The overall annotation strategy for a single source phrase is in the following order:

1. *Candidate Phrase Generation*: Construct a list of candidate phrases based on the common words between all concept instances and the source phrase. If only one candidate phrase remains in the list, return its associated concept name.

2. *Incremental Filtering*: Filter the list of candidate phrases using one or more filters (see Section 3.4). After each filter is applied, if only one candidate phrase remains, its associated concept name is returned. In this work, we apply two filters sequentially. The first filter measures coherence, which is identified by using word order, and the second filter measures coverage (i.e. common words) between a source phrase and a concept instance.

3. *Final Mapping*: If more than one candidate phrase remains in the list of candidate phrases after all filters have been applied, the candidate phrases are passed

to a final stage to perform concept mapping. Final concept mapping finds the best matching candidate phrase among the remaining candidate phrases. Our approach to final concept mapping is to sum the number of candidate phrases belonging to each UMLS concept, and then choose the concept(s) with the largest number of candidate phrases (see Section 3.5).

## 3.2 Domain Resource Pre-processing

Domain resource processing is done once with each new domain resource and then stored externally. Domain resource processing consists of converting UMLS text-based resources into a format usable for fast in-memory access (Section 3.2.1). In addition, calculations which can be performed ahead of time, such as inverse phrase frequency (described in Section 3.2.2) are completed.

### 3.2.1 Table Generation for Fast Lookup

CONANN uses a set of pre-processed hash tables to allow fast in-memory lookup of words and candidate phrases. The use of such in-memory tables has been shown to dramatically increase the response time of concept annotators [6]. Nine hash tables are generated for rapid in-memory lookup, divided into 5 categories: concepts, phrases, words, variants, and language model. The hash tables are built based on the UMLS text-based files, specifically the ones included in the latest MetaMap release, version 2.4.b. These tables are loaded from external storage at CONANN startup as part of CONANN initialization, and remain in main memory until the annotator is shutdown.

### 3.2.2 Inverse Phrase Frequency Calculation

As part of pre-processing, each word in the UMLS is given a weight based on its usage in all concept instances within UMLS. In information retrieval, inverse document frequency value (IDF) uses the frequency of a word across all documents as a way to identify semantically-focused words [11]. Semantically-focused words do not frequently occur across all documents within a collection, and thus are more likely to have more discrimination. TF*IDF is a family of information retrieval algorithms which consider how often a word appears in a document (TF – term frequency) and contrasts with the overall importance of a word in a set of documents (IDF – inverse document frequency). To apply the ideas of TF*IDF [12] to CONANN, each UMLS concept instance is substituted for document, resulting in a weight called Inverse Phrase Frequency (IPF), shown in Figure 2. Each unique UMLS word is assigned its semantic importance based on its inverse phrase frequency value. More semantically important words will be given a higher weight than semantically unimportant words. The idea is to give some indication of the importance of a word in UMLS based on its usage within all UMLS concept instances. Term frequency, typically combined with IDF in information retrieval, is not considered here since we hypothesize it is highly likely the frequency for each word within a phrase will be one, due to the short length of phrases. Existing annotators typically use a binary membership value to measure word coverage, while the IPF value replaces the binary value with a word weight.

$$Inverse\ Phrase\ Frequency = log\frac{N}{n_i} \qquad (1)$$

**Fig. 2.** Inverse Phrase Frequency. $N$ is the number of UMLS phrases, $i$ is a UMLS word, and $n_i$ is the number of phrases word $i$ appears in UMLS.

## 3.3 Candidate Phrase List Generation

The first step in annotating a source phrase is to find a list of all possible candidate phrases in UMLS. This pool of phrases represents possible matches with the source phrase. Finding candidate phrases is done in a series of steps: a) remove stop words from the source phrase; b) map remaining source phrase words to their uninflected base forms; and c) find UMLS concept instances which have one or more words in common with the source phrase. Stop words are removed from the source phrase by removing all words from the source phrase which do not appear in UMLS. Remaining words in the source phrase are mapped to their UMLS base form using available UMLS inflection and word variant information. For example, 'cancers' is mapped to 'cancer.' A candidate phrase list is then generated by finding all concept instances which contain one or more of the base-form words in the source phrase. For example, the phrase *lung cancer* will find candidate phrases having the words *lung* and *cancer*, which will return candidate phrases such as {*lung, chronic obstructive lung disease, lung cancer, liver cancer*}. It is not required that a candidate phrase have all words in common with a given source phrase, since exact mappings between a source phrase and concept instances are expected to be rare. In addition to finding all candidate phrases having one or more of the base words, the same process is repeated for all word variants of the base word. For example, *pulmonary* is a variant of *lung*, so phrases such as *pulmonary carcinoma* will be added to the list of candidate phrases.

## 3.4 Incremental Filtering

The incremental filtering is used to improve computational efficiency by applying successively more computationally complex filters. Two filtering approaches, coverage and coherence filters are used.

### 3.4.1 Coverage Filter

Coverage measures the overlap of common words between a source phrase and a candidate phrase. In contrast to existing systems such as MetaMap [2], SAPHIRE [10], and IndexFinder [6], which consider the count of words in common between a source phrase and a candidate phrase, the scoring of coverage in CONANN considers the contribution of each word in the source phrase, as measured by each word's IPF value. Each candidate phrase is given a score determined by the sum of its words' IPF values, called the PhraseCoverageIPF score, as shown in Figure 3.

$$PhraseCoverageIPF = \sum_{i=1}^{N} IPF_i \qquad (2)$$

**Fig. 3.** Phrase Coverage IPF. $N$ is the number of words in a phrase. *IPF* is the inverse phrase frequency value of *word$_i$*.

Once the PhraseCoverageIPF values are computed for all candidate phrases, a threshold value is chosen as the mean plus one standard deviation of the PhraseCoverageIPF values for the set of candidate phrases. The idea is that a candidate phrase consisting of more semantically-focused words is treated as a better mapping candidate phrase and therefore has a higher possibility of being passed as input to the next filter (e.g., Coherence filter). All candidate phrases whose PhraseCoverageIPF values is greater than or equal to the threshold value are passed to the final concept mapper. There are two exceptions to consider: (a) if there is an exact match between a source phrase and one of the candidate phrases, the candidate concept associated with the candidate phrase is returned; and (b) if no candidate phrases have a PhraseCoverageIPF value greater than or equal to the threshold, the candidate phrases with the highest PhraseCoverageIPF value are passed to the final concept mapper (if there is only one such candidate phrase with a high PhraseCoverageIPF, its candidate concept is returned).

### 3.4.2  Coherence Filter

Coherence is a complimentary filter to coverage. While coverage looks at the commonality of words between a source phrase and a candidate phrase, coherence measures their common word order. The idea is that the common syntactic ordering of source phrase and concept instances will remove candidate phrases which have some words in common but are in a different order, indicating the concept instance may be expressing a different concept than the source phrase. The use of IPF values is not used in Coherence, since we are considering word position rather than word importance. We measure coherence using skip bigrams [13]. A skip bigram is a word pair which allows for an intervening word gaps. A skip bigram list for a phrase is generated by walking the phrase words from start to end, and pairing each word with each word following it, in pairs. An example list of skip bigrams is shown in Table 1.

The advantage of a skip bigram is that it measures word order while allowing intervening words to appear. A candidate phrase is scored by by first generating a skip bigram list for both the candidate phrase and the source phrase, and then calculating recall, as shown in Figure 4, which measures the degree of skip-bigram overlap between the source and the candidate phrases. The performance of skip-bigrams has been evaluated in machine translation evaluation and summary evaluation, and has been shown to perform at or above state-of-the-art measures with less complexity [14]. CONANN uses the recall measure, since it has been shown in machine translation evaluation research that n-gram recall is the biggest factor in evaluations using n-gram measures [15]. Once all candidate phrases have been scored, candidate phrases are pruned based on the same two standard deviation method used in the coverage filter (see Section 3.4.1).

**Table 1.** Skip-bigrams for the phrase *abnormal body temperature elevation*

| Skip Bigram |
|---|
| abnormal, body |
| abnormal, temperature |
| abnormal, elevation |
| body, temperature |
| body, elevation |
| temperature, elevation |

$$Recall = \frac{CommonSkipBigrams(SourcePhrase, CandidatePhrase)}{CountSkipBigrams(CandidatePhrase)} \quad\quad (3)$$

**Fig. 4.** Skip-bigram recall measure [13]

### 3.5  Final Concept Mapping

The final mapping of a source phrase to a UMLS concept is performed after the coverage and coherence filters have been applied to a list of candidate phrases. The remaining candidate phrases are then grouped by the concepts they belong to. Each candidate concept is then scored based on the number of candidate phrases it contains. The highest scoring candidate concept is then output as the concept for the source phrase. In the event of tie scores, multiple candidate concepts can be output. The idea is that the number of candidate phrases per concept after filtering gives an indication of the matching likelihood of a source phrase to a concept.

## 4  Related Work

Most work in semantic annotation for biomedical text is performed to support semantic indexing/retrieval and data mining of biomedical texts [2]. Our work is most closely related to MetaMap [2], KnowledgeMap [4], and SAPHIRE [3]. We focus on scoring candidate phrases, since that is one of the primary differences between systems, SAPHIRE uses simple and partial mapping, and for candidate phrase scoring combines measures of term overlap, term proximity, and length of term matches. KnowledgeMap uses simple and partial matching, and for candidate phrase scoring uses an exact match approach and if no matches are found, performs iterative variant-word-generation and re-matching. KnowledgeMap also offers a disambiguation stage which uses concept co-occurrence information derived from existing medical texts to find a best-matching concept. MetaMap uses simple, partial and complex mapping. MetaMap scores candidate phrases using a mixture of four different scores: a) *Centrality* where a source phrase head term used in concept instance; b) *Variation* how far a source phrase term variant is from concept instance term; c) *Coverage* which measures the overlap between source phrase and concept instance terms, ignoring gaps; and d) *Coherence* which finds term sequence overlaps between source phrase and concept instance. Compared to SAPHIRE, our CONANN uses simple and partial matching, but does not score every candidate phrase for final mapping. Like KnowledgeMap and MetaMap, we incorporate word variants of the source phrase, but we do not incorporate disambiguation or exact matching as KnowledgeMap does or extensive word variants generation as MetaMap does. Our system reduces computational complexity by deferring complex scoring until after most candidate phrases have been eliminated. In addition, we build a language model of each concept's phrases, whereas existing systems consider each candidate phrase as independent of one another, even from the same concept.

Other related systems include SENSE [16], which translates source and concept instance to low-level semantic factors, then performs exact matching of the semantic factors; Concept Locator [17] which simply sub-divides a phrase & looks for exact matches; PhraseX [18] which focuses on phrase identification and performs an exact match with candidate phrases; and IndexFinder [6] which treats the source text as a bag of words and finds all matching words, regardless of their location.

# 5  Evaluation

Evaluation of the annotation system is done using both an intrinsic and an extrinsic method. The intrinsic evaluation is intended to evaluate the speed and accuracy of CONANN against an existing biomedical concept annotator. Four different versions of CONANN are used based on the filtering method(s): a) Coherence only; b) Coverage only; c) Coverage+Coherence, and d) Coherence+Coverage. For (c) and (d), the difference is only in the order of the two filters. The extrinsic evaluation is designed to measure the effect of annotation output on a task. We chose text summarization using concepts as the task, since text summarization can use the phrase, phrase location, and phrase concept mapping output produced by the annotators. This information is combined to identify important areas with a text. The best performing filter in the intrinsic evaluation, Coherence+Coverage, was then used in the CONANN for the extrinsic evaluation.

## 5.1  Intrinsic Evaluation

The intrinsic evaluation is intended to evaluate the speed and accuracy of CONANN against an existing biomedical concept annotator. We use the MetaMap system [2] provided by the United States National Library of Medicine as the baseline system. MetaMap maps biomedical text to concepts stored in the Metathesaurus. The text-to-concept mapping in the MetaMap application is done through a natural language processing approach. Sentences are first identified, and then noun phrases are extracted from each sentence. MetaMap proceeds through several stages to map a noun phrase to one or more concepts. It is possible a noun phrase can map to more than one concept. In this case, no disambiguation step is performed, and MetaMap returns multiple concepts.

The corpus of noun phrases was generated from a citation database of approximately 1,200 oncology clinical trial papers physicians feel are important to the field [19]. Of the 1,200 papers cited, 24 were randomly selected based on the minimum requirements of the ROUGE summary evaluation tool [20]. The PDF versions of these 24 papers were then converted to plain-text format. The resulting text was processed by MetaMap to find all noun phrases and their corresponding concept annotations in the 24 papers, resulting in a corpus of 4,410 unique phrases. The corpus was pruned to retain only those phrases which MetaMap annotated with a single concept, allowing for meaningful mapping comparisons between the two systems. There were 1,628 phrases with a single MetaMap concept annotation. This set of phrases was used to perform the evaluation. Therefore we assume that the precision of MetaMap concept annotation is 100%.

To measure the amount of time it takes for MetaMap to annotate the test corpus of phrases, MetaMap was executed using the 1,628 phrases as input. We used the MetaMap API [21] to annotate each phrase. MetaMap provides various APIs to annotate different text chunk sizes, including document, document section, sentence, or term. We used to the term method so that MetaMap would not need to expend effort finding phrase boundaries, as it would do if passed a document, document section, or sentence to annotate. CONANN was then executed against the same set of 1,628 phrases and its annotation time measured. CONANN also produced concept annotations for the list of phrases. These mappings were then compared to MetaMap, producing the annotation precision metric described in the following paragraph. Three back-to-back runs of each system were performed, and the system restarted after each run to remove variations caused by the operating environment, such as file system caching.

Accuracy is measured by comparing CONANN's annotation of each phrase to the MetaMap's annotation output for each phrase. There are two measures for the intrinsic evaluation: (a) precision, and (b) phrase annotation time. The first measure looks at the accuracy of the concept annotation, and the second measure looks at the speed of the concept annotation. The Annotation Precision measure uses the same idea as in the precision measure in information retrieval, but adapted to fit concept mapping [22]. Annotation Precision is defined as the fraction of mapped concepts which are correct. In this evaluation, we used *Single Concept* matching, where a correct match is counted only if CONANN directly generates a single concept which matches the MetaMap single concept, and *Relaxed Matching* where CONANN generates five top concepts. A correct match is counted if any of the five concepts generated by CONANN match the MetaMap single concept. Recall is not considered because the source phrase corpus that is correctly annotated by MetaMap is only provided to CONANN to annotate, and so recall is not meaningful for this evaluation. For measuring speed, the average time to annotate a phrase is used, which is calculated by taking the total annotation time divided by the total number of phrases annotated. Annotation time is defined as the time it takes to annotate a single phrase, and does not include the annotator initialization, which can be significant, as shown in Figure 5. Total annotation time is the time it takes to annotate all phrases in the corpus, excluding annotator initialization.

## 5.2 Extrinsic Evaluation

The output of a concept annotator is a list of phrases and their associated domain-specific concepts. This output is an intermediate format, not directly useable by an end-user. The extrinsic evaluation is a complimentary evaluation to the intrinsic, designed to show the usefulness of the concept output in some task. We selected text summarization as the end-user task. We used two probabilistic summarizers, FreqDist [5] and a version of SumBasic [23] modified to use concepts rather than terms. Both summarizers only use concept frequency as the sole feature to select salient sentences. The output of a concept annotator is used for the input to the summarizers. Both summarizers' performance is entirely reliant on the frequency of concepts identified in the texts. It is expected if the concept annotation is accurate, summarization performance will improve because the concepts will have identified important areas within a text. Conversely, if the concept annotation is not accurate, text summarization performance will degrade.

The corpus of 24 texts (see Section 5.1) is annotated using both CONANN and MetaMap. The FreqDist and modified version of the SumBasic summarizers are then used to generate a summary of each of the 24 texts using the concept output from both annotators. The summary output from both summarizers is evaluated against manual summaries generated from domain experts using the ROUGE tool (Recall-Oriented Understudy for Gisting Evaluation) [24]. ROUGE is an automated evaluation tool which compares a system-generated summary from an automated system with one or more ideal summaries produced by people. ROUGE uses n-gram co-occurrence to determine the overlap between a summary and the models. An n-gram can be considered as one or more consecutive words. The ROUGE parameters from the DUC 2005 conference [25] are used to evaluate system summarizer performance. Two recall scores are extracted from the output of ROUGE to measure each summarizer: ROUGE-2 and ROUGE-SU4, which are also the measures used by DUC 2005. ROUGE-2 evaluates bigram co-occurrence while ROUGE-SU4 evaluates skip-bigrams with a maximum distance of 4 words. The ROUGE scores indicate the n-gram overlap between the source text and the model summaries.

# 6   Results

The intrinsic evaluation compares the speed and accuracy of our CONANN versus the MetaMap baseline system. The extrinsic evaluation compares the use of the generated concept annotations on a text summarization task.

## 6.1   Intrinsic Evaluation Results

The first measurement is annotator initialization time, which is the time to load domain-specific resources into memory and prepare for annotation. Figure 5 shows the initialization time for each run of both annotators. For MetaMap, initialization time ranged from 1.3 to 1.6 minutes, while for our CONANN, initialization time ranged from 17 to 20 seconds. Both systems exhibit stable initialization behavior.

Figure 6 presents the total time to annotate all 1,628 phrases in the evaluation corpus. MetaMap total annotation time was consistent across all three runs at 5.7 minutes, while CONANN ranged from 14.5 to 16.5 seconds on all three runs. Figure 7 shows the average time to annotate each phrase for each run of the annotator. Average Phrase Annotation Time is calculated by taking the total annotation time and dividing it by 1,628, which is the total number of phrases annotated. MetaMap average time to annotate a phrase was 208 milliseconds, while CONANN ranged from 9 to 10 milliseconds per phrase across all three runs.

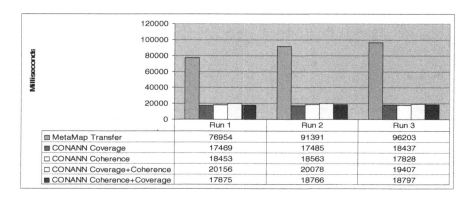

| | Run 1 | Run 2 | Run 3 |
|---|---|---|---|
| ▨ MetaMap Transfer | 76954 | 91391 | 96203 |
| ▥ CONANN Coverage | 17469 | 17485 | 18437 |
| ▢ CONANN Coherence | 18453 | 18563 | 17828 |
| ▢ CONANN Coverage+Coherence | 20156 | 20078 | 19407 |
| ■ CONANN Coherence+Coverage | 17875 | 18766 | 18797 |

**Fig. 5.** Annotator initialization time

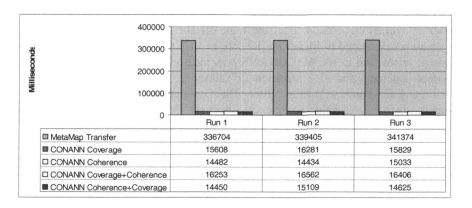

| | Run 1 | Run 2 | Run 3 |
|---|---|---|---|
| ▨ MetaMap Transfer | 336704 | 339405 | 341374 |
| ▥ CONANN Coverage | 15608 | 16281 | 15829 |
| ▢ CONANN Coherence | 14482 | 14434 | 15033 |
| ▢ CONANN Coverage+Coherence | 16253 | 16562 | 16406 |
| ■ CONANN Coherence+Coverage | 14450 | 15109 | 14625 |

**Fig. 6.** Total annotation time

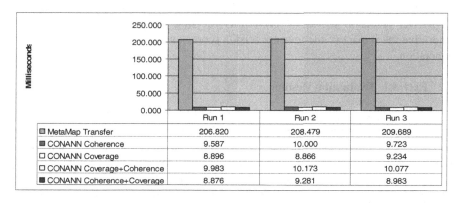

| | Run 1 | Run 2 | Run 3 |
|---|---|---|---|
| ▨ MetaMap Transfer | 206.820 | 208.479 | 209.689 |
| ▥ CONANN Coherence | 9.587 | 10.000 | 9.723 |
| ▢ CONANN Coverage | 8.896 | 8.866 | 9.234 |
| ▢ CONANN Coverage+Coherence | 9.983 | 10.173 | 10.077 |
| ■ CONANN Coherence+Coverage | 8.876 | 9.281 | 8.983 |

**Fig. 7.** Average phrase annotation time

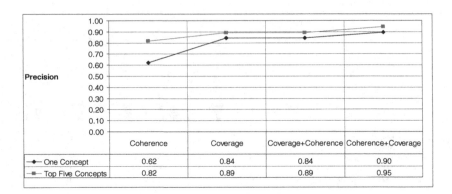

**Fig. 8.** Annotator Precision

While CONANN is over four times faster in initialization and over twenty times faster in average annotation time, the trade-off for the faster performance is less precision as compared to MetaMap (i.e., we assume that the precision of MetaMap concept annotation is 100%). CONANN was measured at 90% precision for exact concept matching, and 95% precision for relaxed concept matching using the best performing Coherence+Coverage filter. The Coverage+Coherence filter had seven percent worse precision than Coherence+Coverage, indicating that the filtering order is important. The worst performing filter is Coherence alone when selecting a single concept, but jumps significantly higher when relaxed matching is used, indicating the correct concepts are available but candidate phrase order alone is not enough to achieve final best mapping. The Coverage filter alone had a precision equal to Coverage+Coherence, indicating the Coverage filter is a strong filter which removes candidate phrases which would have been selected by the Coherence filter. To counter this effect, placing the Coherence filter before the Coverage filter results in candidate phrases with strong ordering being selected, which are then further refined by using the semantic focusing of the Coverage filter. Figure 8 summarizes the CONANN precision scores for each filter. Section 6.2 presents an evaluation to determine the impact of the lower precision scores on a task which uses the concept annotation.

## 6.2 Extrinsic Evaluation Results

Table 2 shows the ROUGE-2 and ROUGE-SU4 scores for evaluating text summarization performance using the CONANN (Coherence+Coverage) and MetaMap annotator output. For the ROUGE-2 metric, MetaMap slightly outperforms CONANN using FreqDist (1% difference), while CONANN outperforms MetaMap using SumBasic by 7%. The results are similar for the ROUGE-SU4 scores. FreqDist using MetaMap has an approximately 2% advantage over FreqDist using CONANN. SumBasic using MetaMap has an approximately 5% advantage over SumBasic with CONANN. We conclude that CONANN performs very closely to MetaMap in the extrinsic text summarization task. In addition, CONANN has a time advantage of performing annotation over twenty times faster than a state-of-the-art system, facilitating its use in online environments.

**Table 2.** Text summarization task performance using Coherence+Coverage version of CONANN (see Section 5.2 for more information on ROUGE)

| Summarization Method | ROUGE-2 Score | ROUGE-SU4 Score |
|---|---|---|
| FreqDist using MetaMap | 0.1207 | 0.2200 |
| FreqDist using CONANN | 0.1192 | 0.2161 |
| | | |
| SumBasic using CONANN | 0.1178 | 0.2098 |
| SumBasic using MetaMap | 0.1094 | 0.2003 |

## 7 Conclusion

We presented an online biomedical concept annotator, CONANN, which takes a source phrase, identifies potential matching concepts and phrases in a domain-specific thesaurus, uses an incremental filter approach to remove candidate phrases using a variation of inverse document frequency, and maps the source phrase to best-matching concepts.

An intrinsic evaluation was performed to compare CONANN's concept output to MetaMap's output. In addition, an extrinsic evaluation was performed to measure the usefulness of the concept output of each annotator. CONANN initialization time is four times faster and average annotation time per phrase is twenty times faster than a state-of-the-art concept annotator. The speed advantage is at some cost in accuracy, as the single concept mapping precision compared to MetaMap is 90%. However, this loss of accuracy did not significantly impact the use of the CONANN's output in a text summarization task.

Future work includes finding methods to reduce the size of the initial candidate list, incorporating concept disambiguation, and using other parts of UMLS to improve precision, such as preferred terms or most frequent vocabulary sources. In addition, while CONANN has been evaluated with UMLS as the domain resource, we would like to explore the result of using CONANN with other domain resources. Our eventual goal is to provide a biomedical concept annotator operating at the phrase level which has high accuracy compared to existing systems, and which can operate in an online environment. Such a system would be useful for ad-hoc physician and biomedical research tasks such as summarizing texts and semantic search.

## References

1. United States National Library of Medicine,PubMed (2006) http:// www.ncbi.nlm.nih.gov/ entrez/query.fcgi
2. Aronson, A.R.: Effective mapping of biomedical text to the UMLS metathesaurus: The MetaMap program. In: Proceedings of the AMIA Symposium, 2001, pp. 17–21 (2001)

3. Hersh, W.R., Greenes, R.A.: SAPHIRE–an information retrieval system featuring concept matching, automatic indexing, probabilistic retrieval, and hierarchical relationships. Comput. Biomed. Res. 23, 410–425 (1990)
4. Denny, J.C., Irani, P.R., Wehbe, F.H., Smithers, J.D., Spickard 3rd, A.: The KnowledgeMap project: development of a concept-based medical school curriculum database. In: Proceedings of the Annual AMIA Symposium, pp. 195–199 (2003)
5. Reeve, L., Han, H., Nagori, S.V., Yang, J., Schwimmer, T., Brooks, A.D.: Concept frequency distribution in biomedical text summarization. In: Proceedings of the ACM Fifteenth Conference on Information and Knowledge Management (CIKM'06), pp. 604–611 (2006)
6. Zou, Q., Chu, W.W., Morioka, C., Leazer, G.H., Kangarloo, H.: IndexFinder: A method of extracting key concepts from clinical texts for indexing. In: Proceedings of the AMIA Annual Symposium, pp. 763–767 (2003)
7. Handschuh, S., Staab, S., Volz, R.: On deep annotation. In: International WWW Conference, pp. 431–438 (2003)
8. Reeve, L., Han, H.: A comparison of semantic annotation systems for text-based web documents. In Web Semantics and Ontology, 1st ed., vol. 1, Taniar, D., Rahayu, J. W., (eds.) Hershey, PA USA: Idea Group (2006)
9. United States National Library of Medicine. UMLS metathesaurus fact sheet (2006) http://www.nlm.nih.gov/pubs/factsheets/umlsmeta.html
10. Hersh, W., Leone, T.J.: The SAPHIRE server: a new algorithm and implementation. Proc. Annu. Symp. Comput. Appl. Med. Care. pp. 858-862 (1995)
11. Manning, C.D., Schutze, H.: Foundations of Statistical Natural Language Processing, 1st edn., p. 620. The MIT Press, Cambridge, Massachusetts (1999)
12. Jones, K.S.: A statistical interpretation of term specificity and its application in retrieval. Journal of Documentation 28, 11–21 (1972)
13. Lin, C.Y., Och, F.J.: Automatic evaluation of machine translation quality using longest common subsequence and skip-bigram statistics. In: Proceedings of the 42nd Annual Meeting of the Association for Computational Linguistics, pp. 605–612 (2004)
14. Lin, C.Y., Och, F.J.: Orange: A method for evaluating automatic evaluation metrics for machine translation. In: Proceedings of the 20Th International Conference on Computational Linguistics, pp. 501–507 (2004)
15. Lavie, A., Sagae, K., Jayaraman, S.: The significance of recall in automatic metrics for MT evaluation. In: Proceedings of the 6th Conference of the Association for Machine Translation in the Americas (2004)
16. Zieman, Y.L., Bleich, H.L.: Conceptual mapping of user's queries to medical subject headings. In: Proc. AMIA. Annu. Fall. Symp., pp. 519–522 (1997)
17. Nadkarni, P.M.: Concept locator: a client-server application for retrieval of UMLS metathesaurus concepts through complex boolean query. Comput. Biomed. Res. 30, 323–336 (1997)
18. Srinivasan, S., Rindflesch, T.C., Hole, W.T., Aronson, A.R., Mork, J.G.: Finding UMLS Metathesaurus concepts in MEDLINE. In: Proc. AMIA. Symp., pp. 727–731 (2002)
19. Brooks, A.D., Sulimanoff, I.: Evidence-based oncology project. In Surgical Oncology Clinics of North America, Anonymous 11, 3–10 (2002)
20. Lin, C.: Looking for a few good metrics: Automatic summarization evaluation - how many samples are enough? In: Proceedings of the NTCIR Workshop 4 (2004)
21. Devita, G.: MMTx API documentation for release 2.3 (2006)
22. Hersh, W.R., Mailhot, M., Arnott-Smith, C., Lowe, H.J.: Selective Automated Indexing of Findings and Diagnoses in Radiology Reports. J. Biomed. Inform. 34, 262–273 (2001)

23. Nenkova, A., Vanderwende, L.: The impact of frequency on summarization. Microsoft Research, Redmond, Washington, Tech. Rep. MSR-TR-2005-101 (2005)
24. Lin, C., Hovy, E.H.: Automatic evaluation of summaries using N-gram co-occurrence statistics. In: Proceedings of, Language Technology Conference (HLT-NAACL 2003), 2003, pp. 71–78 (2003)
25. National Institute of Standards and Technology (NIST), Document Understanding Conferences (2006), http://duc.nist.gov

# Author Index

# Lecture Notes in Bioinformatics

Vol. 2994: E. Rahm (Ed.), Data Integration in the Life Sciences. X, 221 pages. 2004.

Vol. 2983: S. Istrail, M.S. Waterman, A. Clark (Eds.), Computational Methods for SNPs and Haplotype Inference. IX, 153 pages. 2004.

Vol. 2812: G. Benson, R.D.M. Page (Eds.), Algorithms in Bioinformatics. X, 528 pages. 2003.

Vol. 2666: C. Guerra, S. Istrail (Eds.), Mathematical Methods for Protein Structure Analysis and Design. XI, 157 pages. 2003.